STUDIES IN ANALYTICAL CHEMISTRY 6

SOLVATION, IONIC AND COMPLEX FORMATION REACTIONS IN NON-AQUEOUS SOLVENTS

Experimental Methods for their Investigation

STUDIES IN ANALYTICAL CHEMISTRY

STUDIES IN ANALYTICAL CHEMISTRY 6

SOLVATION, IONIC AND COMPLEX FORMATION REACTIONS IN NON-AQUEOUS SOLVENTS

Experimental Methods for their Investigation

K. BURGER

Institute of Inorganic and Analytical Chemistry
Eötvös Loránd University, Budapest, Hungary

ELSEVIER SCIENTIFIC PUBLISHING COMPANY
Amsterdam—Oxford—New York 1983

The distribution of this book is being handled by the following publishers

for the U.S.A. and Canada
Elsevier Science Publishing Company, Inc.
52 Vanderbilt Avenue
New York, New York 10017, U.S.A.

for the East European countries, Democratic People's Republic of Korea, People's Republic of Mongolia, Republic of Cuba, Socialist Republic of Vietnam
Kultura, Hungarian Foreign Trading Company,
P.O.B. 149, H-1389 Budapest 62, Hungary

for all remaining areas
Elsevier Scientific Publishing Company
Molenwerf 1
P.O. Box 211, 1000 AE Amsterdam, The Netherlands

Library of Congress Cataloging in Publication Data

Burger, K. (Kálmán)

Solvation, ionic and complex formation reactions in non-aqueous solvents.

(Studies in analytical chemistry; 6)

Translated from Hungarian ms. by David Durham.

Includes bibliographical references and indexes.

1. Solvation. 2. Non-aqueous solvents. 3. Electron donor-acceptor complexes. I. Title. II. Series.

QD541.B79 541.3'423 81-22091
ISBN 0-444-99697-4 AACR2

ISBN 0-444-99697-4 (Vol. 6)
ISBN 0-444-41941-1 (Series)

Joint edition published by
Elsevier Scientific Publishing Company, Amsterdam, The Netherlands
and Akadémiai Kiadó, The Publishing House of the Hungarian Academy of Sciences, Budapest, Hungary

Printed in Hungary

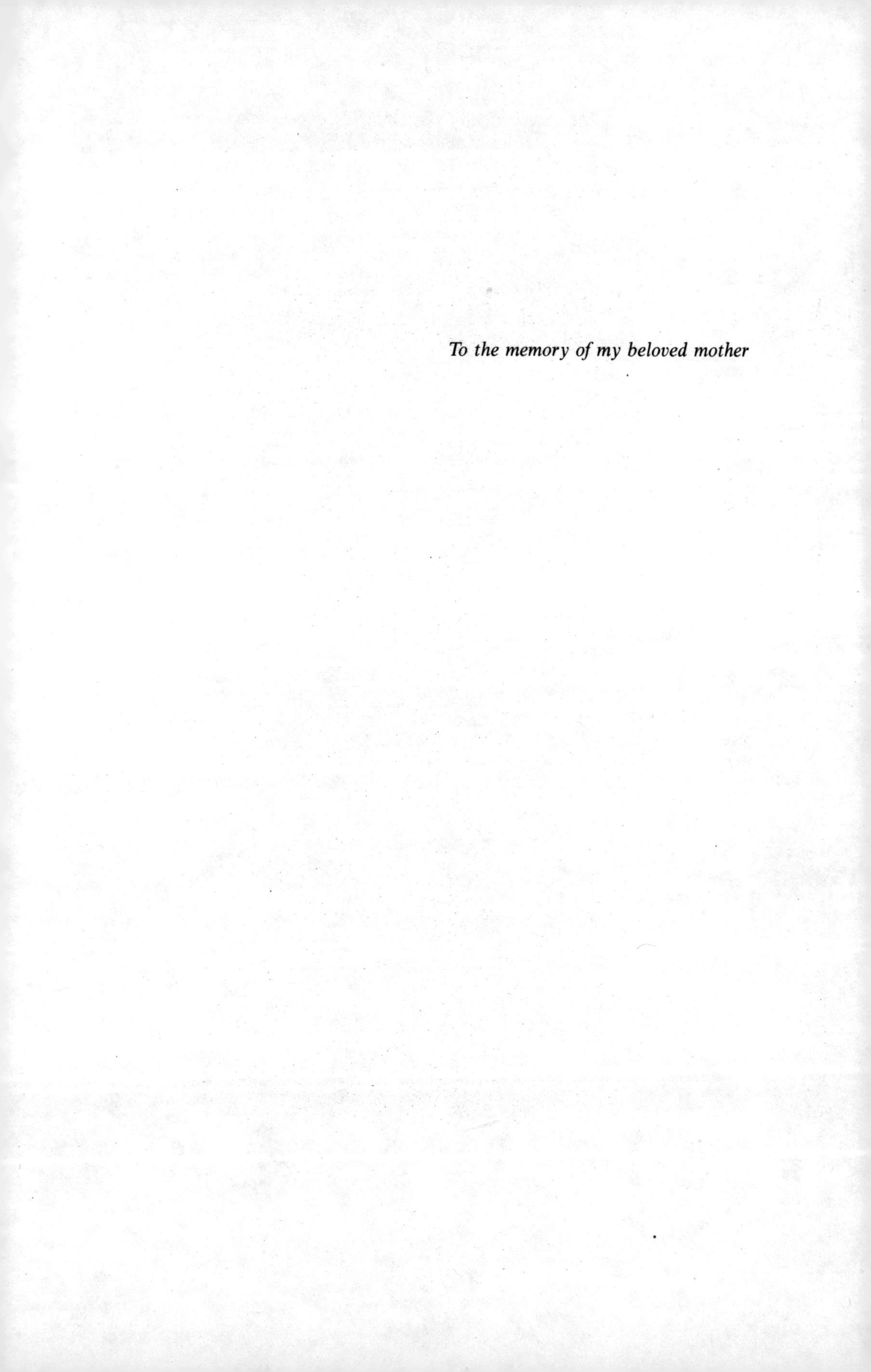

To the memory of my beloved mother

CONTENTS

PREFACE

Rapid developments in the natural sciences and technology during the past two decades have led to an almost incredible expansion in the field of analytical chemistry. Whereas earlier analytical chemistry consisted almost exclusively of the qualitative detection and quantitative determination of elements, functional groups or molecules, recognition of the fundamental importance of the connections between structure (electronic structure, conformation, stereochemistry) and function, and also between the matrix (chemical and physical medium) and function, together with the increasingly extensive investigation of these connections, made necessary the analytical application of a whole series of complex spectroscopic and diffraction methods. An increasing number of new analytical problems have arisen, the solution of which cannot be expected from the traditional methods of chemistry or physical chemistry involving simple instrumentation. Indeed, it may be stated with almost general validity that complicated analytical tasks can be solved only by the combined application of a number of methods that supplement one another. These tendencies are well reflected by the field presented in this volume.

It has long been known that in processes taking place in solution the solvent behaves not only as a medium, but also as a reactant. Its role is, however, so complex that even today no generally valid theory is available for its description. The understanding of the regularities of the solvent effect required the study of a very large number of systems from many aspects. Procedures of investigation in physical chemistry that have not found application in the course of this extensive work are rare. A critical, review-like presentation of the results therefore also provides a good reflection of the modern methods involved in a complex analytical approach.

Accordingly, in this book an attempt has been made to present a literature survey of research dealing with solvent effects, also demonstrating the type of information to be expected from the various methods and the way in which these supplement one another. When discussing the factors governing the solvent effect and the other interactions in the various systems, emphasis has also been placed on the methods of analysis and structural examination which helped to shed light on the problem.

In addition, separate sections deal with the modes of employment of the more important experimental methods in this field, and with the results obtained.

In the above sense, the object of this volume is both to present the available methods and to consider their particular value in investigating solutions composed of non-aqueous solvents; these studies afford information on the interactions between the solvent and the dissolved components, the solvation processes and the solvent-induced changes in the equilibrium and symmetry conditions, in the compositions and electron structures of the species and in the kinetics and mechanisms of the reactions of these species in solution.

A classification of the solvents and empirical solvent strength (donor and acceptor strength) scales are given, based on various experimental parameters, together with various correlations empirically describing the solvent effect; the scope of their use and limitations is discussed. Methods for the purification of solvents and ways of checking their purity are also presented.

As a consequence of the large number of experimental methods employed in this field, and the extremely varied possibilities of application of each procedure, depending on the task to be performed, no attempt could be made to discuss the means of solving individual problems in the detail expected in a reference book. The object has been to draw attention, by means of as many examples as possible, to practically every method that can be used for the study of solutions and the solvent effect, also giving the original literature references on the basis of which the work may be performed. In view of the industrial technological significance of non-aqueous solutions, and hence of the importance of their examination, the material presented attempts to provide information which is indirectly also of industrial analytical interest.

Ideal and regular solutions that can be described by electrostatic theories do not feature in the treatment. A number of other publications deal with these and with the theories elaborated for them. Only passing reference is made to these simple systems in this work, which is concerned primarily with the methods of studying specific solvent effects and their results.

In addition to providing a review of the methodology, we have also attempted, within the limited scope of this volume, to give as complete an account as possible of the individual results achieved during investigations of the solvent effect, and particularly of the general regularities recognized. In this respect, the book also presents a review of the coordination chemistry of non-aqueous solutions. It is hoped that the book will be of interest not only to analytical chemists, but also to those working in inorganic, coordination and physical chemistry, and to university students striving to master these subjects.

Acknowledgements. The sections of the book dealing with positronium chemistry were compiled by Prof. Attila Vértes and Dr. Béla Lévay, whom I should like to thank for their cooperation. The entire manuscript was read and subjected to the closest scrutiny by Prof. E. Pungor, Prof. W. Simon and Prof. J. Inczédy, whose valuable comments contributed to a more complete interpretation of the material. I am extremely grateful to them for their help. I should also like to thank Dr. D. Durham, who translated the Hungarian manuscript into English and Dr. F. Kállay for editing the English translation.

Kálmán Burger

1. INTRODUCTION

Investigations into the structures of liquids and particularly solutions, equilibria in solutions, interactions between the solute and the solvent molecules and interactions between the solvent molecules themselves, represent an extremely wide-ranging group of problems, which are of very great importance from both the theoretical and practical points of view.

The practical significance of these problems is amply demonstrated by the fact that a considerable proportion of industrial processes takes place in the liquid phase, and biological, *in vivo* processes are inseparable from the liquid or solution phase.

The energetics of "*ideal*" solutions, corresponding to Raoult's law, have long been known. However, just as the behaviour of real gases cannot be described with the laws valid for ideal gases, the solutions occurring in practice cannot be characterized on the basis of the laws of ideal solutions.

As early as the beginning of this century, Hildebrand (Hi 16, Hi 62) introduced the concept of *regular* solutions, differing from ideal solutions. Nevertheless, not even this theory takes into account the specific interactions between the solvent and the solute. According to Hildebrand, the Gibbs energies of the formation of ideal and regular solutions differ only in the enthalpy terms; the entropy terms are the same. In such regular solutions the solute and the solvent molecules are randomly distributed; neither polarization nor chemical interactions produce any preferred ordering. The interactions between the different molecules in these solutions cannot be stronger than those between the identical molecules. Such solutions can be prepared only by mixing non-polar and non-interacting molecules.

The interaction between the solvent and solute in these solutions can, in many cases, be described simply in terms of the Van der Waals forces. The corresponding total potential, W, can then be given by the Lennard–Jones function:

$$W = \frac{A}{R^{12}} - \frac{B}{R^{6}}$$

where R is the intermolecular distance and A and B are the repulsion and attraction constants, respectively. A includes the Coulomb and Pauli repulsions, and B takes into account the attractive effects of the Keesom, Debye and London forces. In practice, however, in addition to such Van der Waals forces, other specific interactions also play a part in most systems.

As regards their nature, the interactions between the molecules of the solvent and the solute can be divided into four large groups:

(1) Interactions caused by dispersion forces (London forces);
(2) Dipole–dipole orientation interactions;
(3) Dipole–dipole induction interactions;
(4) Donor–acceptor coordination interactions (the hydrogen-bonded associations feature as a separate sub-group within the last group).

Of these interactions, only those caused by the dispersion (London) forces are taken into consideration by the theory developed for regular solutions. Dipoles, hydrogen bonds and coordination interactions show up as deviations from the theory of regular solutions.

Most real solutions, and virtually all solutions prepared with polar solvents, behave in a different manner to regular solutions. Hildebrand categorized these as *irregular* solutions, but his theory, later modified and further developed by other workers, is not suitable for their description. On the other hand, there is no general theory that can describe quantitatively the conditions prevailing in these solutions. The stronger the interaction between the molecules of the solvent and the solute and the more specific this effect, the less it is to be expected that the system will be described by any of the various solvent theories, which are of limited validity.

In the case of reactions proceeding in solution, fundamental importance is attached to the interaction between the solvent and the solute (various forms of solvation) and to the effect of the solute on the structure of the solvent. Simply by the use of another solvent, basically different molecular and ionic species may be produced and different reactions may occur in systems which otherwise have analogous chemical compositions. It is not surprising, therefore, that numerous research groups all over the world are engaged in investigations of solvent effects, studying the most varied chemical processes in different solvents [see, e.g., ref. Ab 78, Be 81 etc.].

In spite of this, an authentic picture of even the apparently simplest hydration and solvation processes, and especially of their results, and of the compositions and concentrations of the ionic species (solvates, complexes) in solution, can be obtained only by means of individual experiments, even with the most intensively studied aqueous solutions. No general model or theory has been devised that permits the theoretical deduction of an actual solution structure. Some more general conclusions drawn from a large number of experimental facts have proved of value primarily for the interpretation of data obtained experimentally; their prognostic role is at most qualitative.

The complexity of the whole problem is further increased because the interactions possible in solutions are so varied and so intricate that in most cases their separation and individual study are impossible. In addition, their effects may appear in different ways and to different extents, depending on the method of investigation. The experimental data obtained by the different techniques of examination may therefore possibly lead to mutually contradictory interpretations. Hence, the use of one method of examination is suitable at most for the clarification of parts of problems. On the other hand, each different technique may elucidate or throw new light on further valuable details. Even despite the numerous contradictions, comparison of the data provided by the various methods brings us nearer to an understanding of the overall picture.

No matter whether they support one another or are contradictory, the data obtained by the different techniques are all of a supplementary nature; contradictory data contribute to an understanding of the topic by demanding the invention of another approach.

For an understanding of the contradictions or apparent contradictions, it is necessary from the outset to emphasize the differences introduced by the different time scales of the various methods of measurement. It is known, for instance, that the particles present in liquids undergo thermal motion, consisting partly of vibrations about the equilibrium state and partly of changes of position, e.g., diffusion. In water, the average durations of the vibrations and the changes of position are about 2×10^{-13} and 10^{-11} s, respectively. Accordingly, different pictures of the structure of the liquid will be obtained, depending on how the length of time of the recording of the spatial arrangement of the liquid molecules compares with the average durations of the different movements [Ei 69].

The actual distribution of the molecules at a given instant would be obtained if the duration of the recording were shorter than the average vibration time. Unfortunately, no method with such a short exposure time is available. If the duration of the recording is longer than the vibration period of the molecules, but shorter than the time of diffusion, then the positions of each molecule in the picture will reflect the time average of the changes caused by the vibrations, and the effect due to the variations in position will remain unobservable. Such vibration-averaged pictures are provided by the vibration spectra of the systems (infrared and Raman spectroscopy), neutron diffraction, dielectric relaxation, etc.

A picture of the third kind of solution structure is obtained when the exposure time also exceeds the average time of diffusion. Such a picture reflects total disorder at a given spatial point in a liquid, whereas in the coordinate system relating to one selected molecule it displays a certain ordering in the environment of this molecule. Information on this diffusion-averaged structure is provided by thermodynamic data, X-ray diffraction, light scattering, refractive index, etc.

There is naturally a close connection between the different pictures provided by the various time scales. The vibration-averaged structure differs at every position at a given instant, and at every instant at a given position. In both cases, however, its average value gives the diffusion-averaged structure.

Unfortunately, the reconciliation of contradictory experimental results and the drawing of general deductions from partial data are not so simple in all cases. Often a qualitative rationalization of the experimental data has been achieved by taking into account also the comprehensively known macroscopic properties of the system and the components; in other cases it is, even then, impossible. Clarification of such cases may be expected from further research, from the elaboration of new methods of investigation and from the introduction of new theories. For this reason, the recording of experimental data, even though they may be contradictory, is undoubtedly of value. In the compilation of this book, therefore, reference is also made to the results of studies the interpretation of which is not yet clear.

Most reactions occurring in solution are in some respect connected with complex-formation reactions. Even solution itself (whether the solvent acts as a donor or as an acceptor) is a complexing reaction [Gu 68, Gu 78, Ma 75, 79]. Accordingly, coordination chemistry studies, and primarily complex equilibrium studies, have contributed significantly to a deeper understanding of both the structures of solutions and the processes taking place in solution.

Most coordination chemical equilibrium studies have been made in aqueous solutions [Ma 77, Si 65, Si 71]. Although equilibrium studies have also been performed in water–organic solvent mixtures because of the poor water solubility of a ligand or its complexes, comparatively little attention has been paid to the effect of the solvent composition on the dissolved ionic species. In most cases also the solvent extraction procedures yield information about the complexes formed in the aqueous phase only.

The general use and practical importance of water justify unconditionally the investigation of structures occurring in aqueous solutions and hence the complex-formation reactions proceeding in water. Further, the different physical and chemical properties of water compared with those of non-aqueous solvents make such investigations simpler in many respects than those conducted in non-aqueous solutions. Nevertheless, a *full understanding of the complex systems formed in solutions is inconceivable without a study of the effects of the different solvents on the dissolved ionic species.*

Water can by no means be regarded as a typical solvent. It stands out as a consequence of its high polarity, high relative permittivity (dielectric constant), amphiprotic nature, and the physical and chemical properties stemming from these [Ho 72, Lu 74]. Hence, a deeper understanding of the equilibria occurring in aqueous solution also requires the investigation of non-aqueous solutions [Ad 67, Be 81, Gu 68, La 66, Lu 74, Pa 68, Pa 72, Ro 59].

On the above basis, a critical review of the recent results of the more important coordination chemistry research into non-aqueous solutions appears absolutely justified.

References

Ab 78 Abstracts of the IV. International Symposium on Solute-Solute-Solvent Interactions, TU, Vienna 1978.

Ad 67 Addison, C. C.: Internat. Conf. Non-Aqueous Solvents. McMaster University, Hamilton, Canada, 1967.

Be 81 Bertini, I., Lunazzi, L., Dei, A. (eds.): Advances in Solution Chemistry. Plenum Press, New York and London, 1981.

Ei 69 Eisenberg, D., Kauzmann, W.: Structure and Properties of Water. Oxford University Press, Oxford 1969.

Gu 68 Gutmann, V.: Coordination Chemistry in Non-Aqueous Solutions. Springer Verlag, Vienna–New York 1968.

Gu 78 Gutmann, V.: The Donor–Acceptor Approach to Molecular Interactions. Plenum Press, London 1978.

Hi 16 Hildebrand, J. H.: J. Am. Chem. Soc. **38**, 1452 (1916).

Hi 62 Hildebrand, J. H., Scott, R. L.: Regular Solutions. Prentice Hall, Englewood Cliffs, N. J. 1962.

Ho 72 Horne, R. A.: Water and Aqueous Solutions. Wiley-Interscience Publ. Co., New York 1972.

La 66 Lagowski, J. J.: The Chemistry of Non-Aqueous Solvents. Academic Press, New York 1966.

Lu 74 Luck, A. P. (ed.): Structure of Water and Aqueous Solutions. Verlag Chemie, Weinheim 1974.

Ma 75 Mayer, U.: Pure Appl. Chem. **41**, 291 (1975).

Ma 77 Martell, A. E., Smith, R. M.: Critical Stability Constants. Plenum Press, New York–London 1977.

Ma 79 Mayer, U.: Pure Appl. Chem. **51**, 1697 (1979).

Pa 68 Padova, J.: J. Phys. Chem. **72**, 692 (1968).

Pa 72 Padova, J.: Non-Aqueous Electrolyte Solutions (Chapter 4). In: Horne, R. A.: Water and Aqueous Solutions. Wiley-Interscience Publ. Co., New York 1972.

Ro 59 Robinson, R. A., Stokes, R. H.: Electrolyte Solutions. Academic Press, New York 1959.

Si 65 Sillén, L. G., Martell, A. E.: Stability Constants of Metal-Ion Complexes. The Chemical Society, London 1965.

Si 71 Sillén, L. G., Martell, A. E.: Stability Constants of Metal-Ion Complexes, Supplement I. The Chemical Society, London 1971.

2. GENERAL CHARACTERIZATION OF SOLVENTS

Classification of non-aqueous solvents

Non-aqueous solvents are to be found in the most varied classes of chemical compounds, and they may be classified on the basis of their different properties. Since an unambiguous classification taking into account every property is not possible, the individual solvents are differentiated by highlighting the predominant property in the given process, depending on the type of reaction under consideration. Hence, very many variations of the classifications are possible [Br 23, Br 28, Br 30, Co 69, Gu 68, Pa 62, Pa 65, Pa 69].

(a) *Molecular liquids* are used as non-aqueous solvents in an extremely wide range of applications. At room temperature and in a fairly broad interval around this, they are in the liquid state.

(b) *Salt melts* are similarly important class of non-aqueous inorganic solvents and are excellent media for the preparation of numerous compounds. These (usually ionic) melts can be employed over a very wide temperature range, although as a rule substantially above room temperature. They therefore permit the occurrence of many reactions which are appreciably endothermic at room temperature.

(c) The third group of non-aqueous solvents consists of *metals with comparatively low melting points*, such as metallic sodium and mercury.

In this book, only the properties of molecular liquids will be discussed.

Brönsted [Br 23, Br 28, Br 30] classified solvents according to their proton donor and proton acceptor capability, on the basis of three of their properties: relative permittivity, acid strength and basic strength; thus he distinguished between the following eight fundamental solvent classes:

(1) Amphiprotic solvents with high relative permittivity, which are both acidic and basic (e.g. water, methanol);

(2) Protogenic solvents with high relative permittivity, which are acidic (e.g. HF, H_2SO_4);

(3) Protophilic solvents with high relative permittivity, which are basic (e.g. tetramethylurea);

(4) Aprotic solvents with high relative permittivity, which are neutral (e.g. acetonitrile);

(5) Amphiprotic solvents with low relative permittivity, which are both acidic and basic (e.g. higher alcohols);

(6) Protogenic solvents with low relative permittivity, which are acidic (e.g. glacial acetic acid);

(7) Protophilic solvents with low relative permittivity, which are basic (e.g. amines);

(8) Aprotic solvents with low relative permittivity, which are neutral (e.g. benzene, CCl_4).

The Brönsted classification has served as the basis for many, more recently introduced groupings of solvents.

In addition, a series of other solvent classifications have also been devised, based on various properties. For example, solvents are frequently divided into ionizing and non-ionizing solvents.

Ionizing solvents (water, alcohols, acetic acid, etc.) are polar to a considerable extent. They dissolve many ionic and covalent compounds, and the solutions prepared with them behave as conductors of the second kind. The molecules or ions of the solute are appreciably solvated in the solution. In this solvent group the dissolution results as a rule in the formation of compounds containing solvent molecules (solvates).

Non-ionizing solvents (CCl_4, CS_2, hydrocarbons, etc.) are generally apolar. They dissolve ionic compounds to a lesser extent, but covalent molecules more readily. The conductivities of the solutions are usually low. The interaction between the solute and the solvent is as a rule weaker than in solutions prepared with ionizing solvents.

Solvents have also been classified on the basis of their affinities for water, into *hydrophilic* (DMSO, acetone, etc.) and *hydrophobic* (CCl_4, hydrocarbons, etc.) solvents. Although this classification is not particularly favourable from either the theoretical or practical point of view, it clearly shows the prominent role attributed to water among solvents.

Further development of the acid–base concept according to the Lowry–Brönsted theory has led to a division of solvents into *protic* and *aprotic*. In an essentially analogous manner, Parker [Pa 62, Pa 65, Pa 69] distinguishes between protic and dipolar aprotic solvents. Among the former are classified solvents with strong hydrogen donor properties, such as fluoroalcohols, alcohols, formamide, ammonia and hydrogen fluoride. Solvents with relative permittivities higher than 15, but containing no labile hydrogen, are termed dipolar aprotic solvents. These include dimethylformamide, dimethyl sulphoxide, hexamethylphosphoramide, acetone, acetonitrile and even nitromethane.

The above sharp division is based on the different effects of the two types of solvent on the rates of certain reactions; in effect, this reflects the influence of the hydrogen bond on the solvent–solute interaction [Pa 61].

The above classification is generally in agreement with Pearson's acid–base concept. Protic solvents are "hard" in nature, and they solvate small anions with strong hydrogen bonds, whereas dipolar aprotic solvents have a "soft" character, and they interact more strongly with the large, polarizable anions.

On the basis of the conductivity of hydrogen chloride, Janz and Danyluck [Ja 60] classified solvents into groups of *levelling* and *differentiating* solvents. The former group includes methanol, for example, and the latter group acetonitrile, dimethyl sulphoxide and nitrobenzene.

According to Parker [Pa 62], the above classification is based actually on the different degrees of the anion-solvating action of solvents. As a consequence of their strong hydrogen bonding ability, levelling solvents strongly solvate primarily small anions, e.g., chloride, and contribute in this way to the dissociation of hydrogen chloride. Differentiating solvents do not form hydrogen bonds, and therefore their anion-solvating effects are small. In these solvents, the extent of dissociation of hydrogen chloride depends on the relative permittivity and on the ability of the solvent to solvate protons.

In analytical chemistry, where non-aqueous solvents are used in the acidimetric and alkalimetric titration of weak acids and bases, solvents are classified as *acidic* and *basic* [Gy 70]. In the former group the proton-donating and in the latter group the proton-accepting property predominates. Accordingly, bases can be titrated in acidic solvents, which readily transfer their protons to the base dissolved in them, and acids can be titrated in basic solvents, which readily accept protons from the acid dissolved in them.

The more acidic a solvent, the weaker may be the base which can be titrated by acids in it, and the more basic a solvent, the weaker are the acids which can be titrated by bases in it. Hence, it is customary to state that the strengths of bases increase in acidic solvents, and the strengths of acids increase in basic solvents.

In the above classification, those solvents which may behave both as acids and as bases (the most typical example being water) are termed *amphiprotic* solvents.

The most important acidic solvent used in analytical chemistry is glacial acetic acid; some important basic solvents are pyridine, ethylenediamine and dimethylformamide.

As regards the internal structures of liquids, solvents can be classified according to whether they are capable of forming hydrogen bonds or not. The molecules in most protic solvents are considerably associated because of hydrogen bond formation. Association of the molecules in a number of aprotic solvents may also occur in the liquid state; however, in these systems it is not hydrogen bond formation that is responsible for the association, but other structural factors, such as oxygen bridges or halogen bridges.

A much more general classification was made possible by the Lewis acid–base theory, which has also proved of great value in coordination chemistry. In accordance with this theory, a distinction is made between donor and acceptor solvents. *Donor solvents* are Lewis bases, which are capable of donating a lone electron pair, whereas *acceptor solvents* are Lewis acids, which are able to accept an electron pair.

Donor solvents react with acceptor molecules and ions:

$$SbCl_5 + D \rightleftharpoons DSbCl_5 \tag{2.1}$$

$$Co^{2+} + 6D \rightleftharpoons CoD_6^{2+} \tag{2.2}$$

Since most metal ions are electron-pair acceptors (Lewis acids), they are solvated by donor solvents.

Acceptor solvents (A) generally react with donor compounds. Since most anions are electron-pair donors, the interaction of such solvent molecules with anions results in solvated anions:

$$KF + A \rightleftharpoons K^+ + AF^- \tag{2.3}$$

Interactions of this type are generally weaker than the solvation of cations with donor solvents.

Donor and acceptor solvents together form the large group of *coordinating* solvents, which are of outstanding importance in chemistry. Solvents entering into only a weak interaction with the solute, for example by Van der Waals forces, are termed *non-coordinating solvents.*

The donor and acceptor solvent classification is not entirely unambiguous either. There are solvents which may act both as donors and as acceptors [Au 53, Gu 68, Ja 49]. Examples are water and several of the alcohols. Their donor function is ensured by the electron pairs of their oxygen atoms, and the acceptor functions by their hydrogen atoms, which establish connections with the electron-pair donor anions by hydrogen bonding.

Characteristics of solvent properties

The properties of a solvent are determined jointly by its general chemical nature and physical properties. These properties are involved in such a complex interaction that it is difficult to establish unambiguously how they contribute individually to the general behaviour of the solvent. Factors of fundamental importance are the temperature range in which a system is in the liquid state, how it may be purified and how its physical properties influence its applicability.

The most important physical properties contributing to the general solvent nature arc the melting point, boiling point, vapour pressure, refractive index, density, specific conductivity, viscosity, heat of vaporization, dipole moment and relative permittivity. The melting and boiling points are the temperature limits between which the solvent can be utilized. The density and viscosity affect the mobilities of the ions in solution, and hence influence their reactivities. A high heat of vaporization indicates strong association of the solvent molecules. In many cases this feature can be better characterized by the Trouton constant, which is the ratio of the heat of vaporization and the boiling point. Specific conductivity can, in general, be regarded as a criterion of the purity of the solvent. In agreement with the Pearson concept of "hard" and "soft" acids and bases, polarizability is also an important factor. Its significance was recognized even prior to the introduction of Pearson's considerations. A high dipole moment is due to electrostatic interactions between the polar solvent and the ions dissolved in it. With an increase in relative permittivity, the charged dissolved species become increasingly separated, and there is an ever smaller probability that they will be found in the form of ion pairs or ion aggregates in the solution.

If solvation is regarded as the interaction of ions and solvent molecules of dipole nature (i.e., if the specific coordination chemical nature of the solvation is disregarded), ion-pair formation can even be described quantitatively. *To a first approximation*, the dependence of the equilibrium constant (K) of ion-pair formation is given, on the basis of the electrostatic model, by the equation

$$K = A \frac{1}{\varepsilon} \frac{z_+ z_-}{r^2}$$

where ε is the relative permittivity of the solvent, z_+ and z_- are the charges of the ions participating in the reaction, r is the distance between their centres and A is a proportionality factor. This equation correctly reveals that association reactions are favoured if the relative permittivity of the solvent is low, the charges of the dissolved ions are high and their radii are small.

It must be stressed that there is a significant difference between the macroscopic (measurable) relative permittivity of a solution and the microscopic relative permittivity of the solvates present in the solution; the latter cannot be measured. Even if this difference is neglected, the above equation holds only if the electrostatic interaction between the ions does indeed predominate, and other (e.g., polarization, coordination) effects do not play a part.

Efforts have also been made to interpret the effect of the solvent on the solubility, on the basis of the electrostatic properties of the solvent. Thus a number of empirical equations have been published to describe the correlation between the relative permittivity (ε) of the solvent and the solubility (S) of the electrolyte. The most frequently used correlation:

$$\log S = A_1 + \frac{A_2}{\varepsilon}$$

where A_1 and A_2 are material constants, is in fact a simplified form of the Born equation [Bo 20], derived for calculation of the solvation free energy.

Walden [Wa 24] introduced the following empirical correlation between the molar fraction (N) of the solute and the relative permittivity (ε) of the solution:

$$3 \log \varepsilon = B + \log N$$

These and analogous equations have proved suitable for the correct description of numerous experimental data.

On the basis of the electrostatic model only, it usually cannot be understood why certain compounds behave in such different ways in solvents whose relative permittivities (dielectric constants) are nearly identical. No explanation can be given if the solvations of ions with the same charge and the same ionic radius differ considerably in a given solvent owing to the covalent part of the interaction between the dissolved ion and the solvent. The subordinate role of the relative permittivity in certain systems is also clearly demonstrated by the following examples. In anhydrous sulphuric acid ($\varepsilon = 85$) perchloric acid is present in almost undissociated form, whereas in acetic anhydride ($\varepsilon = 7$) it is strongly dissociated; in spite of its high relative permittivity ($\varepsilon = 123$), anhydrous HCN is a poor solvent, whereas tributyl phosphate and pyridine, with relatively small ε values (6.8 and 12.3, respectively), are good solvents for many ionic compounds [Gu 68].

With respect to coordination chemistry, the solvating power of donor solvents is of outstanding importance. This depends primarily on the basicity of the donor atom carrying the free electron pair (i.e., on the electron density on the donor atom), but also on the polarizability and steric properties of the molecule, etc.

Donor solvents which contain oxygen or nitrogen donor atoms are particularly suitable for the solvation of cations. Thus, in solvents such as dimethyl sulphoxide, pyridine N-oxide and dimethylformamide, where a negative charge is localized on the oxygen atoms, the cations are dissolved in a strongly solvated form. In similarly polar solvents, but where the negative charge of the dipole is not localized on an appropriate donor atom, e.g., nitromethane, the cation will be slightly solvated.

A more detailed discussion of the possibilities of characterizing the donor strength will be given later.

The properties discussed interact with one another in a complicated manner [see, e.g., ref. A 78]. For instance, the dissociation occurring in the course of dissolution depends not only on the relative permittivities of the solvent, but also on its solvating ability, which is determined by its donor strength [Ma 75].

References

A 78 Abstracts of the IV. International Symposium on Solute-Solute-Solvent Interactions, Vienna 1978.

Au 53 Audrieth, L. F., Kleinberg, J.: Non-Aqueous Solvents. Wiley, New York 1953.

Bo 20 Born, M.: Z. Physik **1**, 45, 221 (1920).

Br 23 Brönsted, J. N.: Rec. Trav. Chim. **42**, 718 (1923).

Br 28 Brönsted, J. N.: Ber. **61**, 2409 (1928).

Br 30 Brönsted, J. N.: Z. angew. Chemie **43**, 229 (1930).

Co 69 Solute–Solvent Interactions (Coetzie, J. F. and Ritchie, C. D., eds.) Dekker, New York 1969.

Gu 68 Gutmann, V.: Coordination Chemistry in Non-Aqueous Solutions. Springer, Vienna–New York 1968.

Gy 70 Gyenes, I.: Titrationen in nichtwässrigen Medien. Akadémiai Kiadó, Budapest 1970.

Ja 49 Jander, G.: Die Chemie in wasserähnlichen Lösungsmitteln. Springer, Berlin 1949.

Ja 60 Janz, G. J., Danyluck, S. S.: Chem. Rev. **60**, 209 (1960).

Ma 75 Mayer, U.: Pure and Applied Chem. **41**, 291 (1975).

Pa 61 Parker, A. J.: J. Chem. Soc. (London) **1961**, 1328.

Pa 62 Parker, A. J.: Quart. Rev. (London) **16**, 162 (1962).

Pa 65 Parker, A. J.: Inter. Sci. Technol. **28**, Aug. (1965).

Pa 69 Parker, A. J.: Chem. Rev. **69**, 1 (1969).

Wa 24 Walden, P.: Elektrochemie nichtwässriger Lösungen. Barth, Leipzig, 1924.

3. SOLVENT–SOLUTE INTERACTIONS

Theoretical approaches

The first evidence for an interaction between the solute and the solvent is the process of dissolution itself. This interaction may differ in character and may take place to various extents, depending on the natures of the reaction components.

For example, the energy necessary for the dissolution of ionic compounds is supplied by the interaction between the ions and the solvent molecules, i.e., by the energy of solvation of the ions.

If one considers the magnitude of the forces holding together the ions in the crystal lattice, it can be seen that the interaction of the solvent and the solute must entail the release of considerable energy (50–100 kcal/mole) if the ions are to pass into solution. The exact magnitude of this energy cannot be forecast on theoretical grounds for the individual solvents, some empirical regularities may serve, however, as a guide.

For an understanding of the various phenomena characteristic of solutions, a quantitative description of the intermolecular forces would be necessary. The possibility of this arises primarily if the non-specific electrostatic interactions are regarded as predominating. Intermolecular forces of such a type between the solvent and the dissolved molecules (ions) are less sensitive to the distance between the particles and to their orientation than are explicit chemical bonds. To a first approximation, the magnitude of these forces can be regarded as independent of the chemical natures of the participating species.

Three approaches are available for a quantitative description of solvation interactions taking into consideration the above points: (1) the **BBB** model; (2) the chemical model; and (3) the Hamiltonian model [Fr 73]. The last model is also suitable for some considerations of specific interactions.

The **BBB** model is a means of macroscopic approximation to the system on an exclusively electrostatic basis; it describes the solvation effects with the aid of classical field theories, primarily electrostatics and hydrodynamics. The nomenclature **BBB**, an abbreviation of the expression “Brass Balls in a Bathtub”, originates from Frank [Fr 65]. The more important classical theories included in this group have led to a result only for dilute solutions; nevertheless, their refinement has continued up to the present [Ab 79, Be 78, Kr 79, Li 79].

One of the simplest means of a quantitative treatment of the interaction between the solvent and solute by the BBB model is to regard the solvent as a dielectric, situated in the field of the ions, in turn considered as electrically charged spheres. If an ion (an rigid sphere) of radius R and charge q is transferred from vacuum to a solvent of dielectric constant ε, the Gibbs free energy change is

$$\Delta G = \frac{q^2}{2R}\left(1 - \frac{1}{\varepsilon}\right)$$

or, for 1 mole,

$$\Delta G = -\frac{N_A z^2 e^2}{2R}\left(1 - \frac{1}{\varepsilon}\right)$$

where e is the electronic charge, z is the charge on the ion and N_A is Avogadro's number. The corresponding enthalpy change is

$$\Delta H = -\frac{N_A z^2 e^2}{2R}\left[1 - \frac{1}{\varepsilon} - \frac{T}{\varepsilon^2}\cdot\frac{\partial \varepsilon}{\partial T}\right]$$

This theory is naturally only a rough approximation, as the solvent in the immediate environment of the ions can by no means be regarded as a continuous dielectric, not even in those solutions in which it is not necessary to reckon with the coordination of the solvent molecules, i.e., with the formation of solvate complexes. A further uncertainty arises in that the ionic radii are not unequivocally defined in solutions. Hence, the above equations permit at best a rough estimation of the solvation energies of ions with various radii in solvents of various relative permittivities. Some examples are given in Table 3.1. Even from these data it emerges that the solvation energies of small ions may be very high, in agreement with the experimental observation that their salts (e.g., those of the lithium ion) will also dissolve in solvents with low relative permittivity. It is further apparent that the solvation energies increase with increase in the relative permittivity.

If one wishes to take into account the fact that polar molecules polarize their immediate environment "more strongly" than the more distant parts of the liquid, then Onsager's homogeneous dielectric must be replaced with a heterogeneous (possibly inhomogeneous) dielectric. According to such a model [Li 77a, b], the dipole is situated at the centre of a spherical cavity of radius a. The cavity is surrounded by a spherical shell, b–a in thickness, with a relative permittivity ε_1, which in turn is embedded in a dielectric of infinite extent and having a relative permittivity ε_0. Calculations of relative permittivities [Li 77, 78] and that part of the cohesion energy of polar liquids which arises from polar interactions have shown that the layer model gives a better approximation than the Onsager model.

Table 3.1. Effect of the relative permittivity (ε) of the solvent on the solvation energies of ions of various radii (R) (calculated values)

R nm	ε					
	2	4	8	16	32	64
	Solvation energy, kcal/mole*					
0.05	166	249	291	311	322	326
0.1	83	124	145	156	161	163
0.15	55	83	97	104	107	109
0.2	42	62	73	78	80	82
0.3	28	42	48	52	54	54
0.4	21	31	36	39	40	41
0.5	17	25	29	31	32	32
1.0	8	12	15	16	16	16

* 1 kcal/mole = 4.184 kJ/mole

If the two charges of the dipole featuring in the layer model are replaced with a single point charge, then the one-layer continuous model relating to electrolyte solutions is obtained. According to this model, the part of the free energy of solvation due to electrostatic interactions is [Ab 78*b*]

$$\Delta G_e^0 = -\frac{z^2}{2}\left\{\frac{1}{\varepsilon_1}\left(\frac{1}{b}-\frac{1}{a}\right)-\frac{1}{\varepsilon_0 b}+\frac{1}{a}\right\}$$

where z is the charge on the ion. From this equation, the part of the solvation entropy due to electrostatic interactions is

$$\Delta S_e^0 = -\frac{z^2}{2}\left\{\frac{1}{\varepsilon_1^2}\left(\frac{\partial \varepsilon_1}{\partial T}\right)_l\left(\frac{1}{b}-\frac{1}{a}\right)-\frac{1}{\varepsilon_0^2}\left(\frac{\partial \varepsilon_0}{\partial T}\right)_p\frac{1}{b}\right\}$$

where p denotes pressure.

Table 3.2 illustrates the applicability of the model in the case of acetone. The ΔG_e^0 values were calculated with the above equation, with the assumption that the ion takes its immediate environment into dielectric saturation, and thus the relative permittivity of the layer is equal to the square of the internal refractive index ($\varepsilon_1 = n^2$). The thickness of the layer ($b-a$) was selected so that, when every ion is taken into consideration, the difference between the experimental and calculated data should be minimal. The applicability of the equation is demonstrated for 16 solvents in Table 3.3.

Data relating to about 150 ion–solvent systems [Ab 78*c*] have shown that, for monovalent ions, the free energy of solvation can be obtained to a good

Table 3.2. Proportion of solvation free energy arising from electrostatic interactions in acetone at 298 K [Ab 78a]

$\varepsilon_1 = n^2 = 1.95$; $(b-a) = 2.35$ nm

Ion	a nm	ΔG_e^0	
		experimental	calculated
Na^+	0.105	−99.4	−99.7
K^+	0.133	−82.0	−81.8
Rb^+	0.143	−76.9	−76.9
Cs^+	0.166	−68.8	−67.9
Me_4N^+	0.258	−46.8	−47.0
Et_4N^+	0.310	−39.9	−40.2
Cl^-	0.181	−62.7	−63.2
Br^-	0.195	−59.0	−59.4
I^-	0.22	−53.2	−53.7
ClO_4^-	0.245	−50.2	−49.1

Table 3.3. Applicability of the Onsager equation modified by Liszi [Ab 78c]

Solvent	$\varepsilon^{(a)}$	$n^{(b)}$	$\sigma^{(c)}$	$\tau^{(d)}$
1,1-Dichloroethane	9.90	10	0.48	+0.04
1,2-Dichloroethane	10.23	10	0.54	+0.09
Tetrahydrofuran	7.40	5	2.67	−0.20
1,2-Dimethoxyethane	7.30	5	2.41	−0.26
Ammonia	16.90	7	1.33	−0.22
Acetone	20.49	10	0.55	0.00
Acetonitrile	36.02	10	0.77	+0.15
Nitromethane	37.0	8	2.17	+0.22
N,N-Dimethylformamide	36.71	10	1.32	+0.15
Dimethyl sulphoxide	46.68	10	1.47	+0.23
N-Methylformamide	182.4	9	1.17	+0.06
Formamide	109.5	7	1.68	+0.11
1-Propanol	20.45	10	1.73	−0.05
Ethanol	24.33	10	1.51	−0.10
Methanol	32.64	10	1.75	−0.12
Water	78.36	10	2.14	+0.23

(a) Static relative permittivity of the solvent at 25 °C.

(b) Number of ions examined in the given solvent.

(c) $\sigma = \left\{ [\Delta G_e^0\,(\text{calcd.}) - \Delta G_e^0\,(\text{exptl.})]^2 \frac{1}{n-1} \right\}^{1/2}$ where n is the number of ions.

(d) $\tau = \Delta G_e^0$ (calcd., average) $- \Delta G_e^0$ (exptl., average). τ shows the symmetrical difference between the calculated and experimental values.

approximation if the calculations are made with the substitution $\varepsilon_1=2$ and $(b-a)=r$, where r is the radius of the solvent molecule.

Entropy calculations with the above equation show that a two-layer model is required for "structured" solvents, e.g., water and alcohols [Ab 78c]. With the much less polar, "structureless" solvents, the equation can be regarded as a good approximation.

If the dependence of the relative permittivity of the solvent on the electric field strength of the ions is also taken into account, then other thermodynamic parameters of electrolyte solutions (activity coefficient, heat of dilution, partial molar enthalpy content of the solute, etc.) can likewise be calculated in better agreement with the experimental data. Although the introduction of the field-dependent relative permittivity into the ion–ion and ion–solvent interactions is accompanied by very great mathematical difficulties, the problem can be solved successfully by employing various approximations.

One of the greatest difficulties lies in the fact that the superposition laws of linear electrostatics, by which the overall effect of a number of charges or dipoles may be calculated, lose their validity if the relative permittivity is dependent on the field strength; at present, superposition laws with which non-linear electrostatic problems might be treated are not known. Hence, solutions presenting an electrostatic multibody problem because of the many ions dissolved in them can only be modelled in such a way that the ion–solvent interaction is simplified to the interactions between a single ion and its environment; or the interaction of one ion with the others should be reduced to the effects between this ion and a diffuse ionic cloud (Debye–Hückel theory), or between this ion and a regular ionic lattice surrounding it (lattice model). In both cases the correlations are so complicated that the spatial arrangement of the ions is either postulated by theories, or a computer solution is required (Monte Carlo method).

Various approaches are also employed for the modelling of the solvent. Physically, the most reliable model appears to be that in which in the interaction of the ion and the surrounding dipoles the interactions between the dipoles are also taken into consideration as spatial dispersion interactions [Do 72, Do 73, Do 75, Ho 76]. The local relative permittivity in the space surrounding the ion will then be governed by the *relative* magnitude of the field of the ion, referred to the forces acting between the dipoles and determining the characteristic structure of the solvent. With the aid of this model, for example, the hydration energies of alkali metal and halide ions have been calculated and are reported in Table 3.4.

Table 3.4. Calculated hydration energies, kcal/mole [Do 72, 73]

Ion	Li^+	Na^+	K^+	Rb^+	Cs^+	F^-	Cl^-	Br^-	I^-
ΔH_{hydr}	127	101	81	76	67	100	75	68	61

The calculated hydration energies of the salts formed from the individual ions agree with the measured values within 4%. Since the Born calculations, which take the solvent into account with its static, macroscopic relative permittivity, always result in hydration energies substantially larger than those actually observed, the necessity to take the local relative permittivity into consideration is convincingly confirmed by these data.

In the layer with decreased relative permittivity surrounding the ions, the free energy of the solvent is lower than in the absence of the electric field of the ions. The approach of the ions towards one another requires the mutual inter-penetration of the solvate spheres, i.e., the release of a certain amount of solvent from the solvate sphere of the ions. This process needs work, and this work appears as a repulsive force between the ions. (This effect lends stability to electrolyte solutions, for in the absence of such repulsive forces, attraction between the charges would favour the precipitation of the solid salts.) By taking into account such repulsive forces, it was possible to interpret the positive deviation of the average activity coefficients of the ions from the Debye–Hückel limiting law (hypernetted chain equations, HNC, calculation by the Monte Carlo method [Ra 68, Ra 70].

More recently, calculation of the average activity coefficients of most alkali metal halides, alkaline earth metal halides and alkali metal sulphates has succeeded with the assumption of Ruff's regular lattice-like arrangement instead of the Debye–Hückel spherically symmetrical charge distribution, and with the introduction of the space-average of the locally varying relative permittivity, as an average relative permittivity, but dependent on the interionic separation distance (i.e., the concentration). The calculated data are in very good agreement with the measured values over a wide concentration range [Ru 77]. Apart from the approximations mentioned, this theory does not employ fitting parameters. The calculated average relative permittivities are also in good agreement with the experimental values. As an example, the data relating to an aqueous solution of sodium chloride are presented in Table 3.5. The model may also be used for the characterization of non-aqueous solutions.

It is obvious that the major part of the solvation energy is provided by electrostatic ion–dipole interactions only in those systems where the role of the coordinate bonds between the solvent molecules and the dissolved ion is a subordinate one. It follows that it is primarily in the case of anions where agreement can be expected between the experimental solvation energies and those calculated by taking into account exclusively the ion–dipole forces. However, hydrogen bonding may occur here also, giving rise to an increased ion–solvent interaction.

It emerges from the foregoing that good solvents for ionic compounds are those which have large relative permittivities and dipole moments, and which are able to link to cations by coordinate bonds, and to anions by hydrogen bonds.

The chemical models differ from the BBB models primarily in that they distinguish between the solvent situated in the immediate environment of the dissolved molecules and the bulk of the solvent. Models assuming stoichiometric solvate complexes are a limiting case of this concept. Such solvates must naturally be treated by taking into consideration the regularities of coordination chemistry.

Table 3.5. Average relative permittivity and mean activity coefficients of aqueous NaCl solutions, calculated according to Ruff's lattice-like model [Ru 77]

m mole/kg	c mole/dm^3	$\hat{\varepsilon}_{d,\,calc.}$	$\hat{\varepsilon}_{d,\,obs.}$	$\gamma_{calc.}$	$\gamma_{obs.}$
0.1	0.100	77.7	77.7	0.758	0.778
0.2	0.199	76.6	76.7	0.709	0.734
0.3	0.279	75.4	75.6	0.678	0.710
0.5	0.494	72.9	73.2	0.641	0.682
1.0	0.979	66.5	66.7	0.600	0.658
2.0	1.921	55.7	56.5	0.599	0.671
3.0	2.826	46.7	48.1	0.655	0.720
4.0	3.695	39.4	40.2	0.768	0.792
5.0	4.528	33.5	33.9	0.957	—

However, the chemical model may be employed if the solvent spheres surrounding the dissolved ions can be differentiated from the bulk of the solution, although the stoichiometric conditions are unknown. Such are the models of Frank [Fr 45, Fr 57] and Gurney [Gu 62]. These theories have proved successful for the description of the experimentally measurable properties of solutions in only a few cases; nevertheless, they have the merit of having inspired much experimental work.

A more developed variant of the chemical models is the Friedman model, which takes account of the overlapping of the solvate sheaths in the association of solvated ions, and also the solvent displaced from the solvent sheath in the course of this overlapping [Fr 71, Fr 73a, Fr 73b, Ra 71a, Ra 72]. This model permitted good descriptions of the activity coefficient changes, the heats of dilution and the volume changes that occur as a result of dilution in ionic solutions.

One may regard the *Hamiltonian approximations* as the most modern but, because of the complexity of the systems, these are applicable to only a very limited extent even today. A restriction to their successful use is that many simplifications are necessary. Their discussion lies beyond the scope of this book; accordingly, only a few relevant references are made to the literature [Zw 65, Ku 66, Go 65, He 70, Ra 71b, St 72].

Solvation, solvation number

As seen above, as a result of the interaction between a solvent and a solute, the solvent molecules become linked to the dissolved molecules or ions. This phenomenon is *solvation*, and the resulting species containing solvent molecules are *solvates*.

Depending on the reactants, the linkage between the solvent and the solute may be of different kinds (electrostatic, coordination, hydrogen bond formation, etc.) and thus of different strengths, and hence solvates include very varied formations. As a consequence of the subject of this book, but also because of their theoretical and practical importance, we deal here almost exclusively with solvated ions.

Simple ions may differ considerably in their electronic structures, charges and radii, and even larger differences are exhibited by solvent molecules representing different types of chemicals. Hence, the solvated ions may also be very different as regards the number of solvent molecules bound to the ion (the *solvation number*), the forces giving rise to the binding and the resulting physical and chemical properties (spatial requirement, mobility, reaction rate, etc.).

Since ions in solutions always react in the solvated form and, further, since most of these reactions are accompanied by a change in the solvate sheath (release, replacement or even uptake of solvent molecules), a knowledge of the compositions and structures of the solvated ions is one of the fundamental conditions for an understanding of all chemical processes in solution.

There is fairly extensive contradiction and uncertainty in the literature in connection with the solvation number [De 69]. The introduction of the concept of the *primary solvation number* is associated with Bockris [Bo 49]. According to this definition, the solvation number is the number of solvent molecules which are so strongly attached to the dissolved ion that they lose their degree of translational freedom and move together with the dissolved ion in the course of the Brownian movement.

We speak of *secondary solvation* when we take into account the electrostatic interaction between the solvated ion and the solvent molecules surrounding it.

The primary solvation number may be determined by means of various mutually independent methods. However, it must be noted that the different methods do not yield identical values in every case. Padova [Pa 63b, Pa 64a] calculated the solvation numbers (n) of certain electrolytes from the *molar volumes*. He used the assumption that the solute ion gives rise to such a strong electrostatic field that the solvate sheath consisting of solvent molecules bound in the first coordination sphere becomes incompressible. Thus, the molar volume V_s (cm^3/mole), of the solvated electrolyte can be described by the equation

$$V_s = \Phi_v + n_s \cdot \frac{M_0}{d_0}$$

where Φ_v is the apparent molar volume of the solvent, M_0 is its molecular weight and d_0 is its density. Hence, this is suitable for the determination of the solvation number, n_s.

If we assume that the volume of the solvated ion does not decrease as a result of pressure, the molar volume of the solvated ion can be calculated by comparing the *compressibility* of the electrolyte solution with that of the solvent:

$$V_s = \frac{1000(\beta_0 - \beta)}{c\beta_0}$$

where β_0 and β are the compressibilities of the solvent and the solution, respectively, and c is the concentration of the solution. Hence,

$$n_s = \frac{1000 d_0(\beta_0 - \beta)}{c\beta_0 M_0} = \frac{\Phi_v}{V_0}$$

where V_0 is the molar volume of the solvent, and the solvation number can be calculated from

$$n_s = \frac{\Phi_k d_0}{M_0 \beta_0}$$

where d_0 and M_0 are the density and molecular weight, respectively, of the solvent, and Φ_k is the apparent molar compressibility of the electrolyte. In infinitely dilute solutions the equation is modified to

$$n_s = \frac{\Phi_k^0 d_0}{M_0 \beta_0}$$

Numerous investigations have been carried out with this procedure in various alcohols [Ro 67, Ro 68] and in formamide [Mi 64].

Viscosity measurement methods have also been employed for the determination of the solvation number. The viscosity of an electrolyte solution is described by the Jones–Dole equation:

$$\frac{\eta}{\eta_0} = 1 + Ac^{1/2} + Bc$$

where η and η_0 are the viscosities of the solution and the solvent, respectively, A is a calculable constant, which gives the contribution of the inter-ionic forces [Fa 29, Fa 32], B is an empirical constant and c is the molar concentration. It has been demonstrated [Pa 63a, Fe 66] that in solutions more concentrated than 0.1 M the value of A can be neglected, and B is equal to the product $K \cdot V_s$ [St 65], where the

Table 3.6. Solvation numbers determined by various procedures, I

Ion	Viscosity measurements		Mobility measurements	
	Methanol	Formamide	Methanol	Formamide
Cl^-	1.5	1.5	1.5	—
Br^-	1.0	1.0	1.0	1.0
Li^+		4.0		5.4
Na^+		4.0		4.0
K^+	6.0	2.0	4.0	2.5
Rb^+		1.5		2.3
Cs^+		1.5		1.9
NH_4^+		1.0		2.0

value of K (assuming the ions to be spherical) is 2.5. The solvation number, n_s, can then be calculated:

$$n_s = \frac{0.4B - \Phi_v}{V_0}$$

The solvation numbers of ions can also be determined by means of *conductivity measurements*. The method is based on the Stokes equation [Fr 66]:

$$\lambda_i^0 = \frac{z_i F^2}{6\pi\eta_0 r_s}$$

where λ_i^0 is the equivalent conductivity of the infinitely dilute solution, z_i and r_s are the charge and the radius, respectively, of the solvated ion, F is the Faraday constant and η_0 is the viscosity of the solvent. The radius of the solvated ion can be calculated from the equation. With a knowledge of this, and with the aid of the crystal radius (r_c) of the ion, the volume (V_s^s) of the solvent sheath can be calculated from

$$V_s^s = \frac{4\pi}{3}(r_s^3 - r_c^3)$$

The solvation number (n_s) is obtained from this via the equation

$$n_s = \frac{V_s^s}{V_0}$$

A limitation of the above procedure is that it is convenient for the study only of ions of moderate size. Efforts have been made to introduce various empirical [Ni 59, Ro 59] and theoretical [Zw 63, Pa 64a] corrections, with a view to making the method more generally applicable, and in many cases these have been successful.

Perhaps the best method for the determination of the solvation numbers of cations in their non-aqueous solutions is *nuclear magnetic resonance (NMR) spectroscopy* [Sw 62]. The procedure is based on the fact that under optimal conditions the NMR absorption band of the coordinated solvent molecules separates from that of the free solvent, and the ratio of the magnitudes of the areas under the two bands provides direct information on the solvation number of the cation. The conditions for the separation of the two NMR signals are that the rate of exchange between the coordinated solvent and the free solvent molecules should be comparatively low, and the spin relaxation time should be short. Successful studies in dimethyl sulphoxide [Th 66], dimethylformamide [Fr 67a, Fr 67b, Ma 67] and methanol [Al 69b] have been reported. Jackson *et al.* [Ja 60] have developed a method that can also be used in cases where the proton resonance signals of the coordinated solvent molecules and the free solvent are so close to one another that they do not separate [Ch 68, Al 69a].

Robinson and Stokes [Ro 59] calculated the solvation numbers of electrolytes from the difference between the experimentally determined *activity coefficients* and those calculated on the basis of the Debye–Hückel equation. Their method was

Table 3.7. Solvation numbers determined by various procedures, II

Electrolyte	Compressibility measurements		According to Robinson–Stokes	
	Methanol	Ethanol	Methanol	Ethanol
NaCl	4.7	—	3.7	—
NaBr	5.6	2.9	3.2	3.3
NaI	6.2	3.2	2.2	—
KBr	5.2	—	2.7	2.5

modified by Glueckauf [Gl 55]. Longhi *et al.* [Lo 79] used e.m.f. measurements for the determination of the solvation number of electrolytes. It must be noted, however, that these procedures suffer from the error that the solvation number is regarded as independent of the concentration of the solution, which is in contradiction with the experimental facts.

In certain systems, infrared and Raman spectroscopy have also been found suitable for the determination of the solvation number of ions. If the vibrational band of the free solvent molecule is separated from the band of the coordinated solvent molecule, the concentrations of the free and bound solvent can be calculated from the magnitudes of the areas under the bands. With a knowledge of the total amount of solute, the solvation number of the dissolved ion is obtained directly from the above data. By means of such examinations it has been established that in acetone the lithium ion is capable of coordinating a maximum of four solvent molecules.

Other methods, e.g. electron spin echo studies [Na 82a, b] have led also to the determination of solvation numbers.

Solvation numbers obtained *by different methods* for a number of systems in various solvents are listed in Tables 3.6, 3.7 and 3.8. It can be seen that many of the data are in agreement, but at least as many are contradictory.

Table 3.8. Solvation numbers determined by means of mobility measurement

Solvent	Ions					
	Li^+	Na^+	K^+	Cl^-	Br^-	I^-
Acetone	2.9	2.6	2.0	1.0	1.0	—
Ethanol	5.0	4.0	3.0	2.0	2.0	1.0
Methanol	5.0	5.0	4.0	1.5	1.0	1.0
DMF	3.2	3.0	2.0	0.5	0.5	0.5
DMSO	4.3	2.3	2.4	0.6	0.5	0.4

Solvation with donor and acceptor solvents

Donor solvents containing electron-pair donor atoms are suitable for the solvation of electron-pair acceptors, and acceptor solvents are able to solvate electron-pair donor ions or molecules. Donor solvents therefore react primarily with cations and with Lewis acids in general, whereas acceptor solvents react with anions and ligands and with Lewis bases in general [Gu 68].

In the course of the dissolution of salts a donor solvent is thus linked *via* its free electron pair to the positive part (electron-pair acceptor) of the substance to be dissolved (nucleophilic attack):

$$D + M—X \rightleftharpoons D \longrightarrow M\overset{\frown}{—}X \rightleftharpoons D \longrightarrow M^+X^-$$

By increasing the electron density on the molecule, the formation of the coordinate bond D → M promotes the polarization of the M—X bond. If the D—M interaction is so strong that it results in the transfer of an electron from M to X, then ionization of the M—X molecule occurs.

An acceptor solvent is linked to the negative electron-pair donor part of the substance to be dissolved. Most acceptor solvents are capable of forming hydrogen bonds; the above acceptor interaction is, as a rule, the formation of a hydrogen-bonded association, in which the bridging proton plays the part of the acceptor centre:

$$\mathrm{M{-}X + A \rightleftharpoons M{-}X{\longrightarrow}A \rightleftharpoons M^+X^-{\longrightarrow}A}$$

The result is polarization to an extent depending on the strength of the interaction, in the extreme case ionization. Thus, in a medium of high relative permittivity both reactions may lead to complete dissociation:

$$\mathrm{D{\longrightarrow}M^+X^- \rightleftharpoons D{\longrightarrow}M^+ + X^-}$$

$$\mathrm{M^+X^-{\longrightarrow}A \rightleftharpoons M^+ + X^-{\longrightarrow}A}$$

The former can be regarded as a ligand-exchange reaction and the latter as an acceptor-exchange reaction.

It can be seen that the effect of solvation is manifested in two steps:

(1) Formation of a solvated ion pair;
(2) Dissociation of the ion pair.

Equilibrium constants can be written separately for the two processes:

$$K_{\mathrm{ion}} = \frac{[\mathrm{DM^+X^-}]}{[\mathrm{D}][\mathrm{MX}]} \quad \text{or} \quad K_{\mathrm{ion}} = \frac{[\mathrm{M^+X^-A}]}{[\mathrm{MX}][\mathrm{A}]}$$

$$K_{\mathrm{diss.}} = \frac{[\mathrm{DM^+}][\mathrm{X^-}]}{[\mathrm{DM^+X^-}]} \quad \text{or} \quad K_{\mathrm{diss.}} = \frac{[\mathrm{M^+}][\mathrm{X^-A}]}{[\mathrm{M^+X^-A}]}$$

This also reflects the fact that the solvent–solute interaction depends on the various properties of the solvent. The extent of solvation is determined by the strength of the interaction between a donor solvent and the cation [Gu 69a, Gu 69b, Po 79], or between an acceptor solvent and the anion [Gu 76, Ma 70, Ma 75]. The extent of dissociation is a function of the dielectric properties of the solvent.

In a system with a given relative permittivity, ion pair formation will be favoured by an increase in the solvating ability (donor strength or acceptor strength) of the solvent, and by an increase in the polarizability and a decrease in the strength of the M—X bond.

References

Ab 78a Abraham, M., Liszi, J., Mészáros, L.: Unpublished results.

Ab 78b Abraham, M. H., Liszi, J.: J. Chem. Soc. Faraday Trans., I, **74**, 1604 (1978).

Ab 78c Abraham, M., Liszi, J.: J. Chem. Soc. Faraday Trans. I, **74**, 2858 (1978).

Ab 79 Abbaud, J.-L. M., Taft, R. W.: J. Phys. Chem. **83**, 412 (1979).

Al 69a Alger, T. D.: J. Am. Chem. Soc. **91**, 2220 (1969).

Al 69b Al-Baldawi, S. A., Gough, T. E.: Canad. J. Chem. **47**, 1417 (1969).

Bo 49 Bockris, J. O. M.: Quart. Rev. (London) **3**, 173 (1949).

Be 78 Bennetto, H. P., Spitzer, J. J.: J. Chem. Soc. Faraday Trans. I, **74**, 2385 (1978).

Ch 68 Chmelinick, A. M., Fiat, D.: J. Chem. Phys. **49**, 2101 (1968).

De 69 Desnoyers, J. E., Jolicoeur, C.: In: Modern Aspects of Electrochemistry, Vol. 5 (Bockris, J. O. M., Conway, B. A., eds.). Butterworths, London 1969.

Do 72 Dogonadze, R. R., Kornyshev, A. A.: Phys. Stat. Sol., **53b**, 439 (1972).

Do 73 Dogonadze, R. R., Kornyshev, A. A., Kuznetsov, A. M.: Theor. Math. Phys. **15**, 127 (1973).

Do 75 Dogonadze, R. R., Kuznetsov, A. M.: In: Progress in Surface Science, Vol. 6. Pergamon Press, Oxford, 1975.

Fa 29 Falkenhangen, H., Dole, M.: Physik. Z. **30**, 611 (1929).

Fa 32 Falkenhangen, H., Vernon, E. L.: Physik. Z. **33**, 140 (1932).

Fe 66 Feakins, D., Lawrence, K. G.: J. Chem. Soc. **A**, 212 (1966).

Fr 45 Frank, H. S., Evans, M.: J. Chem. Phys. **13**, 506 (1945).

Fr 57 Frank, H. S., Wen, W. Y.: Discuss. Faraday Soc. **24**, 113 (1957).

Fr 66 Frank, H. S.: In: Chemical Physics of Ionic Solutions (Conway, B. A., Barradas, R. G., eds.). Wiley, New York, 1966.

Fr 67a Fratiello, A., Schuster, R.: J. Phys. Chem. **71**, 1948 (1967).

Fr 67b Fratiello, A., Miller, D. Schuster, R.: Mol. Phys. **12**, 111 (1967).

Fr 71 Friedman, H. L.: In: Modern Aspects of Electrochemistry, Vol. 6 (Bockris, J. O. M., Conway, B. A., eds.). Plenum Press, New York 1971.

Fr 73a Friedman, H. L.: Chemistry in Britain **9**, 300 (1973).

Fr 73b Friedman, H. L., Krishnan, C. V., Jolicoeur, C.: Ann. N. Y. Acad. Sci. **204**, 79 (1973).

Gl 55 Glueckauf, E.: Trans. Faraday Soc. **51**, 1235 (1955).

Go 65 Golden, S., Guttman, C.: J. Chem. Phys. **43**, 1894 (1965).

Gu 62 Gurney, R. W.: Ionic Processes in Solution. Dover, New York 1962.

Gu 68 Gutmann, V.: Coordination Chemistry in Non-Aqueous Solutions. Springer, Vienna–New York 1968.

Gu 69a Gutmann, V., Mayer, U.: Monatsh. Chem. **100**, 2048 (1969).

Gu 69b Gutmann, V.: Chimia **23**, 285 (1969).

Gu 76 Gutmann, V.: Electrochim. Acta **21**, 661 (1976).

He 72 Herman, P. T., Alder, B. J.: J. Chem. Phys. **56**, 987 (1972).

Ho 76 Holub, K., Kornyshev, A. A.: Z. Naturforsch. **31a**, 1601 (1976).

Ja 60 Jackson, A., Lemons, J., Taube, H.: J. Chem. Phys. **32**, 533 (1960).

Kr 79 Kruris, P., Poppe, B. E.: Can. J. Chem. **57**, 538 (1979).

Ku 66 Kubo, R.: Rep. Prog. Phys. **29**, 255 (1966).

Li 77a Liszi, J.: Acta Chim. Acad. Sci. Hung. **92**, 409 (1977).

Li 77b Liszi, J., Mészáros, L.: Acta Chim. Acad. Sci. Hung. **93**, 237 (1977); Magyar Kémiai Folyóirat **82**, 546 (1976).

Li 79 Liszi, J., Ruff, I., Szabó, Z. G.: Acta Chim. Acad. Sci. Hung. **100**, 359 (1979).

Lo 79 Longhi, P., Mussini, T., Riva, A., Rondini, S.: Annali di Chimica **69**, 91 (1979).

Ma 67 Matwiyoff, N. A., Movius, W. G.: J. Am. Chem. Soc., **89,** 6077 (1967).
Ma 70 Mayer, U., Gutmann, V.: Monatsh. Chem. **101,** 912 (1970).
Ma 75 Mayer, U.: Pure Appl. Chem. **41,** 291 (1975).
Mi 64 Mikhailov, I. G., Rozina, M. V., Shutilov, V. A.: Akust. Zh. **10,** 213 (1964).
Na 82a Narayan, P. A., Suryanarayana, D., Kevan, L.: J. Am. Chem. Soc., **104**, 3552 (1982).
Na 82b Narayan, P. A., Suryanarayana, D., Kevan, L.: J. Phys. Chem., **86**, 2729 (1982).
Ni 59 Nightingale, E. R.: J. Phys. Chem. **63,** 1381 (1959).
Pa 63a Padova, J.: J. Chem. Phys. **38,** 2635 (1963).
Pa 63b Padova, J.: J. Chem. Phys. **39,** 2599 (1963).
Pa 64a Padova, J.: J. Chem. Phys. **40,** 691 (1964).
Pa 64b Passeron, E. J.: J. Phys. Chem. **68,** 2728 (1964).
Po 79 Popov, A. I.: Pure Appl. Chem. **51,** 101 (1979).
Ra 68 Rasiah, J. C., Friedman, H. L.: J. Phys. Chem. **72,** 3352 (1968).
Ra 70 Rasiah, J. C.: J. Chem. Phys. **52,** 704 (1970).
Ra 71a Ramanathan, P. S., Friedman, H. L.: J. Chem. Phys. **54,** 1086 (1971).
Ra 71b Raman, A., Stillinger, F. H.: J. Chem. Phys. **55,** 3336 (1971).
Ra 72 Ramanathan, P. S., Krishnan, C. V., Friedman, H. L.: J. Solution Chem. **1,** 237 (1972).
Ro 59 Robinson, R. A., Stokes, R. H.: Electrolyte Solutions. Academic Press, New York 1959.
Ro 67 Roshchina, G. P., Kaurova, A. S., Sharapova, S.: Ukr. Fiz. Zh. **12,** 93 (1967).
Ro 68 Roshchina, G. P., Kaurova, A. S., Kosheleva, I. D.: Zh. Strukt. Khim. **9,** 3 (1968).
Ru 77 Ruff, I.: J. Chem. Soc. Faraday Trans. II, **73,** 1858 (1977).
St 65 Stokes, R. H., Mills, R.: Viscosity of Electrolytes and Related Properties. Pergamon Press, New York 1965.
St 72 Stillinger, F. H., Rahman, A.: J. Chem. Phys. **57,** 1281 (1972).
Sw 62 Swinehart, J. H., Taube, H.: J. Chem. Phys. **37,** 1579 (1962).
Th 66 Thomas, S., Reynolds, W. L.: J. Chem. Phys. **44,** 3148 (1966).
Zw 63 Zwanzig, R.: J. Chem. Phys. **38,** 1603, 1605 (1963).
Zw 65 Zwanzig, R.: A. Rev. Phys. Chem. **16,** 67 (1965).
Zw 70 Zwanzig, R.: J. Chem. Phys. **52,** 3625 (1970).

4. DONOR–ACCEPTOR INTERACTIONS. THE DONOR AND ACCEPTOR STRENGTHS OF SOLVENTS

Empirical solvent strength scales

The interactions between the solvent and solute, as discussed above, are the result of a number of different specific (coordination, hydrogen bonding) and non-specific (electrostatic) factors; therefore, it is not possible to find a single *physical* parameter characterizing the solvent, which in itself could rationalize the solvation process. Accordingly, it was necessary to introduce *empirical* parameters serving to characterize the solvent effect.

Empirical solvent strength scales are all based on the assumption that solvents have a *solute-independent property* characterizing their solvation strength. This is, in strict sense, never true [Sn 82], nevertheless surprisingly good correlations have been found between solvent dependent properties and empirical solvation parameters in a vast number of systems [Di 63, Gu 68, Ko 57, La 82, Ma 75 etc.].

Research workers investigating the solvent effect selected model systems with some well measurable property (e.g., light absorption in the UV, visible or IR spectrum, heat of formation, an NMR, Mössbauer or NQR parameter, the redox potential, reaction rate, etc.) which changes appreciably due to the effect of the solvent. Hence, these experimentally measurable data, characteristic of the extent of the interaction between the solvent and the solute, may serve to categorize the solvating powers of solvents. Of course, solvent scales obtained in this way can be compared with one another only if the solvation process in the different model systems is governed by analogous factors.

However, the aim of the research was the introduction of solvent strength scales with as general a validity as possible. Research workers therefore strived to use simple systems in which, if possible, the variation of the experimentally measured parameter is determined by a single property of the solvent. Since the most general of the specific solvent–solute interactions is the Lewis acid–base (or, in the terms of coordination chemistry, the donor–acceptor) interaction, solvent strength scales of general validity were provided by model systems with some experimental parameter that exhibits a variation dependent only on the donor or only on the acceptor properties of the solvent. As many solvents may be both a donor and an acceptor, it was necessary to select the solute so that this should act either as a donor or as an acceptor with regard to the solvent. The better a model corresponded to the above conditions, the more generally valid were the resulting donor or acceptor strength scales. On the other hand, those solvent scales in which the course of the basic reaction is influenced by both the donor and the acceptor strengths of the solvent are of limited validity.

However, donor–acceptor interactions are affected not only by the Lewis acid and base strengths, but also by other, steric and electron structural, factors. Thus, even in systems where either solely the donor or the acceptor property of the solvent is manifested, solvents with different space requirements may interact to different extents because of the steric properties of the reference solute; and a reference acceptor with a tendency for dative π-bonding (back-coordination) will interact more strongly with π-acceptor solvent molecules (e.g., acetonitrile) than would be expected from their basicity. The solvent donicity investigations by Burger *et al.* [Bu 71, 74] with transition metal complex reference acceptor model systems have clearly shown the great extent to which such secondary effects may distort the solvent scale.

Even when the above limitations are taken into consideration, the empirical donor strength scales will remain necessary for characterization of the solvent effect as long as this cannot be described with a theoretical model taking all factors into account. However, it will require a long period before the latter can be attained.

In the following sections the various scales characterizing the donor and acceptor strengths of solvents will be presented.

Characterization of the donor strength on the basis of the ionization energy

Briegleb [Br 61] characterized the donor strengths of solvents by means of their ionization energies. The lower this value, the stronger is the donor property ascribed to the solvent. The resulting donor strength sequence, however, is difficult to reconcile with experience. A number of molecules with different donor properties have very similar ionization energies. For example, the ionization energies of the strong donor ammonia ($I = 10.2$ eV) and the weak donor ethyl acetate ($I = 10.1$ eV) barely differ.

Characterization of the donor strength on the basis of the crystal field splitting

If solvation with donor solvents is regarded as regular complex formation, the experimentally determined stabilities of the solvate complexes formed with a given acceptor can serve as measures of the donor strengths of the solvents.

In most donor solvents, the nickel ion forms solvate complexes with octahedral symmetry. According to the ligand-field theory, the energy levels of the d orbitals of the nickel are split into e_g and t_{2g} orbitals. The magnitude of this splitting depends on the stability of the complex and can be measured spectrophotometrically in the visible spectral region. The value obtained (10 *Dq*) may therefore be regarded as a measure of the donor strength of the solvent [Wa 72].

This characterization has the defect that it can be employed for the comparison only of those solvents which form solvates with analogous compositions and symmetries. It must be considered that in solvents with smaller relative permittivity the solvent molecules do not occupy all six coordination sites of the nickel ion, and anions can also be present in the coordination sphere. Now, the magnitude of 10 *Dq* is determined by the field strength of all of the ligands bound to the nickel. Hence, particularly in solvents with lower donor strengths and relative permittivities, as a result of the effect of the coordinated anions, this method gives much higher values than the actual donor strengths of the solvents.

The *Dq* values measured for the nickel(II) ion in a number of solvents are listed in Table 4.1. The justification of the above critical considerations is well reflected in the

Table 4.1. Crystal field splitting (*Dq*) values of the nickel(II) ion in various solvents

Solvent	*Dq*, cm^{-1}
NH_3	1080
H_2O	860
$(CH_3)_2SO$	773
CH_3OH	850
C_5H_5N	1000
CH_3CN	1026
$CH_3CON(CH_3)_2$	769
$CH_3CONH(CH_3)$	752
C_6H_5CN	970
$HCON(CH_3)_2$	850
$HCONH(CH_3)$	838
CH_3NH_2	993
$C_2H_5NH_2$	987

tabulated data. According to this donor strength scale, for instance, solvents such as dimethyl sulphoxide, dimethylformamide and even water itself, which chemical experience and well tried methods for donor strength determination have shown to have outstanding solvating powers, would be categorized as solvents of low donor strength, whereas acetonitrile, with a much weaker solvating power, appears as a solvent with one of the highest donor strengths, exceeding even that of pyridine..

Characterization of the donor strength on the basis of the heat of solvation

Since the heat of reaction accompanying the solvation process is characteristic of the strength of the solvation in the case of a given acceptor and various donor solvents, calorimetric measurement of the heat of solvation provides a simple possibility for the characterization of the donor strength [Li 60].

The donor strength can therefore be defined as the $-\Delta H$ or, possibly, the $-\Delta G$ value of the interaction between a given reference acceptor and various donors, which can be determined experimentally in an inert solvent [Bo 66, Dr 62a, Gu 66, Li 60, Pe 63].

It appeared obvious to select an acceptor which has only one free coordination site, and with which, therefore, only a 1:1 complex may be formed. In this way the reaction can be well observed and controlled, yet this also means a considerable simplification of reality. Nevertheless, this procedure is very helpful as it provides quantitative data about the donor strengths of solvents.

The first such relative donor strength scale was reported by Lindqvist and Zackrisson [Li 60]. $SbCl_5$ and $SnCl_4$ were employed as acceptors, and the $-\Delta H$ value of the solvation was measured calorimetrically.

A study was also made of the solvation of other acceptor molecules. The $-\Delta H_{DSbCl_5}$ values obtained with antimony(V) chloride were compared with the $-\Delta H_{DA}$ values obtained with other acceptors (where A = antimony(III) chloride, antimony(III) bromide, trimethylin(IV) chloride, phenol and iodine). This work revealed that there is a proportionality between the $-\Delta H_{DSbCl_5}$ and the $-\Delta H_{DA}$ values, and it has become the basis of the currently most widespread and successful donor strength scale: the Gutmann donicity scale [Gu 66, 67, 68, 76].

More recently, Arnett [Ar 74] used the partial molar heat of solution of HSO_3F, determined in various solvents, to characterize the basicity, i.e., the donor strength, of the solvent. This method has not become widespread in practice, although a critical study by Fawcett and Krygowski [Fa 75] indicated that it correctly reflects the base strength.

Gutmann [Gu 67] made systematic examinations with a large number of solvents, which were characterized by their affinities towards antimony(V) pentachloride as the reference acceptor. Gutmann's proposed donicity value is the heat of solvation of antimony(V) chloride by various solvents (in dilute solutions prepared in an inert solvent), determined calorimetrically and expressed in kcal/mole.

The Gutmann donicity is therefore an experimentally easily determined molecular property of the solvent. The donicity expresses the total amount of the donor–acceptor interaction. It includes both the dipole–dipole and the ion–dipole interactions, and even certain steric properties of the solvent molecule. The donicity may therefore be regarded as a semi-quantitative measure of the solvent–solute interaction.

The donicities of some important solvents, together with their relative permittivities, are given in Table 4.2.

The Gutmann concept has been criticized in a number of respects [Dr 62, 65b, 81, Bo 66]. Since the donicity is the $-\Delta H$ value of a solvation reaction, it does not take into account the following:

(a) the contribution of the entropy term to the ΔG value;

(b) the steric factors, which play a role particularly in the interactions of large solvent molecules and small metal ions;

(c) the difference between non-equivalent solvent molecules (bound with different strengths in the coordination sphere, this phenomenon being caused, e.g., by the Jahn–Teller effect.

Even in spite of the above constraints, in many cases the donicities give a picture that agrees well with practice. It is surprising, for instance, that when theoretical

Table 4.2. Gutmann donicities (DN_{SbCl_5}) and relative permittivities (ε) of various solvents [Gu 67, 68]

Solvent	DN_{SbCl_5} (kcal/mole)	ε
Nitromethane	2.7	35.9
Nitrobenzene	4.4	34.8
Acetic anhydride	10.5	20.7
Benzonitrile	11.9	25.2
Acetonitrile	14.1	38.0
Propanediol-1,2 carbonate	15.1	69.0
Benzyl cyanide	15.1	18.4
Ethylene carbonate	16.4	89.1
Methyl acetate	16.5	6.7
n-Butyronitrile	16.6	20.3
Acetone	17.0	20.7
Ethyl acetate	17.1	6.0
Water	18.0	81.0
Diethyl ether	19.2	4.3
Trimethyl phosphate	23.0	20.6
Tributyl phosphate	23.7	6.8
Dimethylformamide	26.6	36.1
N,N-Dimethylacetamide	27.8	38.9
Dimethyl sulphoxide	29.8	45.0
N,N-Diethylformamide	30.9	—
N,N-Diethylacetamide	32.2	—
Pyridine	33.1	12.3
Hexamethylphosphoramide	38.8	30.0

considerations indicate the solvent–solute interaction to be predominantly of an electrostatic nature, the experimental data reflecting this interaction do not show a correlation with the relative permittivities of the solvent (as would be expected on this basis), but rather with the Gutmann donicity. For example, Erlich *et al.* [Er 72] demonstrated that the ^{23}Na NMR chemical shift in sodium perchlorate and sodium tetrafluoroborate, measured in various donor solvents, displays a linear correlation with the donicities of the solvents (Fig. 4.1). A similar correlation was shown by multinuclear NMR studies of other alkali metal ions in non-aqueous solvents [Po 79].

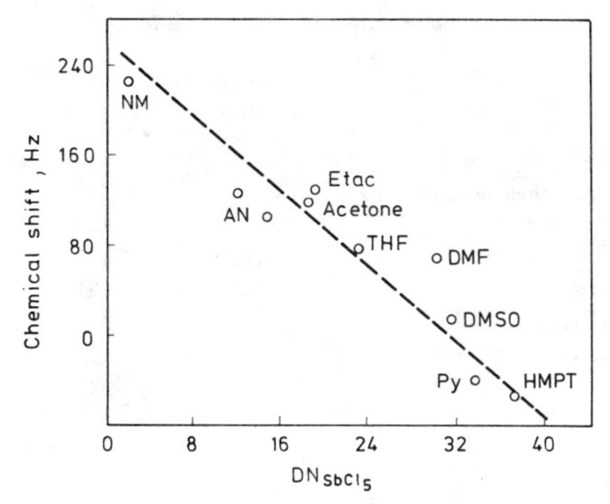

Fig. 4.1. NMR chemical shifts of ^{23}Na in solutions of $NaBF_4$ in various donor solvents [Er 72]

Base strength towards the proton

Their base strengths towards the proton could be regarded as a very good characteristic of the donor strengths of solvents. Unfortunately, the determination of an unambiguous scale of the exact basicities of the various solvent molecules towards the proton is difficult. Benoit and Domain [Be 74] consider the heat of solvation of the proton in the gaseous state to be the most suitable parameter for the characterization of the basicities of solvent molecules. In principle this is undoubtedly correct, but its practical determination is possible only by indirect means.

To solve the problem, Benoit determined the heat of solution of gaseous hydrogen chloride in various solvents experimentally, and then calculated the heat of dissociation of hydrogen chloride from the difference between the heat of the reaction of antimony(V) chloride and hydrogen chloride and the heat of the reaction of antimony(V) chloride and the chloride ion. These data were used to calculate the heats of solvation of the proton in the gaseous state, which are listed in Table 4.3. It can be seen from Table 4.3 that Benoit's data are correlated fairly well with the Gutmann donicities. The only, and rather surprising, difference is the same heat of solvation of the proton in dimethylformamide and dimethyl sulphoxide, which is definitely in contradiction to the different base strengths of these two solvents, as exhibited in numerous chemical reactions.

Table 4.3. Gas-phase proton heats of solvation and Gutmann donicities of solvents [Be 74, Gu 68]

Solvent	Heat of solvation (kcal/mole)	DN_{SbCl_5} (kcal/mole)
Tetramethylene sulphoxide	−245	–
Acetonitrile	−247	14.1
Propanediol-1,2 carbonate	−250	15.1
Water	−261	18.0
Dimethylformamide	−267	26.6
Dimethyl sulphoxide	−267	29.8

1 kcal/mole = 4.184 kJ/mole

Characterization of the donor strength on the basis of conductometric measurements

Conductometric titration curves have also proved suitable for the establishment of the donor strength sequence of solvents. Since the dielectric properties of the system have a great effect on the extent of dissociation of the dissolved salt and hence on the conductivity of the solution, the conductivity data measured in the various solvents naturally do not reflect directly the solvating or donor properties of the solvent.

However, if a reference substance, which does not dissociate in a solvent of low donicity but high relative permittivity (e.g., nitrobenzene or nitromethane), is titrated with a donor solvent, the polarization ensured by coordination of the donor solvent is continued in dissociation in the system with high relative permittivity (see p. 39); conductometric monitoring of this process is suitable for characterization of the donor strength of the titrant solvent. Thus, if trimethyltin(IV) iodide is titrated with various donor solvents in nitrobenzene solution, at any given D: $(CH_3)_3SnI$ ratio the conductivity of the solution depends on the donicity of the solvent (Fig. 4.2). With increasing donicity the conductivity increases [Gu 68].

In the above studies the measurement actually provides information on the extent of dissociation, but since the relative permittivity of the parent solvent changes only slightly or not at all during the titration with the donor solvent, the different extents of dissociation result from the different polarizations caused by the different donor strengths.

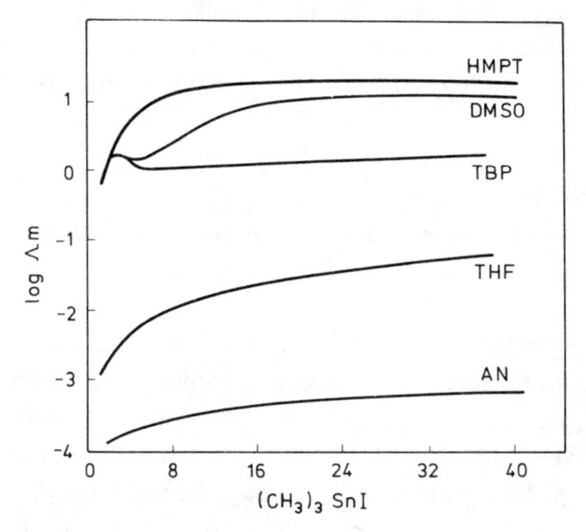

Fig. 4.2. Conductometric titrations curves of $(CH_3)_3$ SnI with various donor solvents in nitrobenzene

Characterization of the donor strength on the basis of polarographic measurements

As has emerged from the foregoing discussion, conclusions may be drawn about the donor strength sequence of solvents from the equilibrium stabilities of the resulting solvates. The exact determination of these is very difficult, however, because of the very laborious and complex nature of equilibrium analysis in non-aqueous solutions. Accordingly, in many cases researchers have been satisfied with the qualitative characterization of the stabilities of the solvates formed. Of the possible electroanalytical procedures, polarography appeared the most readily utilizable for this purpose.

The polarographic half-wave potentials of dissolved cations shift in the more negative direction with an increase in the stabilities of the complexes, including solvate complexes, whereas with anions stronger solvation results in a shift of the half-wave potential in the more positive direction [Ko 57]. However, the comparison of half-wave potentials measured in different solvents is accompanied by extreme difficulties because of the large solvent dependence of the potential of the reference electrode [Sc 55].

In order to avoid this difficulty, researchers sought a reference ion whose polarographic half-wave potential does not depend (or scarcely so) on the solvent, and the rubidium ion proved suitable [Pl 47]. With the aim of characterizing the stability sequence of the solvates, therefore, Gutmann [Gu 68] compared the half-wave potentials, relative to the rubidium ion, of the alkali and alkaline earth metal ions, determined in various solvents. The stability series thus obtained corresponded not only to the sequence of the donicities of the solvents but, as expected, also to the decreasing stability sequence of the complexes formed by the metal ions in the given solvents.

Characterization of the donor strength by means of infrared spectroscopy

As a result of hydrogen bonding or deuterium bonding, O—H and O—D vibrations are known to shift in the direction of lower frequencies. The higher the stability of the hydrogen or deuterium bond, the greater is the shift. For a reference substance with a given —O—H or —O—D content, the stability of the hydrogen or deuterium bond will be a function of the base strength (electron-pair donating ability) of the reaction partner. Hence, if the solvents to be studied are allowed to react with an appropriately selected reference substance capable of hydrogen or deuterium bonding, the sequence of the O—H or O—D vibrations will give the sequence of the donor strengths of the solvents.

The correctness of the above considerations is confirmed by those studies which demonstrate a linear correlation between the O—H vibration and the heat of formation of the corresponding hydrogen-bonded associations [Ba 37, Jo 66, Dr 69, Sh 70].

The shift of the O—D vibration of deuterated methanol in various donor solvents was termed the "electron-donating power" by Kagiya *et al.* [Ka 68], and was denoted by $\Delta\nu_{OD}$. The benzene solution served as a reference. Thus, the Kagiya parameter for the measurement of the donor strength of a solvent is

$$\Delta\nu_{OD} = \nu_{OD(benzene)} - \nu_{OD(solvent)}$$

The B parameter of Koppel and Palm [Ko 72], characterizing the donor strength of a solvent, is analogous with Kagiya's solution, with the difference that the reference is the O—D vibration of deuterated methanol in the gaseous state. Thus,

$$B = \nu_{OD(gas)} - \nu_{OD(solution)}$$

The solvent dependence of the vibration of the O—H involved in hydrogen bonding was used by Allerhand and Schleyer [Al 63] to establish a solvent strength scale. More recently, Arnett *et al.* [Ar 74] attempted to utilize the shift of the O—H vibration of *p*-fluorophenol caused by a solvent to characterize the donor strength of the solvent.

Characterization of the donor strength by means of NMR spectroscopy

The increasingly widespread application of NMR spectroscopy to the study of solvation processes has led to the elaboration of solvent strength scales based on NMR measurements.

The donor strength scale introduced by Bloor and Kidd [Bl 68] is based on the solvent dependence of the NMR shift of ^{23}Na. The concentration-dependent chemical shift of sodium iodide was determined in various donor solvents, and extrapolated to infinitely dilute solutions; the resulting values were regarded as characteristic of the solvent effect.

It was later shown by Erlich and Popov [Er 70, 71] that the NMR shift of the ^{23}Na atom in sodium tetraphenylborate and sodium perchlorate is not concentration dependent in numerous solvents. Accordingly, these salts are better models for characterization of the donor strength.

The solvent dependence of the chemical shift of ^{23}Na in the above systems is governed predominantly by the different extents of solvation of the sodium ion. This is the reason why both the Bloor–Kidd and the Erlich–Popov donor scales are in good agreement with the Gutmann donicity scale.

It was all the more surprising that analogous ^{7}Li NMR studies with lithium perchlorate model systems [Ca 75] could not be correlated with the sodium results. The difference was explained in terms of the different paramagnetic and diamagnetic shieldings of the two nuclei. (Caesium-133 NMR measurements have shown a behaviour analogous with that of the sodium system [Po 79].)

^{19}F NMR studies too have been employed to construct donor strength scales. Gurka and Taft [Gu 69] used *p*-fluorophenol as the reference acceptor. This molecule forms hydrogen-bonded associates (solvates) with donor solvents. The effect of this process on the chemical shift of ^{19}F can serve as a measure of the donor strength of the solvent. The chemical shifts were referred to the value for *p*-fluorophenol measured in carbon tetrachloride solution. This model substance has the disadvantage that not only the phenolic hydrogen, but also the fluorine, may react with the solvent, the halogen naturally as a donor. Hence, the interaction between *p*-fluorophenol and a solvent can be attributed to two reasons: the formation of a hydrogen-bonded associate determined by the donor strength of the solvent, and the interaction between the fluorine atom and a possible acceptor centre of the solvent.

Spaziante and Gutmann [Sp 71] used the chemical shift of ^{19}F to follow the effect of solvation of the model CF_3I by various donor solvents. Abbaud's "generalized solvent polarity scale" [Ab 77, Ab 79, Ka 77], developed for the characterization of aliphatic, aprotic, monofunctional solvents is also based on NMR measurements. The ^{19}F NMR and ^{14}N ESR parameters were formed by Kolling [Ko 77] to reflect the polarity of aprotic solvents in donor–acceptor solvent–solute interactions.

Transition metal complex reference acceptors in the investigation of donor solvents

It appears from the above discussion that the most appropriate of the reference acceptors considered is antimony(V) chloride, used by Gutmann. This compound has the advantage that it possesses only a single free coordination site, and thus only one solvent molecule can be bound in any instance; further, it has no charge, and the coordination reactions are not followed by dissociative side-reactions. The solvent molecules do not replace the chloride ions bound to the antimony. Hence the solvation of antimony(V) chloride corresponds in effect to the coordination of a single solvent molecule. For this reason, the $-\Delta H$ value of the reaction (the Gutmann donicity) may be regarded as a measure of the donor strength of the solvent molecule.

The Gutmann model is naturally a simplified one. In virtually all real solutions more than one solvent molecule is coordinated to each ion, and in many cases heterolytic dissociation resulting in the formation of positively and negatively charged ionic species must also be considered in the course of the reaction. Steric factors may also play a role. These occur to a certain degree in the Gutmann system also, since the originally trigonal bipyramidal symmetry of the $SbCl_5$ reference acceptor is deformed as a consequence of the coordination of the donor molecule, and distorted octahedral $SbCl_5D$ molecule is formed. The extent of this deformation may vary, depending on the steric requirements and donicity of the solvent. It must also be considered that, as will be discussed in greater detail below, an "inert" solvent is not always really inert and its effect cannot always be neglected. Hence, studies with other reference acceptor systems are justified in the investigation of donor solvents.

In the transition metal dioxime systems studied by Burger *et al.* [Bu 71], two solvent molecules are coordinated to the square planar reference acceptor along its z axis. Thus, the reference complex does not undergo distortion on the action of coordination, and in addition to the coordination reaction, but well distinguishable from it, a part is also played by a dissociative equilibrium.

In order to eliminate the use of "inert" solvents employed by Gutmann, the strength of coordination of the solvent molecules was characterized by the equilibrium constant of the displacement reaction of the solvent molecule from the coordination sphere effected by a ligand possessing a greater coordinating power. The requirements were best satisfied by the dimethylglyoxime complex of cobalt(III). In this molecule the central cobalt(III) atom is surrounded, according to square planer symmetry, by the nitrogen donor atoms of two coordinated dimethylglyoximate anions. The complex has a charge of $+1$. If cobalt(II) ion and dimethylglyoxime are allowed to interact in solution under concentration conditions such that the complex formed remains in solution, the cobalt is spontaneously oxidized quantitatively to cobalt(III) by the oxygen dissolved in the reaction mixture, and two solvent molecules are coordinated along the z axis of the low-spin complex. These solvent molecules may be replaced by halide or pseudohalide ions, or possibly by other monodentate ligands which have higher donor strengths than the solvent molecules.

The low-spin cobalt(III) dimethylglyoxime parent complex is inert; even in the presence of only a small excess of dimethylglyoxime it is formed quantitatively, and the dimethylglyoxime cannot be displaced from the complex even by a comparatively large excess of monodentate ligand. At the same time, the displacement of the coordinated solvent molecules by appropriate monodentate ligands is a reversible process leading to equilibrium within a relatively short time.

This model system has the additional advantage that the ligand moiety of the reference acceptor can be modified by the incorporation of substituents without any change in the symmetry of the coordination sphere (nitrogen atoms situated in a square plane).

During the investigation of a number of systems it has emerged that the use of iodide ion is the most advantageous method for displacement of the molecules bound to the parent complex. The stability of the iodide complex is sufficiently high for equilibrium measurements to be carried out with satisfactory accuracy. A well measurable absorption band appears in the absorption spectrum of the diiodo complex. Thus, the two solvent-dependent stepwise stability constants characteristic of iodide coordination may be determined by means of spectrophotometric equilibrium measurements.

A further advantage of the application of the iodide ion is that, as a consequence of its weak hydrogen-bonding tendency, the iodide ion is comparatively little solvated even in protic solvents, and accordingly its effect on the equilibrium constants can be neglected. Hence, these provide direct information on the coordinating powers of donor solvent molecules.

On coordination of the first iodide ion, the originally monopositively charged square-planar parent complex becomes electrically neutral, while coordination of the second iodide ion results in the formation of a mononegatively charged anionic complex. The second iodide dissociates to a lesser or greater extent, depending on the polarity (relative permittivity) of the solvent.

In the absorption spectra of solutions which contain the cobalt ion, dimethylglyoxime and the iodide ion in proportions such that all cobalt is present as the mixed complex bis(dimethylglyoximato)diiodocobalt(III), within the limit of experimental error the position of the absorption band characteristic of the coordination of the iodide does not depend on the solvent. This indicates that the same complex is formed in the different solvents.

The equilibrium constants for the displacement by iodide ions of the two solvent molecules coordinated along the *z* axis of the parent complex (the stability constants of the iodo mixed complexes, measured in various solvents) are presented in Table 4.4, together with the Gutmann donicities and the relative permittivities of the solvents.

It can be seen from the data that, as expected, the equilibrium constants are dependent on the solvent. The ratio of the stepwise stability constants is similarly solvent-dependent. The logarithms of the first stepwise equilibrium constants exhibit a reciprocal correlation with the Gutmann donicities of the solvents; they therefore reflect the stabilities of the solvate complexes governed by the donicity of the solvent. On the other hand, the logarithm of the second stepwise equilibrium constant of the iodide ion association appears to be inversely proportional to the relative permittivity of the solvent.

Table 4.4. Equilibrium constants of coordination of iodide ions to the bis(dimethylglyoximato)cobalt(III) parent complex in various solvents [Bu 71]

Solvent	DN_{SbCl_5}	ε	$\log K_1$	$\log K_2$	$\log \beta_2$
Dimethyl sulphoxide	29.8	45.0	1.23±0.15	0.63±0.06	1.86±0.11
Dimethylformamide	26.6	36.1	3.34±0.15	1.94±0.10	5.28±0.13
Methanol		32.6	3.30±0.10	2.70±0.10	6.00±0.10
Ethanol		24.3	3.74±0.04	3.74±0.04	7.48±0.04
Water	18.0	81.0	3.74±0.10	1.74±0.10	5.48±0.10
Acetonitrile	14.0	38.0	4.35±0.15	2.35±0.10	6.70±0.15
Ethyl acetate	17.1	6.0	K_1 $\ll$	K_2	8.76±0.05
Acetone	17.0	20.7	3.72±0.15	5.12±0.15	8.84±0.15

Burger *et al.* [Bu 74] subsequently investigated the resulting stability changes of the solvates and iodide mixed complexes caused by the introduction of substituents into the dioxime ligand, while keeping the coordination sphere of the reference acceptor complex constant. For this purpose, the equilibrium constants of the complexing reactions between iodide ions and the cobalt(III) parent complexes of cyclohexane-1,2-dionedioxime (nioxime) and α-furyldioxime were determined in various solvents. The inner coordination spheres of these two complexes are completely identical with that of the dimethylglyoxime complex. Hence the process of formation of the iodide mixed complex can also be regarded as analogous.

The equilibrium constants found in non-aqueous solutions are listed in Tables 4.5 and 4.6, together with the donicities and relative permittivities of the solvents.

Table 4.5. Equilibrium constants of coordination of iodide ions to the bis(nioximato)cobalt(III) parent complex in non-aqueous solvents [Bu 74]

Solvent	DN_{SbCl_5}	ε	$\log K_1$		$\log K_2$	$\log \beta_2$
Dimethyl sulphoxide	29.8	45.0	5.10		1.10	6.2 $\pm$0.1
Methanol		38.0	5.70		1.70	7.4 $\pm$0.1
Acetone	17.0	20.7	K_1	<	K_2	8.58$\pm$0.1
Methyl acetate	16.5	6.7	K_1	<	K_2	6.16$\pm$0.1
Ethanol		32.6	5.42		3.42	8.84$\pm$0.1

Table 4.6. Equilibrium constants of coordination of iodide ions to the bis(α-furyldioximato)cobalt(III) parent complex in non-aqueous solvents [Bu 74]

Solvent	DN_{SbCl_5}	ε	$\log K_1$		$\log K_2$	$\log \beta_2$
Dimethyl sulphoxide	29.8	45.0	K_1	<	K_2	1.16$\pm$0.05
Dimethylformamide	26.6	36.1	5.54		2.87	8.41$\pm$0.1
Acetonitrile	14.1	38.0	5.48		3.97	9.45$\pm$0.08

It can be seen that the stability constants do not follow the sequence of the donicities or relative permittivities of the solvents in any of the systems.

The log β_2 values measured in the solvents with the largest (dimethyl sulphoxide) and the smallest (methyl acetate) donicities and also relative permittivities agree within the limit of experimental error in the nioxime complex systems. Only the ratios of the stepwise stability constants differ. (The dissociation favoured by the higher relative permittivity caused a decrease in the value of K_2 in dimethyl sulphoxide.) The largest log β_2 value was measured in ethanol, which has a medium relative permittivity and fairly high donicity.

On the above basis, there can be no doubt that the effect of a solvent on the equilibrium constants depends not only on the donicity and relative permittivity of the solvent, but also on other factors.

A similar conclusion can be drawn from the equilibrium constants measured in the furyldioxime complex systems. Here again the stabilities of the mixed complexes do not follow the sequence to be expected on the basis of the donicities of the solvents. In acetonitrile, with the lowest donicity, for example, a much higher complex stability would be expected than in dimethylformamide. In contrast, it can be seen that the log K_1 value in acetonitrile agrees with the value measured in dimethylformamide, which has a considerably higher donicity. The difference between the donicities of the two solvents (dimethylformamide and acetonitrile) is 12.5. This causes a difference of only one order of magnitude in the log β_2 values. For comparison, the difference of 3.2 between the donicities of dimethyl sulphoxide and dimethylformamide results in a difference of more than seven orders of magnitude in the log β_2 values.

This phenomenon shows that the acetonitrile solvate in the system in question is stabilized by a factor or factors which are not taken into account in the donicity value established on the basis of affinity towards the reference acceptor antimony(V) chloride.

If the π acceptor property of the C≡N bond of acetonitrile is taken into consideration, together with the fact that the low-spin central cobalt(III) atom of the complex is prone to back-coordination, this stabilization can presumably be attributed to an electron shift from cobalt to acetonitrile. A similar effect although to a smaller extent also appears in the analogous dimethylglyoxime systems.

A comparison of the equilibrium constants obtained with the different model systems shows that even a small change in the complex, acting as acceptor, has a significant influence on the stability of the solvate formed with a given solvent molecule. The affinities for solvents even of complexes with identical coordination spheres are changed by this substituent effect to such an extent that the sequences of the stabilities of the solvates, or of the stabilities of the iodide mixed complexes thereby determined, will be different in the analogous systems.

A similar picture is reflected by the stability constants of the mixed complexes formed in various donor solvents between iodide ions and nickel macrocyclic systems with inner coordination spheres analogous to those of the dimethylglyoxime complexes.

With a view to studying the solvent effect, Burger *et al.* [Bu 75] carried out an equilibrium examination of two similar macrocyclic complexes: 1,4,8,11-tetraaza-2,3,9,10-tetramethyl-1,3,8,10-cyclotetraenenickel (II) and bis(diacetylmonoxime-imino)-propane-1,3-nickel (I) (see Table 4.7). The two compounds differ only in that the oxime nitrogens are linked by two propylene bridges in the former, while in the latter one of the propylene bridges is replaced with a hydrogen-bonded O—H—O group. Similarly to the dimethylglyoxime complexes, in a donor solvent two solvent molecules are coordinated along the z axis of these macrocyclic complexes, which have square–planar coordination sphere; these solvent molecules may be replaced with monodentate ligands. The equilibrium constants of these substitution reactions are naturally solvent dependent. Conclusions about the stabilities of the solvate complexes, and indirectly about the donor strengths of the solvents, may be drawn from the solvent dependences of the equilibrium constants.

The stability constants of the iodide mixed complexes of the two macrocyclic systems are given in Table 4.7. The data reveal that the stabilities in the same solvent of the iodide mixed complexes formed with the hydrogen-bond containing macrocyclic system are one order of magnitude smaller than in the case when there are two propylene bridges. This either shows that the solvent molecules are coordinated more strongly by the hydrogen-bond containing macrocyclic complex than by the propylene-bridged complex, or indicates that the affinity of the central atom in the former complex for the monodentate ligands bonded along the z axis is less than in the latter.

This second interpretation is in accordance with the observation that nickel dimethylglyoximate, which is a perfect analogue of the macrocyclic complex with the two propylene bridges, except for the difference that both propylene bridges are replaced with O—H—O groups, is completely unable to coordinate monofunctional ligands [Bu 63].

Analogous investigations were performed by Madeja *et al.* [Ma 76] with the mixed complex consisting of the hydrogen-bond containing nickel(II) macrocyclic complex (I) and thiocyanate ion. The equilibrium constants determined in solutions prepared with various donor solvents are listed in Table 4.8. By comparison with Table 4.7, it can be seen that, with the exception of acetonitrile, the stability of the thiocyanate complex is higher than that of the iodide complex in every solvent; this

Table 4.7. Stability constants of mixed complexes of nickel(II) macrocyclic systems with iodide ions in non-aqueous solvents, and donicities and dielectric constants of the solvents [Bu 75]

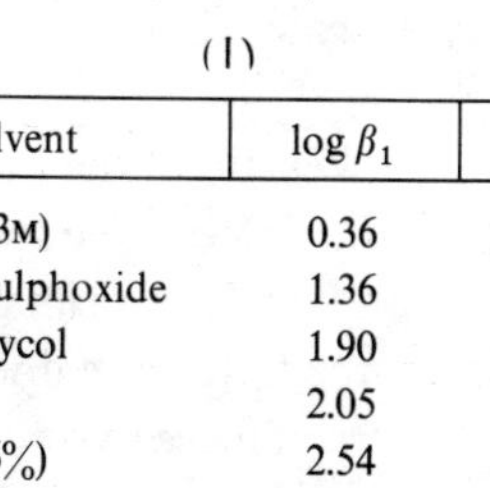

	(I)	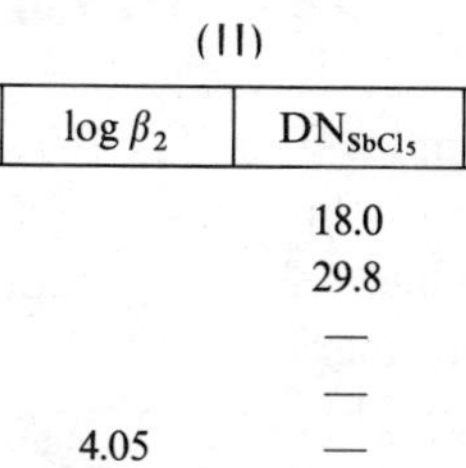	(II)			
Solvent	$\log \beta_1$	$\log \beta_2$	$\log \beta_1$	$\log \beta_2$	DN_{SbCl_5}	ε
Water ($I = 3$M)	0.36		0.68		18.0	81.0
Dimethyl sulphoxide	1.36		2.61		29.8	45.0
Ethylene glycol	1.90		3.28		—	37.7
Methanol	2.05	0.93	3.74		—	31.2
Ethanol (96%)	2.54		3.77	4.05	—	25.0
Ethanol	2.80	2.81	—	—	—	—
Acetonitrile	2.97	4.95	4.81		14.1	38.0
Acetone	4.20		7.00		17.0	20.7

shows that there is a greater affinity between the low-spin "soft" central nickel atom and the similarly "soft" sulphur donor atom of the thiocyanate than between the nickel and the iodide ion, which is less "soft" than the thiocyanate sulphur.

The solvent dependences of the stability constants of the macrocyclic mixed complexes examined are so similar that if the stability constants of any system are plotted as a function of any other, an almost linear correlation is obtained.

Table 4.8. Stability constants of the mixed complex of the nickel(II) macrocyclic system (I) with thiocyanate in non-aqueous solvents [Ma 76]

Solvent	log K
Water	0.80
Dimethyl sulphoxide	1.76
Methanol	2.32
Dimethylformamide	2.50
Acetonitrile	2.98
Ethanol	3.17
Acetone	4.37

However, the sequence of the stability constants does not follow the sequence of either the relative permittivities or the Gutmann donicities, clearly demonstrating that in this system also the stability constants of the solvates, and hence also of the iodide complexes, are determined by combined effects. For example, the small equilibrium constants obtained in ethylene glycol can be ascribed not only to the magnitudes of the donicity and relative permittivity of the solvent, but also to a solvation effect of ethylene glycol on the iodide ion. Ethylene glycol forms hydrogen-bonded chelate-type solvates with halide ions and even, according to the studies of Barcza *et al.* [Ba 74], with the perchlorate ion, which has a much lower basicity than iodide.

The Kamlet–Taft solvent basicity scale

The Kamlet–Taft [Ka 76] scale of solvent basicity makes use of the magnitudes of the solvatochromic shifts of the longest wave band in the UV-visible spectrum of *p*-nitroaniline referred to that of N,N-diethyl-*p*-nitroaniline. The primary amino group of former compound interacts with Lewis bases, the latter one does not. So the difference between the solvatochrom shift in the spectra of these two compounds depends on the strength of the interaction between the Lewis base and *p*-nitroanilin.

In practice $\tilde{\nu}_{max}$-values for *p*-nitroaniline are plotted against $\tilde{\nu}_{max}{}^{ref}$-values (1) of its N-N-diethyl derivative from measurements carried out in nonpolar solvents, leading to the following regression line:

$$\tilde{\nu}_{max} = a_1 \tilde{\nu}_{max}{}^{ref} + b_1 \quad [1]$$

Bathochromic shifts $\Delta\tilde{\nu}$ from this line are observed when measurements are carried out in solvents interacting as Lewis bases. The shift is given by:

$$\Delta\tilde{\nu} = \tilde{\nu}_{max}(\text{calcd.}) - \tilde{\nu}_{max}(\text{obs.}) \quad [2]$$

and its magnitude increases with increase of the solvent's Lewis basicity. This quantity can be taken as a measure of the solvent basicity.

This method was later adapted by Launay *et al.* [La 79] for the study of the effect of substituents on the Lewis acidity of the amino groups in *p*-substituted anilines.

The solvatochrom effect in the visible spectra (between 390–410 nm) of zinc tetraphenyl porphine was also used for the characterization of donor solvents [Ko 78].

Characterization of the acceptor strength on the basis of the solvatochrome effect

The fact that the energy, intensity and sometimes the shape of the absorption band of a dissolved chromophore may change as a result of the action of the solvent has long been known [Ha 22]. In 1922 the phenomenon was named the solvatochrome effect by Hantzsch. Substances displaying this effect are termed solvatochrome substances.

The application of solvatochrome dyes for characterizing the strengths (polarities) of solvents was first proposed by Brooker [Br 51]. His solvent scale is based on the shifts in the absorption bands of the following two merocyanine dyes (I and II):

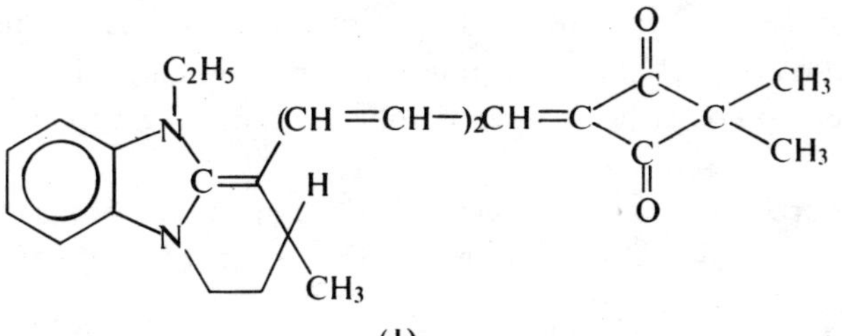

(I)

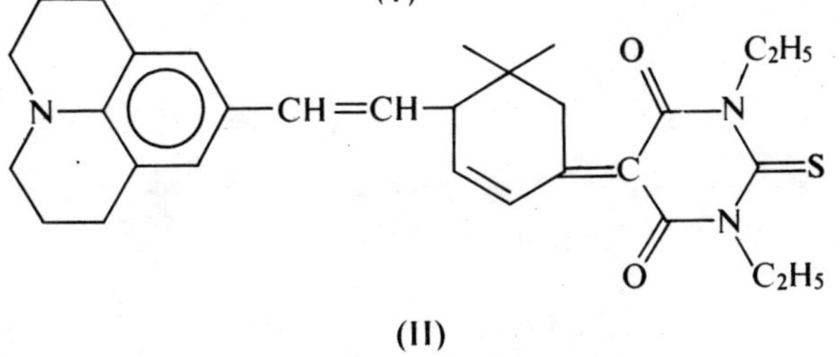

(II)

Solvatochromism of merocyanine dyes is accompanied also with the solvent dependence of their dipole moment [Pa 80]. So dipole moment measurements have also contributed to the explanation of solvatochrom behaviour of these systems. In such systems the electronic ground and excited states of a molecule are described by two resonance structures: non-polar (quinoid) and polar (benzenoid).

An increase in solvent strength leads to a continuous shift in the electron density distribution, making the dye structure closer to its polar resonance form. Indirect evidence of the shift induced by the environment through a conjugated π-electron system from the donor to acceptor end of the dye molecule is the existence not only of strong solvatochromism (both positive and negative) but also of a strong shift in IR vibration, NMR signal of the corresponding groups and also in the change of dipole moments. A strong correlation was shown among the solvent dependence of these parameters.

As a result of increasing solvent strength, the absorption band of molecule **I** undergoes a blue shift, and that of molecule **II** a red shift. These shifts, serving as the basis of the solvent scale, were denoted χ_B and χ_R. The two independent parameters are fairly weakly correlated with each other. This was explained by assuming that the two scales reflect different effects.

Brooker's suggestion did not become widely accepted in practice. A number of successfully employed methods elaborated for the characterization of the solvent strengths of acceptor solvents are nevertheless based on the solvatochrome effect. In these methods solvents are classified with the aid of UV and visible spectroscopic measurements.

The scale frequently used to characterize the solvating power (the acceptor strength) of an acceptor solvent is the Kosower Z scale [Ko 58]. The procedure is based on the fact that the cation of the 1-ethyl-4-carbomethoxypyridinium iodide ion pair used as the model system is not an electron pair acceptor, while the iodide anion is capable of hydrogen bonding. Consequently, the extent of ion pair formation between the 1-ethyl-4-carbomethoxypyridinium cation and the iodide ion in solutions prepared with various solvents depends on the solvation of the iodide ion.

The Kosower Z value is the wavelength of the maximum of the charge-transfer band for the above-mentioned ion pair, given in units of kcal/mole:

CO_2CH_3 / $\overset{+}{N}$ / C_2H_5 $+ I^-$ $\underset{}{\overset{K_{ass}}{\rightleftharpoons}}$ CO_2CH_3 / $\overset{+}{N}$ / C_2H_5 $--I^-$ $\xrightarrow{h\nu}$ CO_2CH_3 / 7π / N / C_2H_5 $--I$

Direction of molecular dipole (vertical arrow)

Direction of molecular dipole (horizontal arrow)

The ion pair is ionic in the ground state (it has a large dipole moment), whereas in the excited state it is much less ionic (the dipole moment is much lower); therefore, solvents of high solvating power (high acceptor strength) depress the energy of the ground state, whereas that of the excited state is scarcely affected (in accordance with the Franck–Condon principle). Hence the Z values of polar solvents of high solvating power (hydrogen-bond forming, strong acceptors) are larger than those of non-polar, weakly solvating solvents.

Determination of the original Kosower Z value was hindered by the insufficient solubility of the model compound in non-polar solvents. Later, therefore, 1-ethyl-4-carbo-*t*-butoxypyridinium iodide, which is readily soluble also in non-polar solvents, was used as the reference donor [Ko 68]. The Z values obtained with the two different references are in excellent agreement.

Kosower initially reported Z values for 21 solvents and 35 solvent mixtures (interpretation of the data relating to the solvent mixtures is, however, often uncertain because of the overlapping of the individual effects of the components). The Kosower scale was later extended to 45 solvents [Le 66, Be 62, Go 65, Fo 69], and Griffiths and Pugh [Gr 77] determined the Kosower Z values of a further 40 solvents.

The solvent with the greatest acceptor strength in the Kosower acceptor scale is water, which exhibits a strong hydrogen bonding tendency. This is followed by the (similarly hydrogen-bond forming) alcohol molecules. At the lower end of the Z scale one finds the non-hydrogen-bond forming, low relative permittivity hydrocarbons, halogenated hydrocarbons, pyridine and acetone. The solvents with a Z value of about 70 in the middle of the scale (dimethyl sulphoxide, acetonitrile) have medium relative permittivities, but do not form hydrogen bonds; their ability to solvate anions can be attributed to electrostatic interactions (see Table 4.9). These few examples indicate that the Kosower scale correctly reflects the strength of the acceptor properties also of those solvents which are strong donors, but weak acceptors.

On the basis of the Kosower data, Brownstein [Br 60] derived a general solvent polarity scale. Analogously to Hammett's linear free energy correlation [Ha 37], he employed the following equation:

$$\log \frac{k_{\mathrm{solv.}}}{k_{\mathrm{EtOH}}} = SR$$

where $k_{\mathrm{solv.}}$ is the measured solvent-dependent parameter in the solvent examined, k_{EtOH} is the analogous parameter determined in absolute ethanol solution, S is a parameter measuring the solvating power of the solvent and R is a constant characteristic of the model system used.

As a model system for confirmation of the correctness of the above correlation Brownstein used 1-ethyl-4-carbomethoxypyridinium iodide, as proposed by Kosower. By definition, the R value relating to this is 1.00, and the S value relating to absolute ethanol is 1.00. The Kosower Z values served as the k values here. On the basis of the experimental data of Kosower, Brownstein calculated S and R for many systems, and found good correlations. Accordingly, he assumed that his equation was of general validity and suitable for the description of the solvent dependence not only of spectral data, but of any solvent-dependent experimental parameter.

In actual fact, this consideration is only true for systems in which one of the many factors influencing the solvent effect is predominant and the others are negligible in comparison. The reason why this correlation was suitable for a description of the Kosower Z values was that the spectral shifts serving as its basis are determined predominantly by the acceptor strength of the solvent. The experiments of Brownstein can actually be regarded as a confirmation of the Kosower acceptor scale.

An acceptor strength parameter more general than the Kosower Z number was introduced by Dimroth *et al.* [Di 63, Re 69, Re 81, Di 73]. Pyridinium-N-phenol-betaine was employed as the reference donor molecule. As can be seen below, the molecule, which is strongly polar in the ground state, becomes much more weakly polar as a result of excitation:

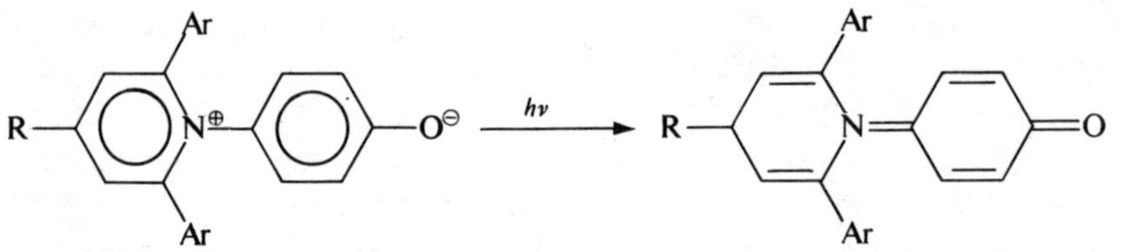

where Ar is phenyl or *p*-methylphenyl. The positive charge shown on the nitrogen in the ground state of the molecule is partially delocalized over the aromatic ring, and is also shielded by the Ar groups; thus the molecule does not react as an acceptor with a donor solvent. On the other hand, the donor oxygen, carrying the negative charge, can interact strongly with acceptor (acidic) solvents. Solvation of the excited molecule proceeds to a much lesser extent because of the decrease in polarization caused by the rearrangement resulting from the excitation. Hence the energy of the π–π^* transitions, which increases with increasing strength of solvation of the ground state and which scarcely depends on the solvation of the excited state, can be regarded as proportional to the acceptor strength of the solvent.

On the basis of these considerations, the wave-lengths of the absorption maxima of the above model, measured in various solvents, were used to calculate [Di 63, Re 69, Re 81, Di 73] the Dimroth–Reichardt E_T values, which were expressed in kcal/mole, these were regarded as the parameters characteristic of the polarity of the solvent molecules; actually they reflect the acceptor strength of the solvent.

The Dimroth E_T values [Di 63] were first determined for 42 solvents, but this was later [Re 65] extended to a total of 62 solvents. This also illustrates well that the Dimroth E_T values unambiguously reflect the acid strengths (acceptor strengths) of the solvents.

A further advantage of the E_T scale is sensitivity. The solvatochrome shift is greatest in the case of the Dimroth model: the absorption band shifts from a wavelength of 453 nm (water) to 810 nm (diphenyl ether). These data also show that the total scale can be obtained from absorptions appearing in the visible spectral range. (The change brought about by replacing a solvent with another is visible to naked eye.)

A limitation to the application of the scale is the temperature dependence of the E_T values, due to the thermochromism of the model compound.

The correctness of the Dimroth–Reichardt concept was clearly proved by subsequent investigations by Krygowski and Fawcett [Kr 75], in the course of which E_T was successfully employed as a parameter to describe the acid strength in their model, named the "complementary Lewis acid–base description of solvent effects" (see also p. 81).

The dependence of solvatochromic behaviour (shift of the absorption maxima) of the Dimroth–Reichardt compound on the composition of the solvent was characterized by Langhals with the help of a simple two-parameter equation in solvent mixtures [La 81a, b].

The Dubois [Du 64] acceptor strength scale also is based on the shift of the electron excitation spectral bands on the action of a solvent. Dubois *et al.* [Du 64] used the solvent dependence of the n–π^* transitions of saturated ketones to characterize the solvent strength. The solvent dependence of the frequency of the absorption band was first described by the equation

$$\nu_s = A_s \nu_H + B_s$$

where ν_s and ν_H are the frequencies measured in the solvent examined and in the reference solvent (n-hexane), respectively, and A_s and B_s are parameters characteristic of the solvent. Further simplification of this correlation led to the following equation, describing the solvent effect with a single parameter, F:

$$\nu_s - \nu_H = F(\nu_H - \nu_I)$$

where $\nu_I = 33\,374\ \text{cm}^{-1}$ is the frequency of a band independent of the solvent. The F values of 13 solvents were determined by this method. The small number of solvents categorized in this way is a limit to the practical application of the method.

The solvatochrome effect may also appear in the infrared spectrum. The acceptor strength scale of Kagiya [Ka 68] is based on the solvent dependence of the C=O vibration band for acetophenone. As for his donor strength scale, the reference solvent is benzene. The acceptor strength is denoted by the band shift, $\Delta\nu_A$:

$$\Delta\nu_A = \nu_{A\,(\mathrm{benzene})} - \nu_{A\,(\mathrm{solvent})}$$

Kagiya's acceptor strength scale has not found wide application.

Many other organic dyes show solvatochromic behaviour. The solvent dependence of the electronic spectra of acridone, N-methylacridone and N-phenylacridone measured at 298 K for 35 solvents of different polarity shows, e.g., a distinct positive solvatochromism [Si 80].

8-Amino-5-ethyl-oxido-6-phenylphenanthridinium betaine was shown recently by Finkentey and Zimmermann [Fi 81] to be a new solvatochromic compound. Its structure was proved by ^{1}H-NMR and mass spectroscopy. In aqueous solutions it is in a pH dependent equilibrium with its conjugated aminophenol and the twice protonated dication. The betaine shows a strong negative solvent effect. Its solutions in aprotic solvents are blue, in amphiprotic solvents red to purple.

In practice, of the acceptor strength scales based on the solvatochrome effect only Kosower's and Dimroth's scales can be considered to have general applicability. Solute–solvent interactions could be rationalized, however, with the help of the solvatochrome effect in several organic systems [Ha 77, 79].

Characterization of the acceptor strength on the basis of NMR measurements

Gutmann's "acceptor number" scale was recently developed to characterize the acceptor strengths of solvents [Gu 75, 76, Ma 75]. The ^{31}P NMR shifts of the adducts formed between the acceptor molecules and triethylphosphine oxide (Et_3PO), as the reference donor, constitute the experimental basis of this scale. With an increase in the acceptor strength of the solvent, the electron density decreases on the phosphorus atom in the solvate, as a result of inductive effect [Gu 75]; this is accompanied by a decrease in the NMR chemical shift (δ). When constructing the acceptor number scale, Gutmann referred the δ values to the δ value of the adduct formed between triethylphosphine oxide and antimony(V) chloride, $Et_3PO \rightarrow SbCl_5$, which was arbitrarily taken as 100. This means that antimony(V) chloride, the reference substance serving as the basis of the donicity scale, is also taken as the reference substance for the acceptor strength scale.

The Gutmann acceptor numbers (AN) of some solvents are listed in Table 4.9, together with the Kosower Z values and the Dimroth–Reichardt E_T values characterizing the acceptor strengths.

Table 4.9. Acceptor strength scales of solvents

Solvent	^{31}P (ppm)	AN	E_T (kcal/mole)	Z (kcal/mole)
n-Hexane (reference solvent)	0	0	30.9	—
Diethyl ether	−1.64	3.9	34.6	—
Tetrahydrofuran	−3.39	8.0	37.4	58.8
Benzene	−3.49	8.2	34.5	54.0
CCl_4	−3.64	8.6	32.5	—
HMPA	−4.50	10.6	40.9	62.8
Dioxane	−4.59	10.8	36.0	—
Acetone	−5.33	12.5	42.2	65.5
Pyridine	−6.04	14.2	40.2	64.0
Nitrobenzene	−6.32	14.8	42.0	—
DMF	−6.82	16.0	43.8	68.5
CH_3CN	−8.04	18.9	46.0	71.3
DMSO	−8.22	19.3	45.0	71.1
CH_2Cl_2	−8.67	20.4	46.1	64.7
$CHCl_3$	−9.83	23.1	39.1	63.2
Isopropanol	−14.26	33.5	48.6	76.3
Ethanol	−15.80	37.1	51.9	79.6
Formamide	−16.95	39.8	56.6	83.3
Methanol	−17.60	41.3	55.5	86.3
Glacial acetic acid	−22.51	52.9	51.2	79.2
Water	−23.55	54.8	63.1	94.6
CF_3COOH	−44.83	105.3	—	—

^{31}P: ^{31}P NMR chemical shift for Et_2PO measured in various solvents and referred to the value measured in n-hexane.
AN: Gutmann acceptor number [Ma 79]
E_T: according to Dimroth and Reichardt [Di 63, 73].
Z: according to Kosower [Ko 68].

It is interesting that, among the solvents with low acceptor numbers, the acceptor strength of the comparatively polar diethyl ether is lower than those of such apolar solvents as benzene and carbon tetrachloride. Polar solvents containing an acidic C—H bond, e.g., dichloromethane, chloroform and formamide, appear with medium acceptor strengths; alcohols, water and acids are solvents with high acceptor numbers.

The data in Table 4.9 indicate that the Gutmann acceptor numbers (AN) follow a sequence similar to E_T and Z values. There is also a close, almost linear, correlation between the heats of solvation of certain anions, e.g., chloride and $Fe(CN)_6^{3-}$, and the acceptor numbers.

It is also obvious from the tabulated data that the acceptor properties of typical donor solvents may differ also. Thus, the heats of solvation of these solvents with antimony(V) chloride as donor are also included in the experimentally determined donicity values. Particularly in solvents of lower donicity, but higher acceptor strength, this may cause the donicity value to differ from the actual value. If the acceptor numbers are known, further refinement of the donicity values will be possible.

NMR measurements likewise led to the Anderson–Symons [An 69] acceptor strength scale. In effect, this is based on the Kosower concept, with the difference that the formation and dissociation of the pyridinium halide model ion pair reflecting the solvent effect are characterized not by UV-visible spectra, but by measurement of the NMR shift of the protons of the pyridine ring. The proton resonance measurement can distinguish between the 2,6- and the 3,5-protons. These two types of proton are affected to different extents by the interaction between the halide and the nitrogen atom of the pyridinium. Hence, the difference in the chemical shifts of the types of proton is a measure of the solvent effect.

A cobalt-59 NMR procedure was elaborated by Laszlo and Stockis [La 80] for the determination of H-bond donation of protic solvent. The ^{59}Co chemical shifts were found to be in linear correlation with the Gutmann acceptor number of the solvents studied.

Characterization of the acceptor strength on the basis of kinetic measurements

Smith *et al.* [Sm 61] utilized kinetic data to characterize the solvent effect. Selecting the rate constant (k_{ion}) (and its logarithm) of the ionization of 2-(*p*-methoxyphenyl)-2-methylpropyl *p*-toluenesulphonate, as described by the following equation:

$$p\text{-}CH_3OC_6H_4\text{-}C(CH_3)_2\text{-}CH_2\text{-}OTs \xrightarrow[75^\circ]{K_{ion}} \left[\text{(bridged phenonium ion)} \; OTs^{\ominus} \right] \longrightarrow p\text{-}CH_3OC_6H_4\text{-}CH{=}C(CH_3)_2 + HOTs$$

where OTs^- is the toluenesulphonate ion. It can be seen that in the transition state the positive charge is delocalized along the π-electron system of the aromatic ring, while the negative charge is localized on the OTs^- anion. Hence, the effect of the solvent on the rate constant, k_{ion}, can be ascribed predominantly to the solvation of

OTs^-. Accordingly, log k_{ion} varies as a function of the strength of solvation of the OTs^- anion, and hence gives a sequence characteristic of the acceptor strengths of acceptor (acidic) solvents capable of solvating the OTs^- anion.

Arrangement of the rate constants in accordance with the Hammett equation was the basis of the Grünwald–Winstein [Gr 48] and the Gielen–Nasielski [Gi 64] concepts. The Y parameter of Grünwald and Winstein [Gr 48], derived from the solvent dependence of the solvolysis of aliphatic compounds following an S_N1 mechanism, can be obtained from the equation

$$\log \frac{k}{k_0} = mY$$

where k and k_0 are the rate contants measured in the solvent under examination and in the reference solvent, respectively, and m is a parameter characteristic of the model system. Grünwald and Winstein used t-butyl chloride as the model compound, and hence for this $m = 1.00$.

From later investigations by Krygowski and Fawcett [Kr 75] it has emerged that the correlation of the Y values relating to the different systems can be described well with a two-parameter approximation, since, in addition to the acidity of the solvent, there is another factor (donor property) playing a slight role in the determination of the reaction rate.

The parameter X proposed for characterization of the solvent effect was also obtained by Gielen and Nasielski [Gi 64] *via* the Hammett equation:

$$\log \frac{k}{k_0} = pX$$

where k and k_0 are the rate constants of the electrophilic aliphatic substitution of organometallic compounds in the solvent under study and in the reference solvent, respectively, and p is a constant characteristic of the reference system.

The value of X has been determined in only very few solvents, hence the Gielen–Nasielski conception is of little importance.

Other solvent strength (polarity) scales

There are solvent strength scales, based on the solvent dependence of various chemical processes or physico-chemical properties of individual model compounds, which cannot be ascribed unambiguously to the donor or acceptor properties of the solvent. The solvent effect may also be manifested in a more complex manner. In a given process, for instance, a solvent may act both as donor and as acceptor, and other factors (electrostatic, steric, etc.) may also play a part. It is rarely possible to separate these effects. All solvent strength scales are of an empirical nature; their justification is their practical usefulness and their contribution to the understanding of the mechanisms of certain processes. The less the utility of some scale, the lower is its importance.

Taft *et al.* [Ta 63] introduced solvent scales based on the ^{19}F NMR chemical shifts of *meta-* and *para-*substituted fluorobenzene derivatives. However, the values of the factors obtained from these measurements are influenced by the interactions of both the substituent and the fluorine atom with the solvent. Later modifications of this method led, however, to the "general solvent polarity scale" of Abboud *et al.* [Ab 77, 79, Ko 77].

Pincock [Pi 64] characterized solvents on the basis of the solvent dependence of the ionic decomposition of *t*-butyl peroxyformate. However, not only the rate, but also the mechanism of decomposition of this compound is solvent-dependent. For example, in chlorobenzene it decomposes in a slow unimolecular reaction in which the peroxide bond is split; in n-butyl ether the decomposition proceeds *via* radical attack on the peroxide oxygen atoms; and in the presence of pyridine a bimolecular elimination reaction occurs, with the formation of *t*-butanol and carbon dioxide. Pincock used the solvent dependence of the rate of this latter reaction to characterize the solvent.

Nagy *et al.* [Na 72, Na 74, Du 75] classified organic solvents into groups on the basis of the solvent dependence of the absorption bands characteristic of the charge-transfer complexes of substituted phthalic anhydrides with substituted naphthenes, naphthalenes and anthracenes, and examined possible correlations between this grouping and other solvent parameters.

The Nagy classification is in a fairly good agreement with practical experience.

A solvent molecule and the usual solute are, in general, of dipolar character, so that the nucleophilic interaction occurs simultaneously with electrophilic interaction at different sites of a molecule depending on the structural characteristics of the solute and solvent. However, a solvated electron is known to be the simplest entity with a unit negative charge held in a cavity created by dilated solvent molecules, so that the interaction of an electron with solvent molecules gives information on the simple electrostatic interaction.

The solvation free energy of an electron in polar solvents can be estimated by electron photoemission (EPE) [Ya 77]. Imai and Yamashita [Im 78] suggested therefore the use of the solvation free energy of the electron for the characterization of the electrophilic properties (acidity) of solvents.

Gurikov [Gu 80] developed semiempirical calculations for the characterization of solvent polarity. His method has, however, a limited use.

Intercorrelation of the various solvent strength scales

A number of authors have dealt with the comparison of the different solvent scales and with the quantitative determination of correlations between them. It is not surprising that a good correlation has been found only when comparing models which characterize the same property of the solvent. Thus the correlation between the Dimroth E_T and the Kosower Z parameters characterizing acceptor strength is very good [Re 65, Gr 77], and then both Z and E_T correlate more or less well with the other acceptor strength parameters: Smith's log k_{ion}, Dubois' F, Berson's Ω, Brooker's χ_B, Mayer's AN, etc. [Br 65, Be 62, Ch 82, Sm 61, Gr 77]. The correlation is similarly good between the various donor strength parameters, such as the Gutmann donicity, Kagiya's ν_D value and the Erlich ^{23}Na NMR data [Gr 77]. Krygowsky *et al.* [Kr 80], for instance, have analyzed the correlation between four basicity parameters. Comparing the Gutmann's donor number, Kagiya's and Koppel's parameters with the extended Kamlet–Taft basicity parameter (B_{KT}), a rough agreement was found between B_{KT} and the other basicity parameters suggesting their equal applicability. In the knowledge of these and similar correlations, it has become possible also to place on the various scales solvents that could not be measured unambiguously by the experimental method used in constructing the given scale.

Naturally, there is not and cannot be any correlation between scales reflecting different solvent properties, e.g., between E_T (typical of the acceptor strength) and the Gutmann donicity (describing the donor strength).

Drago *et al.* [Dr 81] have investigated the correlation between Gutmann's donor numbers and the E and C parameters of his concept. He obtained rather poor fits. The weak adducts (DN < 22) all had positive deviations of the best fit values, and the strong adducts (DN < 25) all had negative ones. On the basis of these investigations new values for some donor numbers were suggested. For the explanation of these anomalies Drago assumed that the increase of the interaction between donor solvent and antimony in the $DSbCl_5$ adduct results in the increased ionicity of the SbCl bonds. This may cause a more extensive solvation of the chlorides in the stronger adduct than in the weaker ones. In this way the solvation of the chlorides also influences the DN values not only the coordination of the solvent by antimony. In spite of the great number of systems showing solvent properties in good correlation with Gutmann's donor numbers Drago questions the significance of such correlations for chemists. According to experience, however, such correlations may help in the better understanding of chemical processes in solution.

Comparisons of the different solvent scales with data reflecting the solvent dependence of a chemical process may reveal the solvent property (donor or acceptor) responsible for the solvent effect; it also may indicate that the solvent effect cannot be attributed to any single such property. The complexity of the solvent effect (and the limitations of simple comparisons of the above type) led to the development of more complicated, several-parameter models. Nevertheless, even the one-parameter models have clearly demonstrated that the most important factor in the solvent effect concerning properties or reactions of ions or polar molecules is the coordinative donor–acceptor (Lewis acid–base) interaction.

Study of Lewis donor–acceptor interactions and related problems

As was seen from the discussion in the preceding sections, the predominant part of the interaction between a solute and a donor or acceptor solvent (solvation) is a Lewis donor–acceptor interaction. The stronger Lewis acid and Lewis base the two reactants are, the more negligible are the effects of other factors influencing the solvation, and the more this process can be regarded as complex formation, in the classical sense, where the stability of the solvate is determined by the stability of the coordinate bond formed between the electron-pair donor and electron-pair acceptor reactants.

The exact means of studying Lewis acid–base interactions would be to determine the reaction heat of adduct formation in the gas phase [Dr 68]. In this process, contributions originating from various side-reactions, such as solvation of the reactant and lattice energies, do not interfere. Such gas-phase reactions can usually be followed by means of pressure-change measurement, which is fairly difficult and time-consuming method. Studies have also been made [Go 65b, Lo 65] in which adduct formation in the gas phase was followed by a spectroscopic method. The errors are so large, however, that general conclusions drawn from such measurements must be considered with great reservation.

In order to eliminate the difficulties associated with gas-phase measurements, Lewis acid–base interactions are investigated in solutions prepared with inert parent solvents. These investigations set out from the assumption that the course of the reaction in an inert solvent corresponds to the gas-phase reaction, hence the various side-effects of solvation, etc., can be neglected.

It is the degree of validity of this assumption which will decide the extent to which the results of the different experimental procedures of measuring the donor and acceptor strengths of solvents are objective and consistent. Below, the factors that determine the assertion of the above condition will be considered.

Effect of the parent solvent

The first and fundamental problem is the correct choice of the inert parent solvent, since different solvents can be regarded as inert in different systems. For instance, it has been shown [Pa 68, No 72a] that adducts of systems involving oxygen donor atoms are solvated by cyclohexane; thus their formation should be examined calorimetrically, e.g., in carbon tetrachloride solution. On the other hand, systems containing nitrogen and sulphur donor atoms interact with carbon tetrachloride [Vo 70], but not with cyclohexane. Accordingly, these may be studied in cyclohexane solution.

The donor–acceptor interaction in solution is described by the equation

$$A_{(\text{solv.})} + B_{(\text{solv.})} \rightleftharpoons AB_{(\text{solv.})}$$

The following thermodynamic cycle may be written for this equation:

$$\begin{array}{ccccc} A_{(\text{gas})} & + & B_{(\text{gas})} & \xrightleftharpoons{\Delta H_4} & AB_{(\text{gas})} \\ \uparrow \Delta H_2 & & \uparrow \Delta H_3 & & \downarrow \Delta H_5 \\ A_{(\text{solv.})} & + & B_{(\text{solv.})} & \xrightleftharpoons[\Delta H_1]{} & AB_{(\text{solv.})} \end{array}$$

where ΔH_1 corresponds to the heat of reaction measured in solution, ΔH_4 to the heat of reaction in the gas phase, ΔH_2 and ΔH_3 to the heats of desolvation of components A and B, and ΔH_5 to the heat of solvation of the product.

It is to be seen that the heat of reaction measured in solution is the resultant of several heats of reaction. In order to establish the heat of reaction of the equilibrium describing the actual extent of the acid–base interaction (ΔH_4), it would be necessary to know the heats of reaction of the other four steps, which is not always possible. Thus, on the basis of various pieces of indirect information, it is usual to regard one or other interaction as negligible, or to assume that the heats or reaction of certain steps just compensate each other. Such simplifications are permissible in many cases without affecting the overall conclusions. However, if any of the assumptions is incorrect, the entire interpretation of the reaction will be faulty.

A good example of the error caused by incorrectly neglecting some step is the method in which the acid–base interaction is studied calorimetrically by using the base simultaneously as the solvent; the enthalpy term due to solvation (process 2) is then corrected by the heats of dissolution of a model compound whose structure is analogous to that of the examined acid, but the acidic functional group of which is blocked. An example is the use of the methyl ester in investigations employing phenol as the Lewis acid [Ar 70]. The fundamental error in this method is that the base, as a polar solvent, may play a role in all of the solvation reactions indicated in the above diagram. The experimentally determined heat of reaction is ΔH_1. The value characteristic of the strength of the actual acid–base interaction, however, is the heat of reaction ΔH_4, which is obtained from ΔH_1 by using for correction ΔH_2, ΔH_3 and ΔH_5. The heat of dissolution used [Ar 70], which was determined with a model system containing a blocked functional group, can be regarded as the ΔH_2 value. Therefore, the method gives correct results only if the heats of reaction ΔH_3 and ΔH_5 exactly compensate each other [Dr 73].

Such a compensation is conceivable when the base used simultaneously as the solvent is not strongly polar, and the solvation energy is small. In polar solvents, however, the probability of this is not very high. For instance, the heat of reaction between *p*-fluorophenol and dimethyl sulphoxide is 7.2 kcal/mole in dimethyl sulphoxide solution, whereas it is only 6.6 kcal/mole if carbon tetrachloride is used as inert solvent.

A number of examples can be found where the heats of solvation ΔH_2, ΔH_3 and ΔH_5 are of the same order of magnitude as ΔH_4, which corresponds to the actual interaction. All this clearly shows the importance of the correct choice of the solvent in investigations of acid–base interactions [Pa 68, Dr 72, No 72b, Sl 72].

The above considerations also reveal the limitations of the Gutmann donicity conception. To illustrate the more general nature of the Gutmann concept, the following reaction was subjected to a quantitative thermodynamic treatment [Gu 74]:

$$[CoCl_3D]^- + Cl^- \rightleftharpoons [CoCl_4]^{2-} + D$$

where D is the donor solvent. A thermodynamic cycle can be written for the reaction (analogously as on page 71, but referred to the ΔG values):

$$\begin{array}{ccc} & [CoCl_3D]^-_{(gas)} + Cl^-_{(gas)} \overset{4}{\rightleftharpoons} [CoCl_4]^{2-}_{(gas)} + D_{(gas)} & \\ \text{solvation } \Delta G & \downarrow 2 \quad \downarrow 3 \qquad\qquad \uparrow 5 & \text{vaporization } \Delta G \\ & [CoCl_3D]^-_{(solv.)} + Cl^-_{(solv.)} \overset{1}{\rightleftharpoons} [CoCl_4]^{2-}_{(solv.)} + D_{(liquid)} & \end{array}$$

The solute–solvent interaction in the reaction is determined by the equilibrium

$$[CoCl_3]^-_{(gas)} + D_{(gas)} \rightleftharpoons [CoCl_3D]^-_{(gas)}$$

the heat of reaction of which is proportional to the Gutmann donicity (it differs from it insofar as the latter was determined in solution prepared with an inert solvent). The term of the above reactions that is independent of the solvent effect is

$$[CoCl_3]^- + Cl^- \rightleftharpoons [CoCl_4]^{2-}$$

It was demonstrated that when the above reaction was examined in two different solvents, one of them being the reference solvent (acetonitrile was actually used), there was a quantitative relation between the difference of the ΔG values measured in the two solutions and the difference of the Gutmann donicities of the solvents.

In the derivation of this correlation, it had to be proved that the difference in the ΔG values ($\Delta\Delta G$) for the solvation of $[CoCl_3D]^-$ and $[CoCl_4]^{2-}$ in the two solvents (steps 2 and 5 of the thermodynamic cycle) was proportional to the difference of the ΔG values for the solvation of Cl^-. Additionally, $\Delta\Delta G$ for the solvation of $[CoCl_3D]^-$ is proportional to the difference in the ΔG values for the vaporization of the two solvents.

In the above system, these correlations proved to hold. Thus, when the $\Delta\Delta G$ values measured in the solutions prepared with solvents of different donicities and referred to the ΔG value measured in acetonitrile were plotted as a function of the donicity values referred to acetonitrile, a linear correlation was obtained (Fig. 4.3). On this basis, it was claimed that it was possible to calculate the equilibrium constants of the reaction in solutions prepared with different solvents, if the donicities of the solvents, the ΔG values for the solvation of the chloride ion in the solvents and the ΔG values for the vaporization of the given solvents are known.

By similar considerations, Gutmann and Mayer [see Ma 79] also carried out calculations on the iodine–triiodide system in different solutions.

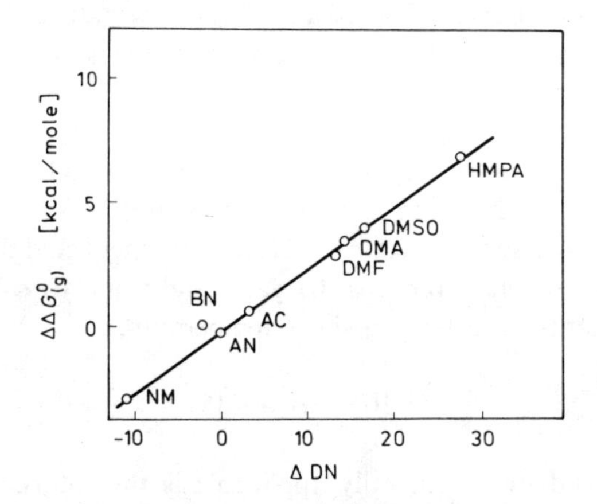

Fig. 4.3. ΔG values, referred to solvation with acetonitrile, for the solvations of $CoCl_3^-$ with various donor solvents ($\Delta\Delta G$), as a function of the corresponding donicity differences [Gu 74]

Notwithstanding the results described above, it must be borne in mind that these considerations lead to a quantitative description of the solvent effect on the basis of the experimental data measured in solution only if conditions dependent on the individual chemical properties of the system are met.

The interaction energies of dissolved ions with the surrounding solvent are large, comparable to the lattice energies of ionic crystals. Changes in these ion-solvent interactions on transfer of electrolytes between solvents are smaller, but are sufficiently large to cause dramatic changes in chemical reactions involving ions. Thermodynamic studies have been reported for a wide range of electrolytes in an enormous number of single component and binary mixed solvent systems, many selected data are listed in [Co 80].

Role of the experimental method

In addition to the solvent, the procedure to be used in the determination of the thermodynamic data must also be selected with great care. If the ΔG, ΔH and ΔS data must be obtained from the temperature dependence of the equilibrium constant, the accuracy of the measurements already automatically considerably limits the accuracy of the data. Particularly in the narrow temperature interval that can be employed for aqueous solutions (from 0 °C to at most 50–60 °C), in many cases the temperature dependence of the equilibrium constant is not much greater than the experimental error. The thermodynamic data calculated from these will thus at best be approximate values of an informatory nature.

A special problem arises if the molecular property measured in the course of the determination of the equilibrium constant is also dependent on the temperature. For example, the temperature dependence of the absorbance may render impossible the use of spectrophotometry in the determination of thermodynamic data from the equilibrium constants relating to various temperatures. In such cases ΔH may be obtained by means of a separate calorimetric measurement, and the entropy term is found in the usual manner from this ΔH value and from ΔG calculated from the equilibrium constants, relating to a single temperature:

$$\Delta G = \Delta H - T\Delta S$$

Even simpler and more generally applicable is the calorimetric equilibrium measurement widely used by Drago *et al.* [Dr 73], which permits the calculation not only of ΔH, but also of the equilibrium constant from calorimetric data. This method makes possible the study of systems where the acid–base interaction is not accompanied by a significant change in the absorption spectrum.

The NMR technique can alco be used to advantage in many cases to follow acid–base interactions [Sl 72]. If the NMR chemical shift of the free acid or the complex formed does not depend on the temperature, in addition to the equilibrium constants fairly accurate thermodynamic data can be obtained from the data measured at different temperatures.

Effect of the reference acceptor or reference donor

In determining the donor or acceptor strengths of molecules, the correct choice of the reaction partner (reference acceptor or reference donor) is also of fundamental importance. It must be borne in mind that both the donor and the acceptor strength sequences inevitably depend on the reaction partner. This fact is well reflected by the heats of reaction of adduct formation for molecules containing oxygen or sulphur donor atoms with either phenol or elemental iodine (Table 4.10) [Ni 64]. It can be seen that compounds with sulphur donor atom systems behave as stronger bases towards iodine than the analogous oxygen donor atom systems. However, the sequence is the reverse when phenol is used as the Lewis acid.

Table 4.10. Heats of reaction of adduct formation of molecules containing oxygen and sulphur donor atoms with elemental iodine and phenol (kcal/mole) [Ni 64]

Donor	Acceptor	
	I_2	C_6H_5OH
$(C_2H_5)_2O$	4.3	7.8
$(C_2H_5)_2S$	5.1	4.6
$CH_3CO.\ N(CH_3)_2$	4.0	6.1
$CH_3CS.\ N(CH_3)_2$	9.5	5.5

The explanation of this effect can be conceived as follows. Phenol has a fairly high dipole moment and has no low-energy acceptor orbitals, whereas iodine has no dipole moment; hence interactions with iodine may be expected to have more covalent character than the analogous reactions with phenol. Accordingly, iodine will react more readily with the better polarizable reaction partners possessing lower ionization potentials. Similar considerations may be employed to interpret, for example, the sequence of basic strengths of primary, secondary and tertiary amines [Dr 63], and the sequence of acid strengths of iodine monochloride, elemental bromine, elemental iodine, phenol and sulphur dioxide [Dr 62].

The above explanation is based on the application of the Mulliken model. In accordance with this, the donor–acceptor bonding can be written as a linear combination of the covalent (charge-transfer) and electrostatic wave functions:

$$\Psi = a\Psi_{\text{cov.}} + b\Psi_{\text{el.}}$$

The phenomenon can also be interpreted qualitatively according to Pearson as "hard–hard" and "soft–soft" interactions. One can similarly make use of the Ahrland concept [Ah 66] of the interaction of reaction partners belonging to groups *A* and *B*.

The last two approaches have the common drawback that at best they may give only a qualitative picture of the interaction, and can explain the phenomenon only subsequently. Their application is particularly difficult in the borderline cases where it is hard to decide the positions of the reaction partners in the above classification.

As even this brief account shows, the Lewis donor–acceptor equilibrium and its thermodynamic data are strongly influenced by the inert solvent, the experimental method and especially the nature of the reaction partner.

The Drago concept

On the basis of the above and analogous considerations, Drago [Dr 73] developed a theory, which could yield information of general validity for the characterization of the strengths of Lewis acid–base interactions.

The essence of Drago's considerations is that the ΔH value of the donor–acceptor reaction is written as the sum of two terms, the electrostatic and the covalent contributions:

$$-\Delta H = E_A E_B + C_A C_B$$

where the subscripts A and B indicate the acceptor and donor molecules, and E and C are empirical parameters typical of the system, which may serve to characterize the electrostatic and covalent contributions, respectively. Drago mainly calculated the E and C values from enthalpy data for the reactions with iodine and phenol. An extremely large number of experimental data were subjected to computer evaluation, to yield E and C values regarded as being of general validity [Dr 65a], the utilizability of which was verified in many systems [Dr 65a, Ha 64, Li 70].

As regards their chemical meaning, the two terms of the Drago equation (formally the electrostatic and the covalent term) might reflect the σ-interaction and the π-interaction also. However, the experimental data are indicative of the former explanation.

The greater complexity of Drago's equation in comparison with earlier correlations could be expected to allow a higher effectiveness, i.e. applicability to the interpretation of a greater number of different reaction types. In reality this has not proved to be the case. The explanation is that the effects of the various factors governing the different donor–acceptor interactions are so complex that even this two-term equation does not lead much closer than the simpler correlations to an understanding of the overall process.

A complete description of a given system can be expected only from an experimental study of the individual reactions in that system. Nevertheless, many efforts have been made to apply a great number of different, more or less general, correlations (possibly elaborated for other purposes) for the description of donor–acceptor interactions.

For example, the Hammett equation, well known in organic chemistry, has also been used for the interpretation of donor–acceptor adduct formation reactions. This equation was originally derived to interpret the effects of electrophilic or nucleophilic substituents on chemical equilibria and reaction rates [Le 63, Sh 63, We 63]. However, it well reflects the Gibbs free energies (ΔG) of numerous reactions, the order of the frequencies of infrared vibrations [La 67, Pi 60] and, particularly important as regards the present considerations, the heats of adduct formation between phenols and molecules containing oxygen, nitrogen and sulphur donor atoms [Dr 69, Vo 70].

A correlation can be demonstrated between the Drago and Hammett equations [Dr 73], starting from the following form of the Hammett equation:

$$-\Delta H = \sigma \varrho$$

where σ is the Hammett substitution constant and ϱ is a constant characteristic of the reaction type.

If the last term of Drago's equation is regarded as the heat of adduct formation between the given base and unsubstituted phenol (considered to be the reference), we can write

$$-\Delta H_B = C_A C_B$$

Hence, C_A is comparable to the σ value of the Hammett equation, and C_B to its ϱ value. With the aid of the appropriate mathematics, Drago [Dr 73] attempted to find a quantitative connection between the Hammett constants and his own constants E and C. However, this proved successful only in a very limited range.

With the utilization of the large number of E and C values tabulated separately for Lewis acids and Lewis bases by Drago, at present the heats of reactions of about 1000 Lewis donor–acceptor interactions can be calculated with Drago's equation. The agreement between the calculated and measured data is not equally good in all systems. This can be attributed in part to the fact that the E and C values cannot be given with the same accuracy for all molecules. The greater the number of reactions whose ΔH values were used in calculating the E and C values characterizing an acid or base, the more reliable are these data, and hence the greater is the probability that other reactions will also be described correctly by them.

Secondly, the occurrence of some effects (e.g., steric hindrance, π back-donation) which did not feature (or appeared in a different manner) in the reactions used for determining the parameters E and C can naturally not be reflected correctly in the calculated data.

When all of these limitations are taken into account, it should be emphasized that the parameters E and C introduced by Drago, in general, faithfully reflect the sequence of acceptor and donor strengths for both acids and bases. The E and C data relating to some of the more important systems are presented in Tables 4.11 and 4.12. These data are conventionally expressed relative to elemental iodine, for which, therefore, $E_A = 1.00$ and $C_A = 1.00$. This means that the absolute values of

Table 4.11. Drago's C_A and E_A parameters for Lewis acids [Dr 73]

Lewis acid	C_A	E_A
Iodine	1.00	1.00
Phenol	0.442	4.33
p-Fluorophenol	0.446	4.17
p-Chlorophenol	0.478	4.34
m-Fluorophenol	0.506	4.42
Pyrrole	0.295	2.54
Sulphur dioxide	0.808	0.92
Antimony(V) chloride	5.130	7.38
Chloroform	0.159	3.02
Methylcobaloxime	1.530	9.14

Table 4.12. Drago's C_B and E_B values for Lewis bases [Dr 73]

Lewis base	C_B	E_B
Pyridine	6.40	1.17
Ammonia	3.46	1.36
Methylamine	5.88	1.30
Dimethylamine	8.73	1.09
Trimethylamine	11.54	0.808
Acetonitrile	1.34	0.886
Dimethylformamide	2.48	1.23
Dimethylacetamide	2.58	1.32
Ethyl acetate	1.74	0.975
Methyl acetate	1.61	0.903
Acetone	2.33	0.987
Diethyl ether	3.25	0.963
p-Dioxane	2.38	1.09
Tetrahydrofurane	4.27	0.978
Tetramethylene sulphoxide	3.16	1.38
Dimethyl sulphoxide	7.46	0.343
Diethyl sulphoxide	7.40	0.339
Pyridine-N-oxide	4.52	1.34
Hexamethylphosphoramide	3.55	1.52

the parameters E and C have no particular significance (if necessary, they can be converted to other scales), and it is their interrelation, sequence, etc., which are suitable for the drawing of chemical conclusions.

The Drago concept and its applicability are also supported by studies which demonstrate that data calculated by means of Drago's equation, and relating separately to the magnitudes of the electrostatic and the covalent contributions, are in good agreement with the data obtained with semi-quantitative calculations on the basis of dipole moment measurements on the given systems [Su 57, Ki 61]. It must be noted, however, that this correlation is not valid for all systems, and Drago's data include a number of examples which are contradictory to chemical knowledge relating to the systems.

On the basis of Drago's concept other two parameter empirical models were suggested [e.g. Sh 70a, Ko 75, 77a, La 81a, b, Ma 79, Ri 80] for the generalization of different type of weak interactions (not only solvation) in different fields of chemistry.

Drago [Dr 73] considered how his concept could be correlated with Pearson's "hard–soft" acid–base theory [Pe 63b, Kl 68]. Setting out from the assumption that the terms $E_A E_B$ and $C_A C_B$ may serve as measures of the electrostatic and covalent interactions, respectively, and identifying the "hard" and "soft" interactions of the Pearson concept with the electrostatic and the covalent interactions, respectively, Drago wished to employ the C/E ratios for the acids or bases examined as a measure of the "soft" nature of the molecules. He considered that the higher the value of C/E, the "softer" is the given acid or base, whereas low value of C/E is indicative of a "hard" character.

However, the sequence of C/E values is not in accordance with practical experience with the solvents. Drago's parameters and his equation also could not be used to describe Pearson's "hard–soft" concept.

The Koppel–Palm concept

Drago's concept has the merit that two specific factors governing the solvent effect are written together in one common equation. As seen above, characterization of the solvent effect by means of this model succeeded well in certain systems, and less well in others. It appeared that replacement of the two-parameter approach by several parameters might lead to a more generally valid solution. For a joint description of non-specific and specific effects, Koppel and Palm [Ko 72] proposed the introduction of the following four-parameter equation:

$$Q = Q_0 + \alpha A + \beta B + \eta Y + \pi P$$

where Q is any physico-chemical parameter varying under the action of the solvent,

A and B are measures of the Lewis acidity and basicity, and Y and P are non-specific parameters characterizing the polarity and polarizability, respectively, of the solvent.

The polarity of the solvent can be described in terms of the relative permittivity (ε), and its polarizability in terms of the refractive index (n), as follows:

$$Y = \frac{\varepsilon - 1}{\varepsilon + 2}$$

$$P = \frac{n^2 - 1}{n^2 + 2}$$

The parameter B, characterizing the base strength, was obtained from infrared measurements (see also p. 50); the shift of the O–D vibration on the action of the solvent referred to that of deuterated methanol in gas phase gave a number that was a measure of the base strength of the solvent:

$$B = \nu_{\mathrm{OD(gas)}} - \nu_{\mathrm{OD(solution)}}$$

The factor A, typical of the acid strength, was calculated by correcting the Dimroth–Reichert parameter E_T (see p. 62) on the basis of the dielectric model:

$$A = E_T - 25.57 - 1.49\left[\frac{\varepsilon - 1}{\varepsilon + 2}\right] - 9.08\left[\frac{n^2 - 1}{n^2 + 2}\right]$$

Koppel and Palm employed the above model to describe the solvent dependences of kinetic and spectroscopic data in numerous systems. However, statistically reliable values for the regression coefficients α, β, η and π were obtained for only relatively few systems.

The model has the fundamental defect that, in the non-specific terms, it uses two macroscopic parameters (ε and n), which clearly differ from the microscopic ε and n values of the solvent bound in the solvate sheath and governing the solvent effect. Unfortunately, there is no means of obtaining these microscopic parameters directly. This defect increased by the fact that the macroscopic parameters are used to "correct" the parameter E_T by which the acid strength is characterized.

Perhaps the most successful application of the Koppel–Palm model to date is to be attributed to Chapman *et al.* [Ch 74], who used it to describe the solvent dependence of the rate of the reaction between benzoic acid and diazophenylmethane. The rate constants determined in 24 different solvents were processed, and a very good correlation was obtained. However, processing of the same data on the basis of the model of Krygowski and Fawcett resulted in an even better correlation

[Fa 75]. The examinations in effect demonstrated that in the description of the solvent effect it is sufficient to take into account the specific solvent–solute (i.e., the Lewis acid–base) interactions, since in any case these also reflect the non-specific interactions. This consideration is accepted in the model of Krygowski and Fawcett [Kr 75] called "the complementary Lewis acid–base description of solvent effects".

The Krygowski–Fawcett concept

Krygowski and Fawcett described the variation of the physico-chemical parameter Q, reflecting the solvent effect, in an analogous manner to that in the first half of the equation of the Koppel–Palm model, but as a linear function of only two independent complementary terms:

$$Q = Q_0 + \alpha E_T + \beta DN$$

where E_T is the Dimroth–Reichardt parameter (see. p. 62) characterizing the Lewis acidity [Di 63], DN is the Gutmann donicity value (see p. 45) characterizing the Lewis base strength, α and β are constants characterizing the acid and base sensitivities of the examined property and Q_0 is the value of the property Q in the solvent-free system (Q_0 can generally only be calculated, but not measured experimentally).

With the aim of bringing the regression coefficients α and β into a common scale, Krygowski and Fawcett calculated the partial regression coefficients α' and β' by the standard method of multi-parameter regression. From these they obtained

$$\bar{\alpha} = \frac{100\alpha'}{\alpha' + \beta'} \quad \text{and} \quad \bar{\beta} = \frac{100\beta'}{\alpha' + \beta'}$$

The values $\bar{\alpha}$ and $\bar{\beta}$ give the acid and base contributions of the solvent to the solvent effect directly as percentages.

Krygowski and Fawcett found their model to be suitable for the description of the variation of the most different solvent-dependent properties (which thus reflect the solvent effect) and of the measurable physico-chemical parameters characterizing these properties. Its advantage is that it simultaneously shows the extents of the effects of both the acid and the base within a given system, if both are reflected in the experimental result. The signs of the coefficients α and β also show whether the proportionality between the examined property and the acid or base strength is direct or inverse.

The practical applicability of the model was tested by evaluating various series of measurements involving solvent effects. In every evaluation the correlation coefficient, R, relating to the system is given correctly. Particularly convincing are the data which, by means of this model, characterize some of the parameters successfully used in practice for a quantitative description of the solvent effect. A number of these are listed in Table 4.13. The parameters examined were derived from measurements of various types. The Kosower Z value originates from the electronic excitation spectrum, and the Koppel–Palm B value and the Kagiya $\Delta\nu_{OD}$ value from the infrared spectrum, the Smith log k_{ion} value is a reaction rate constant, and the Arnett $\Delta\bar{H}$ value is a partial molar heat of dissolution. Two of these parameters (those of Kosower and Smith) characterize the acceptor strength, and the other three the donor strength. Nevertheless, as the final column in Table 4.13 reveals, all of these parameters showed a very good correlation in the Krygowski–Fawcett model.

Table 4.13. Characterization of some solvent effect parameters by means of the Krygowski–Fawcett model [Kr 75, Fa 75]

Parameter	Ref.	α	β	Q_0	$\bar{\alpha}$	$\bar{\beta}$	R
Kosower Z	Ko 58	1.3	−0.02	10.1	98	2	0.97
Koppel–Palm B	Ko 72	−0.1	4.9	41	1	99	0.974
Smith log k_{ion}	Sm 61	0.22	0	−14.4	99	1	0.908
Kagiya ν_{OD}	Ka 68	0.1	4.8	−15.4	0.4	99.6	0.967
Arnett $-\Delta\bar{H}$	Ar 74	−0.1	1.2	0.0	4	96	0.963

Krygowski and Fawcett also utilized their model to process various literature experimental data reflecting the solvent effect [Kr 75, Fa 75, Fa 76]. They found excellent agreement with the solvent-influenced ^{23}Na NMR spectroscopic data of Erlich *et al.* [Er 70, Gr 73], with Parker's solvent activity coefficients for various monovalent ions [Pa 69], with the heat of solvation values of Cox [Co 74], with the solvent dependence of polarographic half-wave potentials taken from the measurements of Fujinaga [Fu 71], etc.

The Krygowski–Fawcett model has the drawback that it does not take into account the entropy term in the solvent effect. Thus, it is primarily suitable for the description of processes in which the role of the entropy change is a subordinate one. For example, it describes heats of solvation excellently ($R \approx 0.91$–0.97), whereas the characterization of free energies of solvation and especially entropy changes is much more uncertain ($R<0.8$). Accordingly, Krygowski and Fawcett suggest the development of an expanded, three-parameter model, which also includes consideration of the entropy change. This would be of importance primarily in the processing of results obtained in solvents possessing a hydrogen-bonded internal structure. The significant role of the entropy change in such systems has been pointed out by Cox *et al.* [Co 74]. However, the data available have not led so far to the elaboration of such a three-parameter model.

The Gutmann–Mayer concept

For characterization of the solvent effect, Mayer [Ma 79] introduced a two-parameter approximation similar to the Krygowski–Fawcett concept. He experimented originally with a three-parameter approximation, using the Gutmann donicity (see p. 45) as a measure of the base strength of the solvent and the acceptor number (see p. 64) as a characteristic of its acid strength; the free energy of vaporization was used as the third parameter, to take into account cavity formation necessary for acceptance of the dissolved ions. With this three-parameter approximation he calculated the standard free energies of dissolution of various salts. The original concept of characterizing the solvent effect may therefore be described with the following equation:

$$\Delta G^S - \Delta G^R = a(\mathrm{DN}^S - \mathrm{DN}^R) + b(\mathrm{AN}^S - \mathrm{AN}^R) + c(\Delta G_{\mathrm{vp}}^{0S} - \Delta G_{\mathrm{vp}}^{0R})$$

where ΔG^S and ΔG^R are the standard free energies of the reaction examined in various solvents S and in the reference solvent R, respectively; DN, AN and ΔG_{vp}^0 are the corresponding donor numbers, acceptor numbers and standard free energies of vaporization. The coefficients a, b and c are constants characteristic of the system in question, and show the weights with which the electron-pair donor ability of the solvent (a), its acceptor ability (b) and the energy of cavity formation in the solvent (c) feature in the solvent effect.

The practical applicability of the Gutmann–Mayer concept was proved by calculation of the standard free energies of dissolution (ΔG values) of sodium chloride and potassium chloride in various solvents (Tables 4.14 and 4.15), of the ΔG values of the coordination reaction $CoCl_3^- + Cl^- \rightarrow CoCl_4^{2-}$ in various organic solvents (Table 4.16) and of the rate constant of the reaction of *p*-nitrofluorobenzene with piperidine (Table 4.17).

Table 4.14 contains ΔG values calculated in two ways, together with the experimental standard ΔG values for dissolution of sodium chloride, determined in various solvents. As regards the calculated values, the first (I) was obtained from the above equation, whereas in the calculation of the second (II) the term proportional to the standard free energy of vaporization, describing the energy of cavity formation in the solvent, was replaced with a constant C.

It can be seen from Table 4.14 that the third term of the Mayer equation may satisfactorily be replaced with the constant C. It is clear that, when the energy of cavity formation is neglected, the agreement of the calculated and measured values is not worse than in the original three-parameter calculation. Mayer found similar results in his investigations of other systems. Thus, his originally three-parameter approximation was changed to a two-parameter model, the term initially relating to the free energy of vaporization being replaced with a constant fitting parameter C, characteristic of the system under examination. Tables 4.15–4.17 list the data calculated with the two-parameter approximation only. The parameters a and b obtained by calculation are given at the foot of each table; these are characteristic of the donor and acceptor effect, respectively. The constant C is also given, together with the correlation coefficient, R, and the mean error, σ.

In all four systems examined, the values of the coefficients of DN and AN show that both the donor and the acceptor properties of the solvent play a role in the reaction. In the dissolution of sodium chloride, for instance, there is scarcely any

Table 4.14. Experimental and calculated standard free energies of dissolution of sodium chloride in non-aqueous solvents at 25 °C [Ma 79]

Solvent	$\Delta G_{(exp)}$	$\Delta G_{(calc)}$ (kcal/mole)	
		I*	II**
Dimethyl sulphoxide	3.57	3.51	3.53
Dimethylformamide	6.41	6.53	6.44
Dimethylacetamide	7.02	6.98	6.96
N-Methylpyrrolidone	7.25	7.25	7.32
Sulpholane	10.01	10.16	10.37
Propylene carbonate	10.69	10.53	10.67
Acetonitrile	11.18	11.17	10.86
Formamide	−0.09	−2.49	−2.44
N-Methylformamide	0.90	−0.04	0.06
Methanol	3.37	−0.87	−1.32

* I: $a=-0.45$; $b=-0.436$; $C=-0.102$; $R=0.999$; $\sigma=0.08$
** II: $a=-0.456$; $b=-0.442$; $C=-0.323$; $R=0.997$; $\sigma=0.13$

Table 4.15. Experimental and calculated ΔG values for the dissolution of potassium chloride in non-aqueous solvents at 25° [Ma 79]

Solvent	$\Delta G_{(exp)}$ (kcal/mole)	$\Delta G_{(calc)}$ (kcal/mole)
Dimethyl sulphoxide	4.70	4.77
Dimethylformamide	7.49	7.28
Dimethylacetamide	7.94	8.04
Propylene carbonate	9.44	9.67
Acetonitrile	9.88	9.70
Formamide	0.56	−2.76
N-Methylformamide	1.65	0.24
Methanol	4.35	−2.28

$a=-0.302$; $b=-0.467$; $C=-0.186$; $R=0.996$; $\sigma=0.16$

difference between the coefficients of DN and AN, which shows well that in dissolution of sodium chloride approximately the same role is played by the interaction between the sodium ion and the solvent (as donor) as by the interaction between the chloride ion and the solvent (as acceptor). As expected on the basis of the weaker solvation of the potassium ion, the acceptor effect was greater with potassium chloride.

Table 4.16. Experimental and calculated ΔG values for the reaction $CoCl_3^- + Cl^- \rightarrow CoCl_4^{2-}$ in various non-aqueous solutions at 25 °C [Ma 79]

Solvent	$\Delta G_{(exp)}$ (kcal/mole)	$\Delta G_{(calc)}$ (kcal/mole)
Nitromethane	−6.59	−6.42
Acetone	−4.69	−4.69
Acetonitrile	−3.92	−4.13
Dimethylacetamide	−2.05	−2.02
Dimethylformamide	−1.64	−1.83
Hexamethylphosphoric triamide	−0.16	−0.08

$a=0.228$; $b=0.191$; $C=-0.209$; $R=0.998$; $\sigma=0.11$

Table 4.17. Experimental and calculated rate constants (k) for the reaction between *p*-nitrofluorobenzene and piperidine in non-aqueous solutions at 25 °C [Ma 79]

Solvent	$RT \ln k_{exp}$	$RT \ln k_{calc}$
	(dm^3 $mole^{-1}$ s^{-1})	
Benzene	−7.39	−7.69
Ethyl acetate	−5.96	−5.69
Dimethoxyethane	−5.84	−5.37
Methyl acetate	−5.71	−5.41
Tetrahydrofuran	−5.67	−5.72
Acetone	−4.69	−4.91
Benzonitrile	−4.68	−4.69
Nitromethane	−4.21	−4.38
Acetonitrile	−4.01	−3.63
Dimethylformamide	−2.91	−3.08
Dimethylacetamide	−2.85	−3.55
Dimethyl sulphoxide	−2.13	−1.94
Isopropanol	−5.28	(0.52)
Methanol	−5.25	(2.41)
Ethanol	−5.25	(1.41)

$a=-0.101$; $b=-0.247$; $C=-0.384$; $R=0.976$; $\sigma=0.27$

In the reaction between trichlorocobaltate and chloride ions, the role of the donor properties of the solvent is more marked, probably because the fourth coordination site of the cobalt atom in the trichloro complex is occupied by a solvent molecule. The higher the donor strength of the solvent, the more difficult it is to replace it with chloride ion. On the other hand, the magnitude of the coefficient of AN shows that the acceptor strength of the solvent also exerts an influence on the system. This is manifested predominantly *via* solvation of the chloride ion, since coordination of the chloride to the cobalt also assumes desolvation of the chloride ion.

The rate of the reaction between *p*-nitrofluorobenzene and piperidine depends more strongly on the acceptor strength of the solvent than on its donicity, although both effects are undoubtedly exerted.

O_2N—F + HN ⟶ O_2N= ($H^{+\delta}$, N^+, F)

The double effect in this system cannot be explained by the solvation of the reactants. The coordination of piperidine increases the acidity of the NH group, and at the same time it decreases the basicity of the partner molecule. Hence, the complex formed is solvated by both nucleophilic and electrophilic solvent attack. The proportions of these show up in the coefficients of DN and AN.

The signs of *a* and *b* provide another source of information, as they reflect the direction of the solvent effect. Thus, in the calculation of the free energies of dissolution (ΔG), increases in the negative values of the coefficients mean an increase in the solvation effect and hence in solubility. In the reaction between *p*-nitrofluorobenzene and piperidine the coefficients *a* and *b* are both negative, showing the effect of increased solvation in the elevation of the reaction rate. The reaction between trichlorocobaltate and chloride ions is inhibited by increasing solvation. Correspondingly, in this system the values of the coefficients of AN and DN are both positive.

By presenting the joint occurrence of the double (donor and acceptor) effects of the solvents, this model also explained why it was not possible to describe solvent effects in these systems by means of a single empirical parameter. It is obvious that the standard ΔG values for dissolution, for example, could not give a linear correlation with parameters predominantly reflecting a single type of effect, e.g., the Kosower Z value, the Reichardt E_T value, the DN value, the AN value, the relative permittivity of the solvent alone. Even in systems where one of the effects is so predominant that the system can be described more or less with a single parameter, only calculations in which the other parameters are also taken into consideration may give information on whether the effect of the latter is present at all in the system and, if so, to what extent.

A good example of this is the solvent dependence of the first-order rate constant of the solvolysis of *t*-butyl chloride. This reaction served as the basis of the determination of the Winstein solvent strength parameter (the logarithm of the rate constant) (see p. 67). It was later shown that the Winstein parameter could be brought into a fairly good linear correlation with the acceptor numbers of the solvents; this result indicates that the rate of the reaction is defined in effect by the strength of solvation of the chloride ion through hydrogen bonding. These investigations suggest that the donor properties of the solvent do not play a part in this reaction. The Mayer two-parameter model, however, pointed to the role of the donicity also, even if the value of coefficient a is small (-0.065) [Ma 79].

The rate of solvolysis of *p*-methoxyphenyl toluenesulphonate, or the logarithm of the rate constant, can also be used for the characterization of the acceptor strengths of solvents. When plotted as a function of the acceptor numbers of the solvents, this parameter similarly varies linearly. Processing of the data on the basis of the Gutmann–Mayer model revealed that consideration of the donor properties of the solvents in this system did not result in a better agreement between the calculated data and the experimental values, i.e., the solvent does not act as a donor in these reactions.

The tables also show that calculations with the two-parameter equation are successful only for certain solvents. The last three entries in Tables 4.14, 4.15 and 4.17 show the results of calculations for some protic solvents for comparison. It can be seen that the calculated values may differ from the experimental values even as regards the sign. It appears that this is due to changes occurring in the structure of the solvent itself in the course of the reaction; these are not taken into account by this model.

All of these investigations show that solvent effects can be described correctly only by means of empirical parameters (with an accuracy corresponding to that of the experimental method), if both the donor and acceptor properties of the solvent can be taken into consideration, i.e., a two-parameter approximation is used. It is certain, however, that agreement with the experimental values of the data obtained by the combined consideration of these two effects shows merely that the donor–acceptor interactions predominate in the given systems, but other properties may also have effects. Attention is drawn to the letter by those systems in which the reactions cannot be described by the two-parameter models either.

It has been mentioned that the authors of the various two-parameter models also experimented with the introduction of models involving three or even more parameters. Different solvent properties were tried as the third parameter. The experiments to date indicate that a property of general validity, similar to the donor–acceptor properties and thus utilizable as the third parameter, has not been found. This means that the introduction of some new empirical solvent scales (based on the experimental determination of some parameter characteristic of the given solvent or dissolved system) may be necessary if solvent effects in a new, individual reaction are to be interpreted. The establishment of such a scale is also an analytical task.

The linear solvation energy relationship (Kamlet–Taft) concept [Ka 80, Ta 81]

Taft and coworkers described the formulation of three scales of solvent properties which were used to unravel and rationalize solvent effects on many types of physico-chemical properties. A π^* scale of polarity/polarizabilities describes the solvent's ability to stabilize a charge or a dipole by virtue of its dielectric effect. The π^* values have been shown to be generally proportional to molecular dipole moments. The α scale of hydrogen bond donor acidities provides a measure of the solvent's ability to donate a proton. The β scale of hydrogen bond acceptor basicities quantifies the solvent's ability to donate an electron pair (accept a proton).

The equations used in the Taft conception are the following:

(*a*) for the π^* scale

$$\mathrm{XYZ} = \mathrm{XYZ}_0 + s\pi^*$$

(*b*) for the α scale

$$\mathrm{XYZ} = \mathrm{XYZ}_0 + s\pi^* + a\alpha$$

(*c*) for the β scale

$$\mathrm{XYZ} = \mathrm{XYZ}_0 + s\pi^* + b\beta$$

(*d*) each of eq *a*–*c* can be extended using a correction term δ, e.g. for eq *a*:

$$\mathrm{XYZ} = \mathrm{XYZ}_0 + s(\pi^* + d\delta)\,.$$

XYZ denotes the property measured, *s*, *a* and *b* are the measures of the response of XYZ to changing the solvent, π^* is the polarity, α the acidity and β the basicity parameter.

This conception was used [Ta 80] to unravel solvent polarity and hydrogen bonding effects on a variety of NMR spectral shifts and coupling constants. The properties analyzed include ^{19}F, ^{1}H, ^{13}C, ^{15}N and ^{29}Si shifts and J(^{13}CH), J(^{119}Sn, C, ^{1}H) and J(^{119}Sn, C, ^{19}F) coupling constants.

The Taft method was also used to examine relationships between Gutmann's solvent donicity (DN) and acceptor number (AN) and the π^*, α, and β [Ta 81]. It is shown that the AN for nonprotonic solvents correlates well with π^* and for protonic solvents with a linear combination of π^* and α. It was therefore concluded that AN, represented as a measure of the solvent's ability to serve as an electron-pair acceptor, is, in fact, a combined measure of solvent polarity/polarizability and hydrogen bond donor ability. It is shown that DN is linear with β for oxygen bases and RCN nitrogen bases but that the correlation breaks down for pyridine. Some of the data are compared in Table 4.18. The linear solvation energy relationship concept proved to be useful in gaining structural and mechanistic information on the solvent effect n several different systems [Ab 82]. Sjöström and Wold [Sj 81] performed a statistical analysis of the π^* scale and showed that for the adequate description of the data a two component model is needed.

Bekarek [Be 81] has shown a good correlation between Taft–Kamlet's π^* parameter and his solvent polarity scale based on the simple function of relative permittivity (ε) and refractive index (n) of the solvent: $(\varepsilon-1)(n^2-1)/(2\varepsilon+1)(2n^2+1)$.

Table 4.18. Comparison of Gutmann's donicity (DN) and acceptor-number (AN) scales with the Kamlet–Taft parameters π^*, α, and β [Ta 81]

Solvent	π^*	α	β	DN	AN
n-Hexane	−0.08	0	0		0
Triethylamine	0.14	0	0.71	30.7	
Carbon tetrachloride	0.29	0			8.6
Diethyl ether	0.27	0	0.47	19.2	3.9
Dioxane	0.55	0	0.37	14.8	10.8
Ethyl acetate	0.55	0	0.45	17.1	
Tetrahydrofuran	0.58	0	0.55	20.0	8.0
Benzene	0.59	0	0.10	0.1	8.2
Acetone	0.68	0.10	0.48	17.0	12.5
1,2-Dichloroethane	0.81	0	0	0	
N,N-Dimethylacetamide	0.88	0	0.76	27.8	13.6
Pyridine	0.87	0	0.64	33.1	14.2
Dimethylformamide	0.88	0	0.69	26.6	16.0
Hexamethylphosphoramide	0.87	0	1.05	38.8	10.6
N-Methylpyrrolidone	0.82	0	0.77	27.3	13.3
Dimethyl sulphoxide	1.00	0	0.76	29.8	19.3
Nitrobenzene	1.01	0	–	4.4	14.8
Nitromethane	0.80	0.29	0	2.7	20.5
Benzonitrile	0.90	0	0.41	11.9	15.5
Acetonitrile	0.76	0.22	0.31	14.1	19.3
Methyl acetate	0.56	0	0.42	16.5	
1,2-Dimethoxyethane	0.53	0	0.41	20.0	
Bis(2-methoxyethyl)-ether	0.64	0			10.2
2-Propanol	0.47	0.77			33.5
Ethanol	0.54	0.85			37.1
Methanol	0.60	0.98			41.3
Water	1.09	1.10			54.8
Formamide	0.85	0.77			39.8

References

Ab 77 Abbaud, J. L., Kamlet, M. J., Taft, R. W.: J. Am. Chem. Soc. **99,** 8325 (1977).

Ab 79 Abbaud, J. L., Taft, R. W.: J. Phys. Chem. **83,** 412 (1979).

Ab 82 Abboud, J. L., Taft, R. W., Kamlet, M. J.: Bull. Chem. Soc. Jpn. **55,** 603 (1982).

Ah 66 Ahrland, S.: Structure and Bonding **1,** 207 (1966).

Al 63 Allerhand, A., Schleyer, P. R.: J. Am. Chem. Soc. **85,** 371 (1963).

An 69 Anderson, R. G., Symons, M. C. R.: Trans. Faraday Soc. **65,** 2537 (1969).

Ar 70 Arnett, E. M., Joris, L., Mitchell, E., Murty, T. S. S. R., Gorrie, T. M., Schleyer, P. R.: J. Am. Chem. Soc. **92,** 2365 (1970).

Ar 74 Arnett, E. M., Mitchell, E. J., Murty, T. S. S. R.: J. Am. Chem. Soc. **96,** 3875 (1974).

Ba 37 Badger, P. M., Bauer, S. H.: J. Chem. Phys. **5,** 839 (1937).

Ba 74 Barcza, L., Pope, M. T.: J. Chem. Phys. **78,** 168 (1974).

Be 62 Berson, J. S., Hamlet, Z., Mueller, W. A.: J. Am. Chem. Soc. **84,** 297 (1962).

Be 74 Benoit, R. L., Domain, R.: Abstracts IV. ICNAS (Gutmann, V. ed.). Vienna 1974, p. 10.

Be 81 Bekárek, V.: J. Phys. Chem. **85,** 722 (1981).

Bl 68 Bloor, E. G., Kidd, R. G.: Canad. J. Chem. **46,** 3425 (1968).

Bo 66 Bolles, T. F., Drago, R. S.: J. Am. Chem. Soc. **88,** 3921 (1966).

Br 51 Brooker, L. G. S., Keyes, G. H., Heseltine, D. W.: J. Am. Chem. Soc. **73,** 5350 (1951).

Br 61 Briegleb, G.: Elektronen-Donor-Akzeptor-Komplexe. Springer, Berlin–Göttingen–Heidelberg 1961.

Br 65 Brooker, L. G. S., Craig, A. C., Heseltine, D. W., Jenkins, P. W., Lincoln, L. L.: J. Am. Chem. Soc. **87,** 2443 (1965).

Bu 63 Burger, K., Ruff, I.: Talanta **10,** 329 (1963).

Bu 71 Burger, K., Zelei, B., Szántó-Horváth, G., Tran, T. B.: J. Inorg. Nucl. Chem. **33,** 2573 (1971).

Bu 74 Burger, K., Gaizer, F., Papp-Molnár, E., Tran, T. B.: J. Inorg. Nucl. Chem. **36,** 863 (1974); Magy. Kém. Foly. **79,** 425 (1973).

Bu 75 Burger, K., Gaizer, F., Márkus, M., Madeja, K.: Unpublished results.

Ca 75 Cahen, Y. M., Hardy, P. R., Roach, E. T., Popov, A. I.: J. Phys. Chem. **79,** 80 (1975).

Ch 74 Chapman, N. B., Dack, M. R. J., Neroman, D. J., Shorter, J., Wilkinson, R.: J. Chem. Soc. Perkin Trans. II, 9622 (1974).

Ch 82 Chastrette, M., Carretto, J.: Tetrahedron **38,** 1615 (1982).

Co 74 Cox, B. G., Hedwigh, G. R., Parker, A. J., Watts, D. W.: Austral. J. Chem. **27,** 477 (1974).

Co 80 Cox, B. G., Waghorne, W. E.: Chem. Soc. Rev. **9,** 381 (1980).

Di 63 Dimroth, K., Reichardt, C., Siepmann, T., Bohlmann, F.: Justus Liebigs Ann. Chem. **661,** 1 (1963).

Di 73 Dimroth, K., Reichardt, C., Schweig, A.: Justus Liebigs Ann. Chem. **669,** 95 (1973).

Dr 62 Drago, R. S., Wenz, D. A.: J. Am. Chem. Soc. **84,** 526 (1962).

Dr 63 Drago, R. S., Meek, D. W., Longhi, R., Joesten, M. D.: Inorg. Chem. **2,** 1056 (1963).

Dr 65a Drago, R. S., Vogel, G. C., Needham, T. E.: J. Am. Chem. Soc. **87,** 3571 (1965).

Dr 65b Drago, R. S., Mode, V. A., Kay, J. G., Lydy, D. L.: J. Am. Chem. Soc. **87,** 5010 (1965).

Dr 68 Drago, R. S., Matwiyoff, N. A.: Acids and Bases. D. C. Heath, Boston, Mass. 1968.

Dr 69 Drago, R. S., Epley, T. D.: J. Am. Chem. Soc. **91,** 2883 (1969).

Dr 72 Drago, R. S., Nozari, M. S., Vogel, G. C.: J. Am. Chem. Soc. **94,** 90 (1972).

Dr 73 Drago, R. S.: Structure and Bonding **15,** 73 (1973).

Dr 81 Drago, R. S., Kroeger, M. K., Stahlbush, J. R.: Inorg. Chem. **20,** 306 (1981).

Du 61 Dubois, J. E., Goetz, E., Bienvenue, A.: Spectrochim. Acta **20,** 1815 (1964).

Du 75 Dupire, S., Mulindab Yuma, J. M., B. Nagy, J., B. Nagy, O.: Tetrahedron **31,** 135 (1975).

Er 70 Erlich, R. H., Roach, E., Popov, A. I.: J. Am. Chem. Soc. **92,** 4989 (1970).
Er 71 Erlich, R. H., Popov, A. I.: J. Am. Chem. Soc. **93,** 5620 (1971).
Er 72 Erlich, R. H., Roach, E., Popov, A. I.: in Gutmann, V.: Topics in Current Chemistry **27,** 59 (1972).
Fa 75 Fawcett, W. R., Krygowski, T. M.: Austral. J. Chem. **28,** 2115 (1975).
Fa 76 Fawcett, W. R., Krygowski, T. M.: Canad. J. Chem. **54,** 3283 (1976).
Fi 81 Finkentey, J. H., Zimmermann, H. W.: Justus Liebigs Ann. Chem. **1,** 1 (1981).
Fo 69 Foster, R.: Organic Charge Transfer Complexes. Academic Press, London, 1969.
Fu 71 Fujinaga, T., Izutsu, K., Nomura, T.: J. Electroanal. Chem. **29,** 203 (1971).
Gi 64 Gielen, M., Nasielski, J.: J. Organomet. Chem. **1,** 173 (1964).
Go 65a Gorden, J. E.: J. Am. Chem. Soc. **87,** 4347 (1965).
Go 65b Goodenow, J. M., Tamres, M.: J. Chem. Phys. **43,** 3393 (1965).
Gr 48 Grünwald, E., Winstein, S.: J. Am. Chem. Soc. **70,** 846 (1948).
Gr 73 Greenberg, M. S., Bochner, R. L., Popov, A. I.: J. Phys. Chem. **77,** 2449 (1973).
Gr 77 Griffiths, T. G.: Personal communication; Pugh, D. C.: Ph.D. Thesis, Univ. of Leeds, 1977.
Gu 66 Gutmann, V., Steiniger, A., Wychera, E.: Mh. Chem. **87,** 460 (1966).
Gu 67–68 Gutmann, V.: Coord. Chem. Rev. **2,** 239 (1967); Coordination Chemistry in Non-Aqueous Solutions. Springer, Vienna–New York 1968.
Gu 69 Gurka, D., Taft, R. W.: J. Am. Chem. Soc. **91,** 4769 (1969).
Gu 74 Gutmann, V.: Proc. 5th Conf. Coord. Chem., Smolenice 1974, p. 81.
Gu 75 Gutmann, V.: Coord. Chem. Rev. **15,** 107 (1975).
Gu 76 Gutmann, V.: Electrochim. Acta **21,** 661 (1976).
Gu 80 Gurikov, Ju. V.: Zh. Phys. Khim. **54,** 1223 (1980).
Ha 37 Hammett, L. P.: J. Am. Chem. Soc. **59,** 96 (1937).
Ha 64 Hamilton, W. C.: Statistics in Physical Science. Ronald Press Co., New York, 1964.
Ha 77 Haberfield, P., Lux, M. S., Rosen, D.: J. Am. Chem. Soc. **99,** 6828 (1977).
Ha 79 Haberfield, P., Lux, M. S., Jasser, I., Rosen, D.: J. Am. Chem. Soc. **101,** 645 (1979).
Im 78 Imai, H., Yamashita, K.: Bull. Chem. Soc. Japan **51,** 3103 (1978).
Jo 66 Joesten, M. D., Drago, R. S.: J. Am. Chem. Soc. **88,** 1617 (1966).
Ka 68 Kagiya, T., Sumida, Y., Inoue, T.: Bull. Chem. Soc. Japan **41,** 767 (1968).
Ka 76 Kamlet, M. J., Taft, R. W.: J. Am. Chem. Soc. **98,** 377 (1976).
Ka 77 Kamlet, M. J., Abboud, J. L., Taft, R. W.: J. Am. Chem. Soc. **99,** 6027 (1977).
Ka 80 Kamlet, M. J., Abboud, J. L., Taft, R. W.: Prog. Phys. Org. Chem. **13,** 485 (1980).
Ki 61 Kimura, K., Fujishiro, R.: Bull. Chem. Soc. Japan **34,** 304 (1961).
Kl 68 Klopman, G.: J. Am. Chem. Soc. **90,** 223 (1968).
Ko 57 Kolthoff, I. M., Coetzee, J. F.: J. Am. Chem. Soc. **79,** 870 (1957).
Ko 58 Kosower, E. M.: J. Am. Chem. Soc. **80,** 3253, 3261, 3267 (1958).
Ko 68 Kosower, E. M.: Physical Organic Chemistry. Wiley, New York, 1968.
Ko 72 Koppel, A. I., Palm, V. A.: The Influence of the Solvent in Organic Reactivity. In: Advances in Linear Free Energy Relationships, Chapter 5 (Chapman, N. B., Shorter, J., eds.). Plenum Press, New York, 1972.
Ko 77 Kolling, O. W.: Anal. Chem. **49,** 591 (1977).
Ko 78 Kolling, O. W.: Anal. Chem. **50,** 1581 (1978).
Kr 75 Krygowski, T. M., Fawcett, W. R.: J. Am. Chem. Soc. **97,** 2143 (1975).
Kr 80 Krygowski, T. M., Milczarek, E., Wrona, P. K.: J. Chem. Soc. Perkin II, 1564 (1980).
La 67 Laurence, C., Bruno, B.: Compt. rend. 264 (1967).
La 79 Launay, G., Wojtkowiak, B., Krygowski, T. M.: Can. J. Chem. **57,** 3065 (1979).
La 80 Laszlo, P., Stockis, A.: J. Am. Chem. Soc. **102,** 7818 (1980).

La 81a Langhals, H.: Z. Phys. Chem. Neue Folge **127,** 45 (1981).
La 81b Langhals, H.: Nouveau Journal de Chimie **5,** 97 (1981).
La 82 Langhals, H.: Nouv. J. Chim. **6,** 285 (1982).
Le 63 Leffler, J. E., Grunwald, E.: Rates and Equilibria of Organic Reactions. Wiley, New York, 1963.
Le 66 Leermakers, P. A., Thomas, H. T., Weiss, L. D., James, F. C.: J. Am. Chem. Soc. **88,** 5075 (1966).
Li 60 Lindqvist, I., Zackrisson, M.: Acta Chem. Scand. **14,** 453 (1960).
Li 70 Lingane, P. J., Hugus, Z. Z., Jr.: Inorg. Chem. **9,** 757 (1970).
Lo 65 Long, F. T., Strong, R. L.: J. Am. Chem. Soc. **87,** 2345 (1965).
Ma 76 Madeja, K.: Personal communication.
Ma 75 Mayer, U., Gutmann, V., Gerger, W.: Mh. Chem. **106,** 1235 (1975).
Ma 79 Mayer, U.: Pure Appl. Chem. **51,** 1697 (1979).
Na 72 B. Nagy, O., B. Nagy, J., Bruylants, A.: J. Chem. Soc. Perkin Trans. II, 968 (1972).
Na 74 B. Nagy, O., B. Nagy, J., Bruylants, A.: Bull. Soc. Chim. Belg. **83,** 163 (1974).
Ni 64 Niedzielski, R. J., Drago, R. S., Middaugh, R. L.: J. Am. Chem. Soc. **86,** 1694 (1964).
No 72a Nozari, M. S., Drago, R. S.: J. Am. Chem. Soc. **94,** 6877 (1972).
No 72b Nozari, M. S., Drago, R. S.: Inorg. Chem. **11,** 280 (1972).
Pa 68 Partenheimer, W., Epley, T. D., Drago, R. S.: J. Am. Chem. Soc. **90,** 3886 (1968).
Pa 69 Parker, A. J.: Chem. Rey. 69, 1 (1969).
Pa 80 Pawelka, Z., Sobczyk, L.: J. Chem. Soc. Faraday **I, 76,** 43 (1980).
Pe 47 Pelskov, V. A.: Uspekhi Khimii **16,** 254 (1947).
Pe 63a Pearson, W. B., Golton, W. C., Popov, A. I.: J. Am. Chem. Soc. **85,** 891 (1963).
Pe 63b Pearson, R. G.: J. Am. Chem. Soc. **85,** 3533 (1963).
Pi 60 Pimental, G. C., McClellan, A. L.: The Hydrogen Bond. Freeman, W. H. and Co., San Francisco, 1960.
Pi 64 Pincock, R. E.: J. Am. Chem. Soc. **86,** 1820 (1964).
Po 79 Popov, A. I.: Pure Appl. Chem. **51,** 101 (1979).
Re 65–69 Reichardt, C.: Angew. Chem. Int. Ed. Eng. **4,** 29 (1965); Liebigs Ann. Chem. **727,** 93 (1969).
Re 81 Reichardt, C., Harbusch, E., Müller, R.: in: Advances in Solution Chemistry. Plenum Press, New York, London, 1981, p. 275.
Ri 80 Rider, P. E.: J. Appl. Polymer Sciences **25,** 2975 (1980).
Sc 55 Schoap, W. E., Messner, A. E., Schmidt, F. C.: J. Am. Chem. Soc. **77,** 2683 (1955).
Sh 59 Shorter, J.: Chem. Brit. **5,** 2969 (1959).
Sh 70 Sherry, A. D., Purcell, K. F.: J. Phys. Chem., **74,** 3535 (1970).
Si 80 Siegmund, M., Bendig, J.: Z. Naturforsch. **35a,** 1076 (1980).
Sj 81 Sjöström, M., Wold, S.: J. Chem. Soc. Perkin II, 104 (1981).
Sl 72 Slejko, F. L., Drago, R. S., Brown, D.: J. Am. Chem. Soc., **94,** 9120 (1972).
Sm 61 Smith, S. G., Fainberg, A. H., Winstein, S.: J. Am. Chem. Soc., **83,** 618 (1961).
Sp 71 Spaziente, P. M., Gutmann, V.: Inorg. Chim. Acta **5,** 273 (1971).
Su 57 Subomura, H. T., Nalzagura, S.: J. Chem. Phys., **27,** 819 (1957).
Su 82 Suppan, P.: Nouv. J. Chim. **6,** 285 (1982).
Ta 63 Taft, R. W., Price, E., Fox, I. R., Lewis, I. C., Anderson, K. K., Davies, G. T.: J. Am. Chem. Soc. **85,** 709, 3146.
Ta 80 Taft, R. W., Kamlet, M. J.: Organic Magnetic Resonance, **14,** 485 (1980).
Ta 81 Taft, R. W., Pienta, N. J., Kamlet, M. J., Arnett, E. M.: Journal of Organic Chemistry **46,** 661 (1981).
Vo 70 Vogel, G. C., Drago, R. S.: J. Am. Chem. Soc., **92,** 5347 (1970).
Wa 72 Waddington, T. C.: Nichtwässrige Lösungsmittel. UTE Hüthig, Heidelberg, 1972.
We 63 Wells, P. R.: Chem. Rev., **63,** 171 (1963).
Ya 77 Yamashita, K., Imai, H.: Bull. Chem. Soc. Japan, **50,** 1066 (1977).

5. EXPERIMENTAL METHODS EMPLOYED IN THE STUDY OF NON-AQUEOUS SOLUTIONS OF COMPLEX SYSTEMS

It is apparent from the foregoing that a large variety of methods have been used for the study of solutions, and often their combinations are advantageous. The most different experimental techniques, ranging from the simplest *electrochemical procedures* (conductometry, polarography) to the most up-to-date electromagnetic and magnetic structural investigation methods (EPR, NMR spectroscopy), have been utilized to characterize the magnitudes of the donor–acceptor interaction serving as the basis of solvation. The solvation number has similarly been studied by many different procedures and with different solutions. Solvent bound in the solvate sheath has been differentiated from the bulk solvent by many methods, from compressibility investigations to NMR spectroscopy. There is virtually no structure determination procedure which has not been put to use to follow the structural and electron structural changes brought about by the solute–solvent interaction. There is hardly any method used in equilibrium chemistry with which attempts have not been made to follow the formation of new species resulting from the action of the solvent.

A barrier to the examination of solvent effects in organic solvents may be the limited solubility of the dissolved model systems. With the spreading of high-performance spectroscopic methods and other modern analytical procedures, increasing importance is attached to the proper selection of the model compounds for these examinations.

The success of the Gutmann donicity concept can be attributed primarily to the fortunate choice of antimony(V) chloride as reference acceptor. The limitations of the empirical solvent strength scales presented in Chapter 4 are due, in almost every system, to various physical or chemical properties of the model. Also in examinations aimed at learning more about special phenomena of less general validity, the primary step is the selection of the appropriate model.

Excellent examples are the reference substances employed by Day [Da 77] and his co-workers for IR [Ho 69] and NMR spectroscopic [Sc 68] studies of the solvent effect, viz., tetraalkylaluminate salts. These compounds are readily soluble in hydrocarbons and other apolar solvents. When a donor solvent is added to a hydrocarbon solution containing the reference substance, the solvates of the cations belonging to the tetraalkylaluminate anion are formed without a substantial alteration in the dielectric properties. The anion itself is inert to donor solvents, and its role is confined to ensuring the dissolution of the cation which is to be studied in the inert parent solution.

With the aid of this model, not only could the solvation of cations be determined (e.g., the solvation number of the sodium ion in various organic solvents), but it was also possible to differentiate between contact, solvent-shared and solvent-separated ion pairs [Na 77, We 72, We 73].

In the course of the study of association interactions [Gr 60b], a particularly large number of workers have dealt with the conditions of ion-pair formation in non-aqueous solutions [Ge 70, 71, 79, Bj 26, De 55, Fu 58, Gi 60, Kr 56, Pa 76a, Pr 66]. Efforts have also been made to clarify the correlations between solvent properties and the constants of the association equilibria [Pr 66, Pe 66]. Far fewer workers, however, have dealt with multi-step complex formation in non-aqueous solutions, and particularly with the study of the effect of the solvent on this process.

In most cases the structures and compositions of solvates and solvated or unsolvated complexes isolated in the solid state from solution, or prepared in the solid phase by the interaction of the substance in question and the solvent molecules, differ from those of the ionic species present in solution. This gives special importance to those procedures which are suitable for the study of the dissolved species in the solution itself.

In systems where a reversible electrode is available that can be used also in non-aqueous solutions to measure the activity of the cation under investigation, unambiguous information relating to complex formation may be obtained directly by means of *potentiometric titrations*. For example, the study of silver complexes is particularly favoured, since the silver electrode is eminently suitable for the exact following of the silver ion activity in most solvents. This has permitted the investigation of the formation of silver complexes in acetonitrile [Al 67, Le 66], dimethylformamide [Pr 70], hexamethylphosphoramide [Al 67], acetone [Le 66], nitromethane [Ba 70], dimethyl sulphoxide [Ru 70, Le 70], sulpholane [Be 69] and propylene carbonate [Bu 67]. A similar method has been used to determine the solubility products of numerous silver complexes in acetonitrile, dimethylformamide and dimethyl sulphoxide [Ch 73a]. Similar measurements have also provided information on the changes in the mean activity coefficients in solutions prepared with non-aqueous solvents. Amalgam electrodes could also be used with success in different non-aqueous solvents affording valuable information on the solvent effect, e.g. in zinc(II), copper(II) and cadmium(II) complex systems [Ah 80, 81].

Potentiometric measurements with a silver–silver chloride and glass electrode pair in *liquid junction-free* systems have proved suitable for equilibrium studies in solutions prepared with various non-aqueous solvents (e.g. dimethyl sulphoxide, acetonitrile, acetone, methanol, ethylene glycol) [Bu 75a].

Reaction kinetic studies [Am 66] have furnished information on the effects of solvents on the rates and mechanisms of reactions in solution.

Valuable data relating to systems in solution may be provided by the various spectroscopic methods. The changes (or at least many of them) caused by solvation in the *electronic excitation spectra* of dissolved species have been attributed to charge-transfer interaction [Bl 70].

Infrared and *Raman spectroscopic* examinations of solutions reveal changes in the vibration spectra of the solvent molecules caused by the action of the solute. The appearance of new bands or the shift of the original bands permits conclusions to be drawn about solvate formation [Ad 68, Ah 78, Am 66, Ke 64, Pe 64, Pe 68, Pu 58]. Solvation may also show up in the change of the vibration spectrum of the solute.

Proton magnetic resonance studies on solutions prepared with proton-containing solvents [Br 81, Bu 68] have yielded valuable data on the extent of solvation, but NMR studies of dissolved species containing other magnetic nuclei are also suitable, for instance, for the determination of the stability sequence of various solvates [Ba 81, Bu 69, Ha 69, Ma 68, Mo 68, Sy 67].

Solvent molecules situated in the neighbourhood of dissolved ions may be investigated with *dielectric measurements* in the microwave range [Gl 64, Ha 58b, Ja 77b].

In the case of paramagnetic ions, the effect of the solvent on the dissolved ion may be followed by studying the effect on the unpaired electrons by means of *electron paramagnetic resonance (EPR) spectroscopy* [Sy 69]. In concentrated solutions, *solution X-ray examinations* may give information on the immediate environment of the dissolved ion [We 69, Ah 78]. *Mössbauer studies* on rapidly frozen solutions [Bu 72, Bu 73, Vé 70a, Vé 70b, Vé 79] permit conclusions to be drawn primarily about the symmetry and covalency of the inner coordination sphere in dissolved species containing a Mössbauer atom.

Many workers have been engaged in the measurement and calculation of *thermodynamic data* relating to solvates [Ha 58a, Ho 72]. Nowadays, the heat of solvation is mainly determined by means of calorimetric measurements, but procedures based on the measurement of the electromotive force, solubility studies and calculations with the modified Born equation [Co 54, Cr 68] are also used. Special mention should be made of the heats of solvation obtained *via* the *mass spectrometric* study of interactions between the ion and solvent molecule in the gas phase [Ke 68, 79].

In the following sections we shall deal in detail with the applications of some of the more important methods of investigating the structure of solvates and of other ionic species and complexes formed in non-aqueous solution.

Conductometry

Of the classical electroanalytical methods, conductometry was used even in the early stages of solvent effect investigations. As the conductivity of a solution depends in part on the number of ions present and in part on their mobilities, conductivity measurements could provide information on changes in the dissociation conditions caused by the solvent, and also on the mobility variations due to the change in the solvate sheath [e.g., Bo 79, Ja 79, On 79].

The practical application of this procedure was facilitated by the fact that dissociation equilibria are repressed in these systems, the relative permittivity being generally lower than that of water; because of the resulting lower conductivity, the method is also suitable for the examination of more concentrated solutions.

The determination of experimental conductometric data is relatively simple. However, the clear interpretation of the measured data is more difficult. The concentration-dependent association and dissociation reactions occurring in solution give rise to changes in the ionic strength and hence in the activity coefficients, and this shows up as apparent ion concentration or mobility changes in the conductivity data. The errors arising in this way cannot be eliminated by adjustment of the system to constant ionic strength, since the salts (inert in other examination methods) added for this purpose contribute to the conductivity; accordingly, the measured data may be distorted so much that it is very difficult (if possible at all) to draw conclusions from them as to the original effect which was to be studied. Conductivity examinations must therefore be carried out in solutions containing only the components under investigation. Even then, a quantitatively evaluable result is obtained only if the solute consists of no more than two or three components. Even in such systems, where generally only ion-pair formation takes place, exact results can be obtained only by fairly complex theoretical calculations. Employment of the procedure as a qualitative method may be the source of many errors. Therefore qualitative studies are not discussed here.

A number of adequate equations are available for the quantitative description of the conductivities of totally dissociated electrolyte solutions [Ba 68, Ba 76, Eb 78, Fa 71, Fe 73]. Within these, short-range forces may be taken into consideration with various models, such as the model involving rigid charged spheres [Fu 59, Fu 62, Fu 65], or by a statistical-mechanical treatment [Eb 78, Fa 71]. In view of the generally low relative permittivity of non-aqueous solvents in comparison with that of water, complete dissociation of the electrolytes can be assumed in comparatively few systems and only in very dilute solutions. A condition for the success of investigations aiming at interpreting the solvent effect is therefore the consideration of association between the ions, i.e., ion-pair formation. This is made possible by the following conductivity equation, elaborated on the basis of recent studies [Ba 76, Ba 77, Eb 78, Ju 77]:

$$\alpha^{-1}\Lambda = \Lambda^{\infty} - S(\alpha c)^{1/2} + E\alpha c \log \alpha c + J_1(\alpha c) + J_2(\alpha c)^{3/2} \tag{5.1}$$

where Λ is the equivalent conductivity in a solution of a symmetrical electrolyte $M^{z+}A^{z-}$, c is the concentration, α is the degree of dissociation, S is the Onsager boundary law coefficient, and E, J_1 and J_2 are constants (for details, see ref. [Ba 76]). It should be noted that, as regards the data relating to the ions, S and E contain only the charges of the ions, whereas J_1 and J_2 additionally depend on the ionic distances.

The concentration range in which the above equation is valid depends on the dielectric properties of the solvent. Usually the upper concentration limit of its successful application is a few tenths of 1 mole/dm^3.

Naturally, only free ions take part in the charge transport. The concentration of these is αc, and the concentration of the undissociated MA species is $(1-\alpha)c$. The following relationship between the equilibrium constant K_A of ion-pair formation and the α value in Eq. (5.1) holds:

$$K_A = \frac{1-\alpha}{\alpha^2 c} \cdot \frac{y'_{MA}}{y'^2_{\pm}} \tag{5.2}$$

where $y'_{\pm}$ is the mean activity coefficient of the ions and y'_{MA} is the activity coefficient of the undissociated electrolyte (ion pair). Equation (5.2) is in effect independent of whether the process occurs in one or more steps. For a one-step process, however, the relationship can be developed further:

$$K_A = \frac{Q_{MA}}{Q_M Q_A} \exp\left[-\Delta E_A/RT\right] \tag{5.3}$$

where Q_{MA}, Q_M and Q_A are the reduced state sums of the reactants and ΔE_A is the energy difference between the lowest energy level of the ion pair MA and the lowest energy levels of the component ions:

$$\Delta E_A = E_{MA} - E_M - E_A \tag{5.4}$$

On the basis of the correlations outlined briefly above, quantitative information on the association and dissociation conditions could be obtained from the concentration dependences (determined at various, but constant, temperatures) of the electric conductivities of simple non-aqueous solution systems. By consideration of the terms, relating to the activity coefficients, the following system of equations can be derived from Eqs (5.1) and (5.2) for numerical evaluation of the data:

$$\alpha^{-1}\Lambda = \Lambda^{\infty} - S(\alpha c)^{1/2} + E\alpha c \log \alpha c + J_1(R_1)\alpha c + J_2(R_2)(\alpha c)^{3/2} \tag{5.5a}$$

$$K_A \alpha^2 c y'^2_{\pm} + \alpha - 1 = 0 \tag{5.5b}$$

$$y'_{\pm} = \exp\left[-K q(1 + K R_y)^{-1}\right] \tag{5.5c}$$

where

$$q=\frac{z^2e^2}{8\pi\varepsilon_0\varepsilon kT} \quad \text{and} \quad K=16\pi q N_A \alpha c \cdot 10^n$$

where q is the distance parameter calculated in accordance with the Bjerrum theory and K is the reciprocal of the radius of the Debye ion cloud.

Perhaps the greatest uncertainty in the evaluation is caused by the selection of the appropriate ionic radii, R. In accordance with the introduction of Bjerrum's association constant K_A [Bj 26], Justice [Ju 71b, Ju 75a, Ju 75b] assumes the equation $R_y=R=q$. From chemical considerations, Barthel [Ba 78a] considers it more correct to describe the radius R_y as the sum of the contact distance, a, of the ions and the size, s, of the solvent molecule: $R_y=a+s$.

By studying a number of model systems (generally alcoholic solutions of alkylammonium salts), Barthel has also proved experimentally that the latter consideration gives a better approximation to reality. Important results of these studies are the Gibbs free energies, calculated from the temperature dependence of the association constant K_A with the following equation:

$$\Delta G_A^0=-RT\ln K_A \tag{5.6}$$

For a description of the short-range (non-Coulombic) interactions between the ions in the ion pair, Barthel [Ba 78b] introduced the equation

$$W_{ij}^0(r)=-\frac{e_0^2z^2}{4\pi\varepsilon_0\varepsilon}\cdot\frac{1}{r}+U^* \tag{5.7}$$

where $W_{ij}^0(r)$ is the energy of interaction between the ions and U^* is the potential induced by the short-range forces. On the basis of the equation $\Delta G_A^*=N_A U^*$, the ΔG_A values calculated in accordance with Eq. (5.6) from the temperature dependence of the K_A values can be resolved into the sum of terms originating from the Coulombic (ΔG_A^C) and non-Coulombic (ΔG_A^*) interactions: $\Delta G_A^0=\Delta G_A^C+\Delta G_A^*$. Thus, the enthalpy and entropy terms may be calculated in the usual way for both types of interaction.

The above treatment has led to a deeper understanding of the solvent effect. Whereas alkali metal salts and tetraalkylammonium salts displayed analogous behaviour on the basis of thermodynamic data calculated from the temperature dependence of the constant K_A for ion-pair formation, the non-Coulombic part of the Gibbs energy reflected different behaviours of these two kinds of cation. Whereas in solutions of alkali metal salts the alkali metal ions coordinate solvent molecules, and hence ion-pair formation is accompanied by displacement of solvent molecules from the solvate sphere, the interactions of the tetraalkylammonium salts, as they do not coordinate solvent, can be ascribed primarily to dispersion forces. This different behaviour is manifest in the entropy and enthalpy terms of the non-Coulombic interactions; in solutions of alkali metal salts $\Delta S^*_A > 0$ and $\Delta H^*_A > 0$, whereas in solutions of tetraalkylammonium salts $\Delta S^*_A \approx 0$ and $\Delta H^*_A < 0$.

It can be seen that such examinations in simple systems may lead to a greater understanding of the intricacy of the solvent effect. However, the more complicated the system, the less suitable are conductometric methods for its understanding and characterization. In the investigation of the solvent dependence of several-step complex equilibria or processes associated with solvent substitution in systems with complicated solvate spheres, conductometry may at best be a source of supplementary qualitative information. Examples of this may be seen in the section dealing with the use of electrochemical methods for characterization of the donor strengths of solvents.

Dielectrometry

As seen in the preceding sections, for a quantitative description of the electrostatic part of the solvent effect, or even for its qualitative evaluation, it is necessary to know the dielectric constant (relative permittivity) of the solvent (or possibly that of the solution). Accordingly, even in the early stages of research, the study of the dielectric properties of solvents and solutions served as the basis of attempts at elucidating the solvent effect.

If an electrically non-conducting substance (dielectric) is placed in the field of a condenser, the result will be polarization in the dielectric. Consequently, the capacity of the condenser will increase, and the relative permittivity (the dielectric constant) of the dielectric can readily be calculated from this capacity increase.

If the capacity of the condenser *in vacuo* is C_0 and it increases to C as a result of the introduction of the dielectric, then, to a first approximation, the relative permittivity (ε) will be

$$\varepsilon = \frac{C}{C_0}$$

Various instruments have been constructed for measurement of the capacity or capacity increase. In the early research, primarily instruments based on electrostatic methods were employed, e.g., those utilizing Coulomb's law [Fü 24] or static capacity measurement [Op 33]. Later, instruments based on alternating current capacity measurement gained preference. Among these can be listed the instruments based on the classical bridge method [Ad 64, Ga 50], the resonance methods [Im 65, So 64] and the heterodyne method [Ba 51, We 51]. All of these can be used only for relative measurements. The dielectric data sought are obtained by comparison with known calibration substances. Instruments based on the measurement of electromagnetic wave phenomena are suitable for absolute measurements also [Hu 54, Ra 59]. As regards the details, the reader is referred to the original literature cited above, a review [Br 71] and books [Pu 65, Na 70c, Oe 62].

The relative permittivity of liquids, including solvents and solutions, are usually determined by a calibration method. Many reliable standard substances are available for this purpose, particularly in the range of low relative permittivities. Examples are various hydrocarbons and halogenated hydrocarbons. For higher relative permittivities, not only does the number of available reliable calibration liquids become smaller, but also the errors in the measurement increase. If the relative permittivity is between 2 and 5, the accuracy of the measurement in the optimal case varies between ± 0.0001 and ± 0.0005, whereas in systems with high relative permittivities the error may attain a value of ± 0.01 or even ± 0.1. In spite of this, investigations have been carried out to measure the dielectric properties not only of solvents with high relative permittivities, but also of numerous electrolyte solutions prepared with them [Bo 59, Ge 64, Ha 54, Ha 58, Ma 58b, Se 56]. Reliable dielectric data have been obtained partly by suitable reconstruction of the measurement cell, and partly by the selection of an appropriate frequency range (microwave range, etc.), possibly using a varying frequency, and also in cells with variable capacities. In the reports of these investigations, the description of the special experimental technique is often even more interesting and valuable than the conclusions drawn.

Apart from the construction of electrostatic solvent theories, the dielectric properties of solutions are of limited value in drawing conclusions on the solvent effect, because the experiments provide macroscopic data relating to the entire solution, and hence microscopic data characteristic of the solvent molecules bound in the coordination sphere cannot be obtained. Investigations carried out in concentrated solutions to follow the macroscopic changes occurring in the solvent in the course of solution formation are more promising, as is research aimed at obtaining a full understanding of the interactions between the components in solvent mixtures [Hi 69, Li 75]. We shall deal with some examples of this approach in the relevant sections.

In spite of the difficulties mentioned, an increasing number of workers have made use of dielectrometric methods in the investigation of solution structures [Ja 77b, Wi 82a, b]. All of these methods are based on the fact that intermolecular interactions between the solvent and solute also change the dielectric properties of the system. In the studies of the solvent effect, changes in the dipole moment resulting from the formation of hydrogen-bonded associates and other electron-pair donor-acceptor complexes are of particular importance [So 76].

Complex formation is always accompanied by a change in the charge distribution of the components. In dielectrometric measurements, this shows up in the fact that the dipole moment of the product formed will differ from the vectorial sum of the dipole moments of the reactants. The component of the dipole moment due to the interaction is naturally also a vector which, in hydrogen-bonded associates, points in the direction of the more strongly basic central atom.

Analogous effects can, of course, be expected if one of the reactants participating in the hydrogen bond is the solvent itself. However, a solvent effect also arises in solutions containing hydrogen-bonded complexes, or other associates formed as a consequence of donor–acceptor interactions. Complexes of moderate polarity are particularly sensitive to the effect of the solvent [Ja 72, Pa 75].

The effect of a solvent can actually be considered as the resultant of two components: a macroscopic effect depending on the dielectric permittivity of the solvent as an environmental continuum, and specific interactions related to the formation of definite associates. The former effect predominates in solvents of low relative permittivity, containing no donor or acceptor groups. It can be taken into account with appropriate model calculations, e.g., on the basis of the Onsager reaction space model.

On the other hand, the specific donor–acceptor interactions arising in dipolar solvents cannot be calculated with such a simple, essentially electrostatic approximation. If such complexes containing coordinated solvent molecules are regarded as "supermolecules", attempts may be made to calculate the solvation effect using the Monte Carlo method [Ko 72]. Even this comparatively intricate method, with considerable computer requirements, has proved successful only for the quantitative description of relatively simple systems. Such specific solvent effects, however, can be well described by means of some empirical solvent parameter or an experimental (e.g., spectroscopic) parameter characteristic of the stability or possibly of the electronic structure of the complex.

For instance, by plotting the dipole moment of *p*-nitrophenol–triethylamine complex measured in solution as a function of the Onsager parameters of the solvents, Pawelka and Sobczyk [Pa 75] obtained a virtually completely random distribution, whereas plots of the dipole moment as functions of the charge-transfer absorption band of the complex, its O—H infrared vibration or the Reichardt E_T parameter of the solvent led to linear correlations. Similarly, a plot of the dipole moment of the iodine–pyridine charge-transfer complex measured in solution against the Onsager parameter of the solvent gave a random scatter, whereas when it was plotted against empirical solvent parameters (such as the log k_M value of the Menshutkin reaction), linear correlations resulted. The above concept is supported by the correlation of the dipole moment change with the heat of reaction of complex formation for electron-pair donor–acceptor complexes [Gu 73, Ra 73].

Dielectric relaxation spectroscopy

If the permittivity of a dielectric material is measured over a frequency range extending from very low values up to the infrared, a number of absorptions will be observed. Frequencies at which these absorptions occur can yield information about the structure of the molecules of the material, and of the molecular dynamics of the system [Ca 80, Le 74].

Isotropic liquids usually show only one absorption maximum in their relaxation spectrum. Broadening of such lines may be sufficient to allow the relaxation to be resolved into multiple processes corresponding to whole molecule rotation and the internal rotation of a flexible group within the molecule.

The examination of the frequency spectrum can be used to investigate the behaviour of aggregates of molecules of the micelle type. Molecular dynamics can also be studied by the rather unusual technique of non-linear dielectric spectroscopy, when materials are polarized by fields of high strength.

Interpretation of dielectric relaxation spectroscopic measurements can be aided by the construction of models which corresponded in some analogous way with the physical reality. Although in principle the construction of such models might be fairly straightforward, the analysis of the model to determine the behaviour which it predicts is usually a matter of considerable technical complexity. Computer simulation is liable to be expensive in computer time, but reliable values of complex permittivity of highly polar fluids could be obtained by a two-dimensional simulation involving a few hundred molecules. This is an encouraging advance in a field in which progress is not easy.

Studies of dielectric relaxation processes [Le 74] are more promising than classical dielectrometry. By this means, phenomena with characteristic durations of 10^{-10}–10^{-9} s can be examined even in electrolyte solutions. With this method, for instance, it is possible to follow the effects of the ions in a solution on the motion of the solvent molecules. Thus, the dielectric examinations permit conclusions to be drawn about the solvation numbers of ions [Ba 67, Ba 71], and they may also give information on other orientation processes.

Both solvent–solvent and solvent–solute interactions can result in associates with more or less definite compositions. In some of these, the mutual orientation of the components has a reasonably long lifetime, whereas in others it is only momentary [Ko 72]. Information on the dynamics of these interactions is provided by the dielectric relaxation of the system and the relaxation time.

Good examples of such examinations can be found in studies of the association of various solvents, primarily alcohols. A number of authors have dealt with primary and secondary alcohols [Da 70, Ga 67, Mi 69], with solutions of an alcohol in apolar solvents [Cr 71a, Cr 71b, Ge 72] and with mixtures of alcohol and polar solvents [Da 71, He 71]. It was found that the systems could be described by several (in general, three) relaxation times. Chemical information relating to the systems was obtained from the dependence of these times on the various parameters.

Interpretation of the experimental facts was possible on the basis of different concepts. In systems displaying a weaker interaction, the dielectric properties could be attributed to short-range ordering of a dynamic nature. In systems involving a stronger interaction, assumption of the joint presence of monomers and of oligomers with various, defined compositions and lifetimes was necessary for the interpretation.

In short-chain alcohols the dielectric dispersion is described by three relaxation times (three dispersion ranges), the predominant of which is the low-frequency Debye-type process. The experimental results may be interpreted in accordance with various assumptions.

Smyth *et al.* [Cr 70, Ga 65, Ga 67, Jo 69] assume the formation of long-chain oligomers in the liquid, consisting of molecules interconnected with O—H. . . O hydrogen bonds. This results in a dynamic short-range ordering. In addition to oligomers, the presence of free monomers in the system is also assumed. According to this model, the first of the three dispersion ranges reflects the rotation associated with the partial splitting of the long-chain alcohol oligomers, monomers being liberated as a result. The relaxation time here is fairly long, since it is governed by the rate of cleavage of the hydrogen bond. The second dispersion range reflects the reorientation of the monomers and smaller oligomers. Finally, the third dispersion range can be ascribed to reorientation of the O—H groups of the monomers around the C—O bond.

In the model of McDuffic and Litovitz [Mc 62, Mc 63] the first dispersion range can be interpreted by the assumption of short-range partial ordering. The ordered ranges and the extent of ordering depend on the temperature, pressure and chemical properties of the liquid. The stronger the interactions between the solvent molecules, the larger will be these ordered ranges, yet the dynamic character is retained; at one instant the individual molecules belong to one associate, and in the next instant they become members of another associate. Thus the associates undergo continuous break-down and reformation. The structural relaxation times characterize the kinetics of these processes. The dielectric reorientation process is manifested as part of the process accompanied by previous chain rupture and reconstruction. Reorientation can occur after the chain ruptures proceeding by cleavage of the hydrogen bonds have taken place and the "liquid lattice" structure has disintegrated.

According to the above picture, the dielectric relaxation time is longer than the time of structural relaxation. Hence, this theory gives a better description of systems possessing two dispersion ranges.

A model differing fundamentally from the previous ones is that of Dannhauser [Da 65, Da 67], who took into account the formation of monomers, cyclic dimers and short-chain oligomers in alcohols. In the case of octanols, for instance, the first dispersion range was explained by the cyclic rearrangement of the linear alcohol chains; the second dispersion range was attributed to the reorientation of the OR groups and the third to the rearrangement of the monomers.

Similarly to Dannhauser, Bordewijk *et al.* [Bo 69a] refuted the role of cleavage of the hydrogen bonds in the first dispersion range. On the basis of investigations on heptanol isomers, they assumed the joint presence of polar cyclic tetramers, dimers and monomers in the liquid. The first dielectric absorption range was explained by the reorientation of the tetramers.

From these examples, it can be seen that given experimental data can sometimes be described well on the basis of different models. In addition, even within such analogous systems as the alcohols, individual chemical properties (e.g., length of the alkyl chain) may not only cause a change in the dielectric properties, but may also demand the introduction of contradictory model systems for the interpretation of the experimental data.

Most dielectric relaxation investigations reported in the literature deal with simple systems, and attempts have also been made to describe the dielectric behaviour of these with theoretical calculations. For the methods and results of these studies, reference may be made to several reviews [Ch 73b, Le 74, Va 73].

Even this brief survey shows that although the relative permittivity is one of the most important characteristic properties of a solvent, serving as the basis of the qualitative interpretation of numerous phenomena, the use of dielectrometry has limited use in understanding the specific interactions of the solvent effect.

UV and visible spectroscopy

The "solvatochrome effect" [Ha 22], changes in the energies, intensities and shapes of the absorption bands of chromophores due to the effect of the solvent, has long been known. The early work in this field was reviewed by Sheppard [Sh 42]. Even from a survey of these data it follows that the UV and visible spectra of solutions may afford valuable information on the solvent–solute interaction and, indirectly, on the effect of solvation on the structure of the solution.

The cause of the solvatochrome effect is that, depending on their solvating power, different solvents stabilize the ground and excited states of the chromophore in different ways, thereby giving rise to changes in the excitation energies and hence in the spectrum [Ha 79].

The effect of a solvent on the absorption spectrum is particularly large in conjugated systems where the charge transfer accompanying excitation causes significant changes in the magnitude and direction of the dipole moment of the molecule; examples include the merocyanine dyes [Br 65] and pyridinium-N-phenol betaines [Di 63].

A high solvent dependence is also often displayed by the charge-transfer bands between the components of ion pairs [Ko 56], since the different degrees of solvation change the dissociation conditions of the ion pair and hence the polarity of the solute. If this brings about a substantial alteration in the magnitude and direction of the dipole moment of the molecule in the course of excitation, the solvatochrome effect will be large. The absorption bands reflecting the formation of ion pairs are definitely solvent dependent, although to different extents [Cr 70].

Similarly to the use of other physico-chemical parameters in characterizing the solvent effect, models based on an electrostatic approach have been produced for the description of the solvent dependence of the absorption spectrum; these models describe the shifts of the absorption bands in terms of the relative permittivity and refractive index of the solvent [Oo 54, Ba 54]. These have fairly limited validity, however, since specific coordinative interactions cannot be neglected when considering the factors that determine the stabilities of solvates.

Nevertheless, the solvatochrome effect has served as the basis of a number of scales characterizing the solvating powers of solvents. These include the Kosower Z scale and the Dimroth–Reichardt E_T scales, dealt with in Chapter 4.

It is a more difficult task to interpret the solvent dependences of the electronic excitation spectra of metal complexes, particularly transition metal complexes [Ma 79a].

It is known that the spectra of the latter have absorption bands corresponding to the following three types of transition: (1) *d–d* transitions; (2) charge-transfer transitions; and (3) transitions within the ligand. The *d–d* transitions are in fact forbidden, and appear only as a result of the perturbing effect of the ligands; this is the explanation of their low intensities [Gr 60a]. In accordance with the energies of the *d–d* transitions, these bands are generally found in the visible or near-ultraviolet ranges. Their positions (energies) depend on the donor strength of the ligand. The *d–d* absorption bands of aquo complexes, for instance, usually appear in the visible spectral range. If water in the coordination sphere of a transition metal is replaced with a stronger donor solvent (or ligand), the *d–d* bands are shifted in the direction of the UV range; substitution with a weaker donor leads to a shift in the opposite direction. If the substitution of the ligand causes a change in the symmetry of the coordination sphere, the intensities of the *d–d* bands also usually change. A symmetry decrease is generally accompanied by an increase in the molar absorptivity.

On the basis of the above considerations, the energies of the *d–d* bands of solvate complexes would give the spectrochemical series of the solvents [Gr 56], and with their help the nephelauxetic series of the solvents too could be calculated in a similar manner to other ligands. In fact, however, this is not so simple. Solvent exchange is not merely ligand exchange. In the course of solvent exchange the relative permittivity of the system changes, which is accompanied by changes in the dissociation equilibria. In contrast with other ligands, the solvent may interact not only with the cation, but also with the anion. Solvent exchange may cause alterations in the solvations of both ions, which again results in changes in the dissociation conditions. This is the reason why in analogous systems, naturally containing anions in addition to the metal ion, parent and mixed solvates of different compositions are formed depending on the donor–acceptor and dielectric properties of the solvent. In aqueous solutions of divalent transition metal halides, for instance, the metal is present in the form of its hexaquo complex. If the water is replaced with a solvent with lower relative permittivity (e.g., acetonitrile), in addition to the solvent, halide ion also is bound in the first coordination sphere of the transition metal ion. The magnitude of the *d–d* splitting will then be determined by the two types of ligand (halide and acetonitrile) together; thus the energy of the *d–d* band will be characteristic not of the solvent, but of this mixed ligand complex.

The metal → ligand or ligand → metal charge-transfer bands generally appear in the near-ultraviolet range, but sometimes in the visible spectral range. Their intensities are about 100 times larger than those of the *d–d* bands.

A charge-transfer band between the metal ion and the coordinated solvent in the solvate complex is rare. However, by changing the dissociation conditions in the system, and by promoting or inhibiting entry of the anion into the coordination sphere of the transition metal ion, solvent exchange may give rise to or eliminate charge-transfer between the metal and the given anion. This is the explanation, for example, of the solvent dependence of the colour of iron(III) thiocyanate.

Charge-transfer bands are of outstanding importance in the study of molecular complexes (charge-transfer complexes) [Br 61, Kö 75]. The formation of a molecular complex can be recognized primarily by the spectrum containing an absorption band that is not to be found in the spectra of the components. Since many solvents are capable of forming molecular complexes with the most varied dissolved molecules, charge-transfer bands may play a considerable role in the study of the solvent effect.

Typical examples of charge-transfer complexes formed with a solvent are the complexes of elemental halogens [Be 49, Br 61, Ha 54, Kö 75]. Figure 5.1 presents absorption curves for iodine in a number of solvents in the range 280–600 nm. In carbon tetrachloride, which has an extremely low solvating ability, the single absorption band of iodine appears at about 520 nm. If the carbon tetrachloride is replaced with a solvent able to form a charge-transfer complex (benzene, mesitylene, diethyl ether, etc.), this band is shifted slightly towards the UV range, and in addition a new band of higher intensity appears at about 300 nm, the charge-transfer band of the iodine–solvent molecular complex. The energy of this depends on the ionization potential of the donor (solvent), on the electron affinity of the acceptor (iodine) and possibly on the magnitude of the interaction. Such charge-transfer bands have large line-widths. Certain assumptions suggest that the line-width increases with the decrease in the stability of the complex [Ju 71, Ju 73].

The bands of electron transitions within the ligand are the least sensitive to the solvent effect. The absorption bands of most solvents used in practice are located in the ultraviolet range. As a result of the coordination of the solvent to the metal ion, these bands shift at most only slightly in the direction of higher energies (shorter

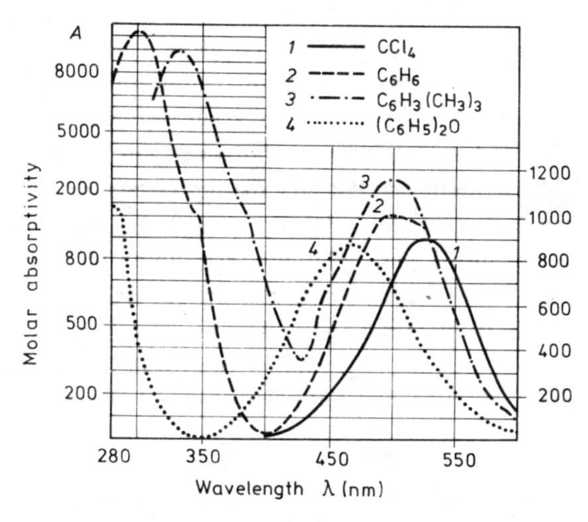

Fig. 5.1. Solvent dependence of the absorption spectrum of iodine [Kö 75]

wavelengths). The change resulting in the dissociation conditions on the action of a solvent may cause a change similar to that due to the coordination of another ligand (or organic anion) to the metal ion.

In studies of the solvent effect in non-aqueous solutions, electronic excitation (UV and visible) spectroscopy is most frequently used as a method for measuring equilibrium. The utilization of this methodology is not limited by solvent exchange. The potentiometric and other electroanalytical procedures used most often for the study of complex equilibria in aqueous solutions cannot be employed (or to only a very limited extent) for the determination of the compositions and/or stabilities of complexes in aprotic systems and systems with low relative permittivities. The results obtained by their means in the various solutions cannot always be compared, as they refer to different standard states. Spectrophotometric equilibrium measurements are not influenced by the dielectric properties of the solution or by the protic or aprotic nature of the solvent. All processes which are accompanied by a change in the light absorption of the system (whatever the solvent may be in which the process takes place) may be studied with the aid of this method. Since the introduction of computers for the evaluation of complex equilibrium measurements, spectrophotometry can be used much more easily for the successful study even of complicated, multi-step equilibrium processes.

The absorbance (A) of monochromatic light in a solution may be written as the sum of the absorbances of the species in the solution. For example, for the absorbance of a solution in which n complexes are formed in a stepwise manner from a metal ion M and a ligand A, we have the relationship

$$A = \varepsilon_A[A] + \varepsilon_M[M] + \varepsilon_1[MA] + \varepsilon_2[MA_2] + \ldots + \varepsilon_n[MA_n] \qquad (5.1)$$

where $\varepsilon_A, \varepsilon_M, \varepsilon_1, \varepsilon_2, \ldots, \varepsilon_n$ are the extinction coefficients (molar absorptivities) of the free ligand, the free metal ion and the stepwise-formed complexes, respectively, and the terms in brackets are the concentrations of the corresponding species. If the latter are expressed in terms of the equilibrium constants ($K_1, K_2, \ldots, K_n$) of complex formation, we have

$$A = \varepsilon_A[A] + [M](\varepsilon_M + \varepsilon_1 K_1[A] + \varepsilon_2 K_1 K_2[A]^2 + \\ + \ldots + \varepsilon_n K_1 K_2 \ldots K_n[A]^n) \qquad (5.2)$$

If the analytical composition of the solution, i.e. the total concentrations of the metal and the ligand (C_M and C_A) are known, the material balance equations may be written:

$$C_M = [M] + [MA] + [MA_2] + \ldots + [MA_n] \qquad (5.3)$$
$$C_A = [A] + [MA] + 2[MA_2] + \ldots + n[MA_n] \qquad (5.4)$$

When the concentrations of the complexes are expressed in terms of the equilibrium constants also in these equations, we have

$$C_M = [M](1 + K_1[A] + K_1K_2[A]^2 + \ldots + K_1K_2 \ldots K_n[A]^n) \qquad (5.5)$$
$$C_A = [A] + [M](1 + K_1[A] + 2K_1K_2[A]^2 + \ldots + nK_1K_2 \ldots K_n[A]^n) \qquad (5.6)$$

The equilibrium constants $K_1, \ldots, K_n$ can readily be calculated by computer on the basis of Eqs (5.2), (5.5) and (5.6), from the spectral data of solutions with various analytical compositions (variations of C_M and C_A). When the equilibrium constants are known, it is possible to calculate the compositions of the complexes (the possible values of n) formed in solutions of given analytical compositions, and how the concentrations of the individual complex species vary with the change of the total concentrations.

If such examinations are carried out with series of solutions prepared with different solvents, the effect of the solvent exchange on the equilibria can be established. Depending on the donor and acceptor properties and the relative permittivity of the solvent, the values of the constants K and possibly also n, will be different in the systems made with different solvents. With a knowledge of these data, quantitative information is obtained with regard to the solvent effect.

The above example generalizes the simplest case. If the solubility conditions permit the study of the concentration dependence of the spectrum over a wide concentration range, and the light absorption conditions allow the study of a broad spectral range, then also equilibria much more complicated than the above (e.g., the formation of mixed ligand and polynuclear complexes) can be investigated effectively by this means. Naturally, the more complicated the equilibrium system, the larger is the number of terms in the analogous correlations corresponding to Eqs (5.2), (5.5) and (5.6), and the larger is the number of unknown molar absorptivities and equilibrium constants. Accordingly, the solution of the equations demands a larger number of experimental data (concentration-dependent extinction values).

The simplicity of the experimental method makes it relatively easy to collect the necessary large number of data, and the bulk of the work consists of the calculations. For this reason, spectrophotometric methods could not be used for the solution of more complicated problems in the early stages of equilibrium chemistry. Even today, the calculations required for this would be inconceivable without a computer.

The evaluation and checking of the objectivity of the results are to a certain extent facilitated by the fact that whereas the molar absorptivity values are dependent on the wavelength, the equilibrium constants naturally are not. Hence, comparison of the equilibrium constants calculated from the concentration dependence of the absorbances measured at the different wavelengths is of assistance in the calculation of the correct constants and correct molar absorptivities. When the real equilibrium data are known, the distribution curves describing the system can be calculated and from these one can read off directly how the concentrations of the various species in solution vary with the change of the analytical concentrations (cf., Figs 8.1 and 8.2).

Mutatis mutandis, the above correlations are also applicable to other spectroscopic methods. In the study of complex equilibria, however, it is the electronic excitation spectra which have found the most widespread use.

It is obvious from what has been said that a condition for practical application is the elaboration of suitable computer programs which permit the simultaneous processing of a huge mass of experimental data containing a large number of unknown constants [Ga 79].

Experimental data obtained by equilibrium measurements and quantitatively reflecting the solvent effect will be reported in the sections that deal with the individual factors determining the solvent effect and present the special features of the individual solvents.

Circular dichroism (CD) spectroscopy

It is known that molecules which do not contain a second-order symmetry element (mirror plane, inversion centre of symmetry and gyroid) exhibit optical activity. Among the optically active compounds there are low-symmetry organic molecules as well as various kinetically inert metal complexes. All reactions that are accompanied by a change in the symmetry of these molecules give rise to a change in the optical activity. Thus it is not surprising that, as a result of the symmetry changes caused by solvation, in the case of optically active systems the solvent effect shows up in the optical rotatory dispersion (ORD) spectrum, and more markedly in the circular dichroism (CD) spectrum. Accordingly, CD spectroscopy has become a valuable method in the investigation of the solvent effect.

The phenomenon of CD is based on the fact that in optically active systems the molar absorptivities of radiation circularly polarized to the right and to the left are different. If the resulting difference in the absorptions of the two light beams is plotted as a function of the wavelength of the radiation employed, the CD spectrum is obtained. In accordance with this, two methods, direct and indirect, may be used for the measurement of CD.

In the direct procedures, light beams circularly polarized to the right and to the left are passed alternately through the solution under examination, and the difference between the absorptions of the two beams is recorded. Mason [Ma 62] and Velluz [Ve 65] have constructed instruments operating on this principle.

As a result of the difference between the molar absorptivities of the light beams circularly polarized in the two directions, the resultant of the electric vectors of the two beams, meeting after having passed through the optically active substance, describes an ellipse; the major axis defines the plane of the resulting polarized beam, and the length of the minor axis corresponds to the difference in absorption of the beams circularly polarized in the two directions.

In indirect procedures for measurement of the CD spectrum, therefore, the CD-induced elliptically polarized beam emerging from the optically active solution is transformed to plane-polarized light by means of an anisotropic crystal with known optical properties. From the angle of the polarization plane it is then possible to calculate the data relating to the given ellipse and necessary for construction of the CD spectrum. Instruments operating on this principle were constructed, e.g., by Jeffard [Je 48] and Arvedson [Ar 66]. Instruments based on these two principles are currently commercially available.

The solvent dependence of CD

A number of detailed studies have been concerned with the solvent dependences of the optical activity and CD of organic compounds; the reader is referred to the excellent book by Legrand [Le 73].

The solvent dependence of the CD spectra of inorganic compounds and metal complexes is more difficult to rationalize. Attention was drawn to this in particular by the investigations of Bosnich and Harrowfield [Bo 72] on the mixed complexes *trans*-CoN_4Cl_2 (where N denotes monodentate nitrogen bases). It was established that the solvent dependence of the ${}^1A_{1g}-{}^1A_{2g}(D_{4h})$ *d–d* transition in this group of compounds, as manifested in the CD spectrum, is so large that it may even lead to inversion of the sign of the Cotton effect. Consequently, the solvent effect was shown to give rise to possible uncertainties in the structural conclusions drawn from the CD spectrum, since it was just this transition that served as the basis of the application of the sector rule [Ha 65] in the determination of the configuration.

The studies by Hawkins *et al.* [Ha 77] permitted interpretation of the solvent dependence of the CD spectra of cobalt(III) mixed complexes. In effect it was as a result of their investigations that this spectroscopic method became applicable to the experimental monitoring of the solvent effect in inorganic and coordination chemistry.

The solvent dependence of the CD spectra of solute molecules (including complexes in solution) in an achiral solvent can be attributed to the following three main causes:

(1) Formation of an association between the solvent and the solute [Mo 63]. The energy conditions of the solvated and unsolvated molecule, and the rotatory powers depending on these, may be so different that essential changes may appear in the CD spectrum owing to a change in the extent of solvation, or because of the different structures of the associates (solvates) formed with the various solvents. Hydrogen-bonded associates, for instance, may be accompanied by the formation of new dissymmetry centres near to the chromophore group, and this is naturally manifested in the CD spectrum.

(2) The difference between the free energies of the various conformers of the complex, or of some other dissolved substance, may be so small that the conformers may undergo interconversion as a result of solvation. This phenomenon may arise if, for example, one of the conformers is more prone to form hydrogen bonds than the other, and hence, in protic solvents, the solvation may give rise to a conformational change *via* the formation of hydrogen bonds connecting solute and solvent [We 65].

The interconversion of different conformers may also occur on the action of the solvent as a dielectric, for instance if the existence of the different conformers is due to the different orientations of two neighbouring dipoles in the molecule [Ku 67].

The dipole–dipole interaction is a function of the relative permittivity of the medium (solvent): an increase in the relative permittivity leads to a decrease in this interaction, and vice versa.

(3) Ion-pair formation in the system. In systems containing complex cations, for example, the solvating effect and relative permittivity of the solvent control the extent of ion-pair formation between the complex cation and the associated anion; similarly to the effect of hydrogen bond formation, this may lead to a change in, or the formation of, a dissymmetry centre.

Hawkins *et al.* [Ha 77] studied the solvent dependence of the CD spectra of mixed complexes of bis(R-propane-1,2-diamine)cobalt(III) with monodentate ligands, and established that the exchange of the solvent leads to changes in the higher-energy $^1A_{1g}-{}^1A_{2g}$ Cotton effect, in accordance with the following sequence: py > Me_2SO > dma > dmf > thf > mf > $HCONH_2$ > Me_2CO > MeOH > MeCN > sul, where py = pyridine, dma = dimethylacetamide, dmf = dimethylformamide, thf = tetrahydrofuran, mf = methylformamide and sul = sulpholane. As an example, Fig. 5.2 illustrates the CD spectra of the tetrafluoroborate salt of the complex cation *trans*-bis(*R*-propane-1,2-diamine)dichlorocobaltate(III), measured in various solvents.

Proton resonance studies reflecting the effect of solvation of the NH proton of the ligand gave a similar sequence for the solvents. This showed that the solvent dependence of the CD spectrum of the Hawkins system can be interpreted by the different abilities for hydrogen bond formation of the solvents. In accordance with point (2), the stereoselective solvation, involving hydrogen bonding, of the NH group of the ligand influences the CD spectrum by causing a change in the asymmetric nitrogen donor group. This explanation is in agreement with the picture acquired in earlier examinations, according to which the system is labile from a conformational aspect [Go 71, Ha 76], and the NH proton of the ligand bound in the complex is accessible to hydrogen bonding [Mi 67].

With a view to studying the effect of the presumed ion-pair formation on the CD spectrum of the complex, a comparative investigation was made of salts of the bis(R-propane-1,2-diamine)cobalt(III) cationic complexes formed with different anions. It may be assumed that, in the solvents studied, the tetraphenylborate anion does not form an ion pair with the complex cation in the concentration range investigated, and halides display a decreasing tendency towards ion-pair formation in the sequence chloride > bromide > iodide. In conformity with these considerations, the experimental data revealed the effect of ion-pair formation on the CD spectrum only in the presence of chloride anion. The solvent dependences of the CD spectra of the complex cations in the presence of tetraphenylborate or bromide anion were identical; this suggests that the solvent effect presented in Fig. 5.2 can be attributed exclusively to the above-discussed different participations in hydrogen bonding of

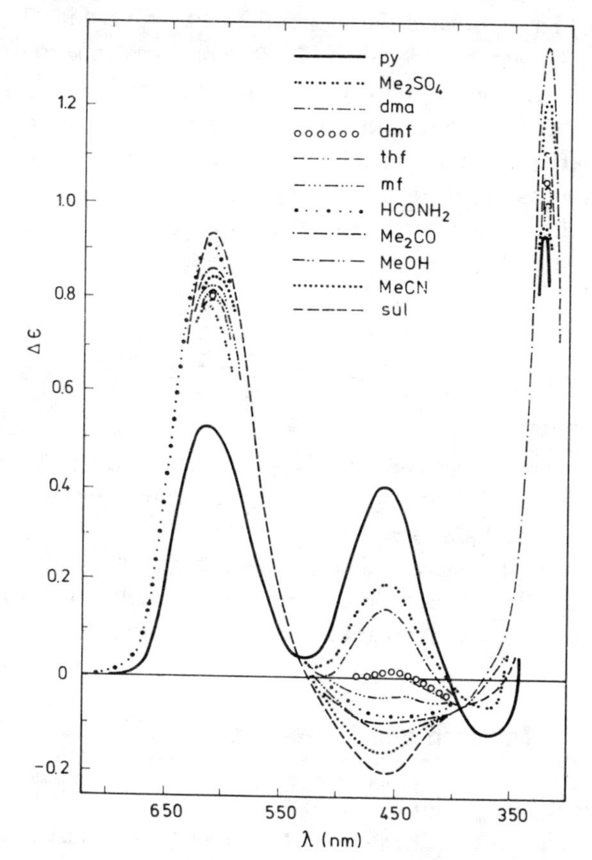

Fig. 5.2. CD spectra of the tetrafluoroborate salt of the *trans*-bis (*R*-propane-1,2-diamine) dichlorocobaltate(III) complex, determined in solutions prepared with various non-aqueous solvents [Ha 77]

the ligand bound in the complex. In the system containing chloride as the anion, an additional contribution is made by the effect of ion-pair formation.

The determination of the absolute configuration of a chiral substance is a very important part of the characterization of that molecule. That is also true for chiral solvents. Circular dichroisms induced in the UV spectra of metal complexes solvated by a chiral solvent can be used for the determination of the absolute configuration of the solvent [Br 80]. This method is attractive in that the experiments are easy to carry out.

When a chiral molecule is also capable of luminescence after being excited by UV light, it is often possible to observe the excited-state analogue of circular dichroism, circulary polarized luminescence (CPL) [Ri 77].

Brittain [Br 80] has correlated the sign of the CPL induced in $Tb(dpm)_3$ (dpm =2,2,6,6-tetra-methylheptane-3,5-dione) with the absolute configuration of several chiral solvents. It was found that the magnitude of total luminescence (TL) and CPL depended strongly on the nature of the solvent used, and reflected its configuration. In general the Tb(III) emission was at least an order of magnitude more intense in the amine solvents than in the alcohol solvents.

The CPL results were placed on a quantitative basis by calculating the luminescence dissymmetry factor g_{lum} as defined by Richardson and Riehl [Ri 77]:

$$g_{\mathrm{lum}} = \frac{2(\Delta I)}{I} = \frac{2(I_{\mathrm{L}} - I_{\mathrm{R}})}{(I_{\mathrm{L}} + I_{\mathrm{R}})}$$

where I_{L} and I_{R} refer, respectively, to the intensities of left and right circulary polarized emission, ΔI is the differential emission of left and right circulary polarized light, and I is the mean light intensity.

The method seems to offer a fairly convenient way to predict the absolute configuration of a substrate capable of forming an adduct with $Tb(dpm)_3$, and is particularly suited to the study of amine solvents.

Infrared and Raman spectroscopy

Infrared spectroscopy has long been successfully used to study the structures of solutions and liquids. Since the spreading of laser-Raman instruments, Raman examinations, which were earlier employed to only a limited extent, have become particularly useful supplements of infrared examinations (e.g., ref. [Me 80a]).

It is known that complex formation (the coordination of a free electron pair of a donor atom of the ligand to the acceptor) changes the electronic structures, energy states and symmetry conditions of both coordinating ligand and the acceptor ion or molecule; this results in change in their vibrational spectra, in the force constants determined from these, etc. Thus, in the event of the donor–acceptor interaction of a solvent and solute, the infrared or Raman bands of both the solute and the solvent may provide information on this process.

The new species, the solvate complexes, formed in the course of solvation will display new vibrations, that are not observed for the free reactants. The structure and symmetry of the molecule, the strengths of the bonds and the interactions of the molecule with its environment (solvent, ions bound in the outer sphere, other molecules, etc.) all influence the vibrational spectrum. This is the reason why infrared and Raman spectroscopy, experimental procedures involving the recording and study of vibrational spectra, have become important tools for the investigation of the structures of solutions [Yu 78a, b].

Theoretical analysis of the vibrational spectra of complex molecules is a difficult [Pa 76b, 79a, b], and in certain cases an insoluble task. However, the knowledge of the spectra of many different types of molecules has made possible the recognition of qualitative correlations between molecular structure and the vibrational spectrum. The study of the structures of complicated molecules and metal complexes can thus be performed with the aid of the vibrational spectra.

One of the most important advantages of these methods over other procedures for investigation of structures (X-ray diffraction, nuclear magnetic resonance, electron spin resonance, etc.) is that they give certain information about the structure of a molecule within a short time and without laborious methods of evaluation.

An additional great advantage is that the examinations may be carried out in the liquid state, and hence the picture is not distorted by the changes caused by the liquid–solid transformation.

With solvent molecules containing several potential donor atoms, the infrared spectrum may also yield information as to which of the donor atoms are linked to the different metal ions. In this way it has been possible to demonstrate, for instance, that the solvent dimethyl sulphoxide is bonded *via* its oxygen donor atom to the ions of the first transition metal series, but *via* its sulphur donor atom to the palladium ion [Co 60, Co 61a, Co 61b, Gr 79, Me 60, Dr 61].

The direct proof for the ambidentate character of dimethyl sulphoxide is, however, the single-crystal X-ray diffraction study of some of its complexes as, e.g., that of tetrakis (propionato)- and tetrakis(trifluoroacetatodirhodium(II)) adducts [Co 80].

The application of infrared spectroscopy to the solution of such solvation problems is hampered to a certain extent by the fact that complex formation leads to the appearance of new, skeletal vibrations, the coupling of which with the vibrations of the original molecules makes the vibrational spectrum more complicated. The situation may similarly be complicated by other effects, such as changes in the orbital hybridization, back-coordination, etc. In the coordination of the solvent acetonitrile to metal ions, for example, if only the effect of coordination causing a decrease in the electron density on the nitrogen is taken into consideration, the frequency of the C—N vibration would be expected to decrease. In fact, in the course of this process the coordination increases the σ character of the C—N bond, which is accompanied by an increase in the frequency of the C—N vibration [Be 61]. In many cases, it is difficult to assign the infrared bands to the corresponding vibrations; the conclusions drawn from the spectral data may therefore be uncertain.

In spite of the above difficulties, much valuable information has been obtained from the infrared study of the solvation of both cations and anions.

It is obvious that measurements in the most varied regions of the infrared and Raman spectral ranges may yield information adding to the knowledge of the structures of non-aqueous solutions.

The near-infrared region (between 4000 and 12 500 cm^{-1}) contains the combination bands connected with the hydrogen atom and the harmonic overtone bands, e.g., the first overtones of the O—H vibration at around 7140 cm^{-1}, and the N—N vibration at 6667 cm^{-1}. These bands are sensitively affected by the formation of hydrogen bonds, which strongly influence the vibrations, and by coordination of the O or N donor atom.

For practical reasons, the mid-infrared region (650–4000 cm^{-1}) may be subdivided into two parts. The region of the "group frequencies" (1300–4000 cm^{-1}) contains the relatively easily identifiable vibrations characteristic of the various functional groups, e.g., valence vibrations of the O—H, N—H and C—H groups lie between 2500 and 4000 cm^{-1}, the vibrations of triple bonds between 2000 and 2500 cm^{-1} and the vibrations of most double bonds between 1540 and 2000 cm^{-1}. Even within these groups, the bands of the different vibrations can be readily distinguished in most cases, e.g., the C═C, C═N, N═O and S═O bands in the last group.

The energies of the vibrations assigned to the individual functional groups are naturally also influenced by the other structural elements in the molecule, but they depend on the various intermolecular interactions, such as solvation, complex formation and other association equilibria. It is well known, for example, that the O—H vibration of alcohols in the free (non-associated) state appears as a sharp, low-intensity band at 3600 cm^{-1}. The formation of a hydrogen bond shifts this band in the direction of lower frequencies, and increases its intensity significantly. As will be discussed later, such changes are suitable for the characterization of equilibria of hydrogen-bonded associates. The group frequencies are similarly influenced by other interactions: for example, the N—H vibration of a metal ammine complex has a lower frequency than that of free ammonia. With increase in the strength of the metal–nitrogen bond, the frequency of the N—H vibration decreases. The frequency of the C≡N vibration in cyano complexes is higher than that of the free cyanide ion. On the other hand, the C≡O vibration of metal carbonyl complexes has a lower frequency than that of free carbon monoxide. With cyanide complexes, an increase in stability of the complex is accompanied by an increase in frequency of the C≡N vibration, while the higher stability of the carbonyl complex causes a decrease in frequency of the C≡O vibration [Bu 73]. As will be seen below, the vibrations of the NH, CN and CS groups of the solvent molecules are no less sensitive to association interactions caused by the various solvation changes. Hence, changes in the group frequencies may give information on the association conditions between the solvent molecules themselves, on solute–

solvent associations and on what functional groups of the solvent and the solute participate in the bonding, etc.

In the lower part of the mid-infrared region (650–1300 cm^{-1}) one finds the vibrations of the C—C, C—O, C—N single bonds, etc. and the bands of numerous deformational vibrations and skeletal vibrations. In this spectral region it is no longer easy to assign the individual bands to the corresponding vibrations. Nevertheless, on the basis of a comparison of the vibrational spectra of the pure solvent and the solution, the changes appearing in this spectral region may yield information on the system.

The formation of solvate complexes is accompanied by the occurrence of new skeletal vibrations. The attachment of the solvent molecule to the solute gives rise to changes (generally a decrease) in the symmetry of both reactants. This alters (usually increases) the number of infrared-active vibrations. It is known that only those vibrations which are associated with changes in the dipole moments of the molecules are infrared-active. It is understandable that a decrease in the symmetry of the molecules increases the number of such vibrations. Further, a decrease in the molecular symmetry can also cause the separation of degenerate vibrations which originally had identical energies. The two effects may be simply illustrated by comparing, e.g., the infrared spectra of the carbonate ion and carbonate complexes. The free carbonate ion belongs to the $\mathbf{D}_{3h}$ point group. Of its four vibrations (Fig. 5.3), the ν_1 vibration is not infrared-active and the ν_3 and ν_4 vibrations are doubly degenerate vibrations. Thus, the infrared spectrum of the free carbonate ion has two bands. Carbonate bound in a complex has $\mathbf{C}_3$ or $\mathbf{C}_{2v}$ symmetry. As a result of the decrease in the symmetry, the ν_1 vibration will be infrared-active, and the ν_3 and ν_4 vibrations will split into two bands.

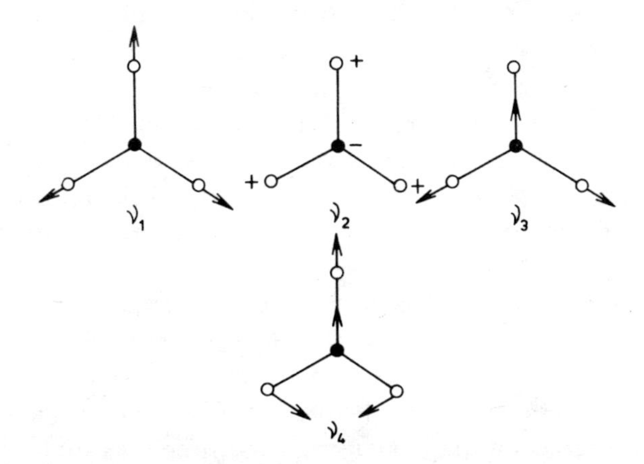

Fig. 5.3. Vibrations of the carbonate ion [Bu 73]
● — Carbon atom; ○ — oxygen atom

The far-infrared region (10–650 cm^{-1}) contains a series of deformational vibrations and the vibrations of the coordinate bonds of metal complexes (metal solvates). In the study of solutions, assignment of the latter vibrations is of particular importance, since these give direct information on the stabilities of the metal solvates (or other metal complexes).

In the following, some examples are presented of the results obtained from infrared investigations on non-aqueous solutions.

Dimethyl sulphoxide (DMSO) is known to be able to coordinate in its metal complexes in two ways. It is bound by its oxygen donor atom to "hard" transitional metal ions, and *via* its sulphur donor atom to markedly "soft" acceptors. The simplest means of deciding the mode of coordination is the application of infrared spectroscopy. Depending on the mode of coordination, there is a variation in the vibration of the sulphur–oxygen bond of DMSO. Coordination of the oxygen causes a decrease in the order of this bond, which is seen from the decrease in the S=O vibration. Coordination of the sulphur atom, in contrast, results in an increase in the S=O bond order and hence in the vibration.

The above effect was demonstrated by the study of the DMSO complexes of many metals, primarily those belonging to the first transition metal series [Cu 67, Se 61, Me 60, Ed 67]. The conclusions drawn from the infrared studies have been confirmed by X-ray structural examinations on a number of complexes [Be 67, Kr 73].

Whereas the infrared spectra of DMSO complexes linked *via* the sulphur atom can be interpreted fairly unambiguously, in the spectra of solvates containing solvent bound through the oxygen the S=O vibration and the deformational (rocking) vibration of the methyl group are probably coupled [Dr 61, Co 61]. Some authors [Se 61, Me 60] assign the band appearing at 1000 cm^{-1} to the S=O vibration and that at 930 cm^{-1} to the deformational vibration of the methyl group. Other authors [Co 60] prefer the opposite assignments.

If the infrared spectrum of the pure solvent is compared with that of deuterated DMSO, the former assignment seems correct. As a result of deuteration, the DMSO band at 950 cm^{-1} shifts to 800 cm^{-1}, while there is barely any change in the position of the band above 1000 cm^{-1}. It is obvious that deuteration will drastically change the vibration of the methyl group, while it will scarcely influence that of the SO group.

The above example shows well that, with appropriate circumspection, an infrared examination may reveal the mode of coordination between a solvent molecule and a metal more simply and more quickly than any other method.

In acetone solutions of silver [Pu 58], lithium and magnesium [Pe 64] perchlorates, an interaction was demonstrated between the cation and the donor oxygen of the solvent. In the corresponding iodides, splitting of the C—H vibrations of the methyl groups of the solvent even revealed the formation of comparatively very weak hydrogen bonds between the iodide ion and acetone [Pe 64]. Raman studies of acetone solutions of sodium, lithium and barium perchlorates led to results similar to the above, and also gave the sequence of stabilities of the acetone solvates of the cations [Mi 63b].

Various research groups have reported analogous results from studies of the vibrational (infrared and Raman) spectra of solutions of electrolytes in alcohol [Ha 67, Ke 62, Mi 63a, Pe 62a] and in acetonitrile [Ba 65, Ev 65, Pe 62b, Pu 66, Ro 70].

The vibrational spectrum gives information not only on the structures and structural changes of the dissolved ionic species and the solvent molecules attached to these, but also on the changes occurring in the interactions between the solvent molecules themselves.

Raman examinations have demonstrated [Ha 67b, Ke 62, Ke 64] that anions exert a stronger effect than cations on the OH bonds of alcohols. The perchlorate ion proved to be the most effective in splitting the hydrogen bonds between the methanol molecules in methanol solution, thereby leading to the formation of monomeric solvent molecules.

Studies of acetonitrile solutions showed that cations caused shifts of the C≡N and the C—C bands [Ke 62, Ke 64, Pe 62b], whereas anions changed the C—H vibrations.

Raman [Ke 62, Ke 64] and infrared [Gu 65, Ke 70] investigations in acetonitrile solutions of metal perchlorates, nitrates and chlorides indicated the splitting of the C≡N and C—C acetonitrile bands due to the solute–solvent interaction. The first lines in these line pairs formed were found to be independent of the nature of the dissolved salt, but the second one exhibited a definite dependence on the cation. This was explained in that the first bands originate from the free solvent (not bound in the solvate sheath). The solvated ions act to such a slight extent on these molecules that the resulting shift in the infrared bands is not greater than the experimental error. The second line in the line pair could be assigned to the solvent belonging to the solvate sheath.

A comparison of the results obtained in acetonitrile and acetone solutions has shown [Po 64] that in acetonitrile the vibrations of the solvent are almost independent of the anion, whereas in acetone they change considerably on the action of the anion. This can be attributed to the much weaker interactions of the anions with acetonitrile than with acetone.

Kecki *et al.* [Ke 73] carried out quantum chemical (semi-empirical) calculations to interpret the data obtained from vibrational spectra. By this means they attempted to determine the rearrangement of the σ and π electrons as a result of the formation of the solvate complex. It was shown, for example, that in an acetonitrile solution of cobalt perchlorate the π-interaction between the cobalt and the solvent was primarily responsible for the change in intensity of the C≡N bands. In contrast, in acetone solution the change in the C=O band was caused by both the σ- and the π-interactions. The examinations also showed that acetonitrile is bound to the dissolved cation generally *via* the free electron pair of the nitrogen.

In formamide and dimethylformamide solutions the cation is primarily bound to the oxygen of the amide group [Pe 68], but in formamide there is also an interaction with the nitrogen atom of the amide. On the other hand, the hydrogen atoms of the amide group of formamide may form hydrogen bonds with the electronegative anions.

Far-infrared studies of sodium tetraphenylborate in pyridine, dioxane and tetrahydrofuran solutions have shown that unsolvated ion pairs are also present in these solutions [Fr 68b]. Other investigations point to the solvation of the sodium ion by tetrahydrofuran [Hö 69].

Infrared examinations have revealed that dimethyl sulphoxide [Ma 67, Wu 70] and pyridine [Mc 70, Th 64] are solvents with such high donor strengths that even alkali metal ions form solvate complexes when dissolved in them.

Infrared examinations in solution also indicated the extent of ionic dissociation of alkali metal salts of organic acids in various organic solvents, and the structures of the associates in solution, could also be determined. Regis *et al.* [Re 74, 75], for instance, demonstrated the presence of the following two (asymmetric and symmetric) complexes in addition to the free ligand ion in an acetonitrile solution of lithium trifluoroacetate:

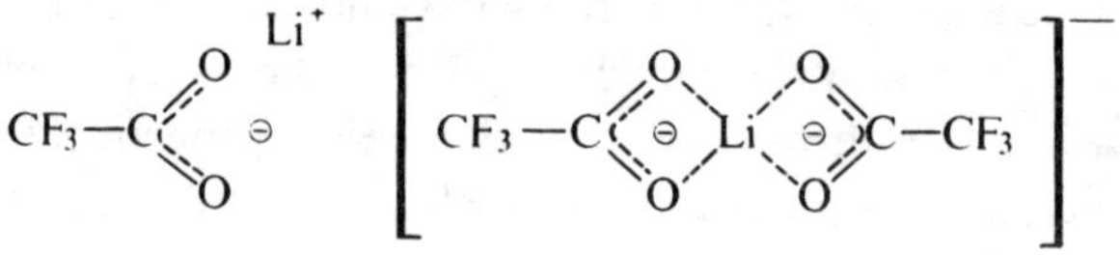

$$CF_3{-}C\langle^{O}_{O}\rangle^{\ominus}\ Li^+ \qquad \left[CF_3{-}C\langle^{O}_{O}\rangle^{\ominus} Li\ {}^{\ominus}\langle^{O}_{O}\rangle C{-}CF_3\right]^-$$

Mention should be made of infrared spectroscopic studies performed to determine the strength of hydrogen-bonded solvation; these studies characterized the strength of the interaction by means of the shift in the X—H vibration (where X may be oxygen, nitrogen or fluorine) and by the changes in width and intensity of the band [Bu 59, Bu 61, Ha 72, Hy 62, Lu 58, Sc 59].

As a result of hydrogen bonding, the X—H vibrations (both fundamental and overtones) shift towards lower frequencies, the extent of the shift being about 30–100 cm^{-1}. In contrast, the deformational vibrations shift towards higher frequencies, although to a smaller extent. The reason for these changes is that the force constants of the valence vibrations decrease owing to the formation of the hydrogen bond, whereas those of the deformational vibrations increase.

A band shifted as a result of hydrogen bonding also changes its shape; a considerable broadening can be observed, which can be interpreted in two ways. If several types of hydrogen bond can form in a given system, the broadening may be due to superposition of the absorption bands. In a system proved to contain only a single type of hydrogen-bonded species, the cause of the band broadening is that the proton may oscillate between two fairly close-lying potential minima, and this frequency is combined with the fundamental vibration [Ha 72, So 68]. This theory at the same time explains the appearance of the submaxima observed in the case of strong hydrogen bonds.

The band broadening caused by hydrogen bonding is always accompanied by an increase in the intensities of the IR bands. Between certain limits, both phenomena are proportional to the extent of the frequency shift. The strengths of the hydrogen bonds, however, are characterized most sensitively by the shifts in frequency of the vibrational bands. The magnitude of the shift increases with increasing strength of the hydrogen bond.

In this respect, particular interest is attached to those studies which show a correlation between the O—H vibration frequencies measured in hydrogen-bonded associates of various hydroxyl-containing compounds (phenols, alcohols) with various donor molecules and the ΔH values for formation of the associates.

A linear correlation between the heats of formation of hydrogen-bonded associates and the shifts in the O—H vibrations due to the above effect was first pointed out by Badger and Bauer [Ba 37]. More recently, Drago *et al.* [Dr 63, Dr 69, Dr 70, Ep 67, Jo 66, No 70, Pu 67, Pu 69, Sh 70, Wa 64] made comparisons between the heats of reaction of various phenols and alcohols with different donor molecules and the resulting O—H vibration shifts. From their numerous experimental data they constructed empirical equations by means of which the heat of formation of an associate can be calculated from the O—H vibration shift, which is simply read off the infrared spectrum. Some of these empirical equations are presented in Table 5.1.

Table 5.1. Correlation between the shift in the frequency of the OH vibration ($\Delta\nu_{OH}$) and the heat of formation of hydrogen-bonded associations ($-\Delta H$) with various donors [Dr 70]

Hydroxylic compound	Correlation between $-\Delta H$ and $\Delta\nu_{OH}$
Phenol	$-\Delta H = (0.0105 \pm 0.0007)\Delta\nu_{OH} + 3.0$
t-Butanol	$-\Delta H = (0.0108 \pm 0.0003)\Delta\nu_{OH} + 1.64$
2,2,2-Trifluoroethanol	$-\Delta H = (0.0121 \pm 0.0005)\Delta\nu_{OH} + 2.7$

The method has been criticized [Ar 70], and the authors themselves were aware of its limitations [No 70, Pa 68, Va 70]. Although there can be no doubt that values obtained in this way have a fairly large error, the simplicity of the method (especially when compared with the considerable time- and labour demands of calorimetric measurements) definitely makes it suitable for the acquisition of informatory enthalpy data. Among limitations recognized more recently, it is worth mentioning, for example, that the interactions of bases containing sulphur as the donor atom with phenol cannot be described with the given equation [Vo 70]. This emphasizes that this procedure can be used reliably for the quantitative study of the hydrogen bond formation only of reactants for which the validity of the ΔH *vs.* $\Delta \nu_{OH}$ relation has been proved by analogous reactions of related compounds.

A comparison of the calorimetrically determined heats of reaction with spectral data (IR, Mössbauer, ESCA, etc.) of the products may also yield valuable information about the causes of the changes.

Elaboration of the technique of *low-temperature measurements* significantly widened the scope of application of infrared spectroscopy in the investigation of solvation effects. As a result of the lowering of the temperature, broad, overlapping bands separate, and a spectrum with fine structure is obtained. Spectroscopic measurements at low temperature require the use of an appropriate vacuum technique. A special problem is the protection of the optical window from the condensation of moisture. This difficulty has been overcome by the development of suitable cryostats and connected equipment, and at present commercial instruments are available that permit measurements in systems thermostated to ± 1 K in the range 100–500 K.

A further problem is the consideration of the changes taking place in the system in the course of cooling. For instance, the danger exists that the solute will commence to precipitate during the cooling. This leads to a strong scatter (uncertain background) in the spectrum, which fortunately also draws attention to the presence of the disturbing effect. Various independent methods may also be used to check that phase separation has not occurred in the system.

The low-temperature infrared technique, and within this the study of solutions frozen to amorphous solids, have primarily contributed to the understanding of the structures of solutions prepared with protic solvents.

In the room-temperature infrared spectra of low-molecular-weight alcohols and electrolyte solutions prepared from them, the bands of the O—H vibrations are generally very broad. In most cases the electrolyte causes comparatively small changes only in this part of the spectrum, and it is fairly difficult to draw conclusions from these about the solvation conditions [Ad 71, Wa 70]. Exceptions are perchlorate and tetrafluoroborate salts [Br 70, Wy 70], which give rise in methanolic solutions to the appearance of a new, high-frequency band in the spectrum. According to certain authors [Wa 70], this is a result of the structure-breaking effect of the perchlorate ion, and the new band can in fact be ascribed to the free O—H groups released in this way. Other authors [Be 71b, Br 70b, Sy 75a, Wy 70] consider that these new bands arise from the vibrations of O—H groups bound by weak hydrogen bonds to the perchlorate or tetrafluoroborate anion. Regardless of which view we accept, it is certain that the "bulk" solvent and the solvent bound in the solvate sphere of the cation feature together in the other, broader band in the spectrum. The dependence of the latter band on the cation has been utilized to demonstrate the formation of cation solvates [Sy 75b].

It was shown by Strauss *et al.* [St 77] that in frozen methanol the O—H vibration bands become narrower, and in frozen methanol solutions of alkali metal salts the solute causes the appearance of fine structure in the spectrum. Appropriate assignment of the distinct bands has contributed to the understanding of the solvation processes. With a series of detailed examinations employing this method, Symons *et al.* [St 76, St 77, Sy 75a, Sy 75b, Sy 75c] investigated the structures of alcoholic solutions of alkali metal and tetraalkylammonium halides.

For a long time, the following model was used for the characterization of the solvation of ions [Fr 57]. The dissolved ion is surrounded by a strongly binding primary sheath of solvent molecules. Around this is situated the still ordered secondary sheath, less firmly bound, which is separated from the bulk of the solvent by a non-ordered layer. Symons [Sy 75c] cast doubt on the existence of such a non-ordered, third layer in solutions formed with protic solvents. In his model, the entire system is characterized by continuous hydrogen bonding interactions, the solvent molecules of the ordered solvate spheres being connected directly to the similarly ordered bulk solvent.

Using methanolic electrolyte solutions frozen to amorphous solids, Jackson *et al.* [Ja 77a] assigned the various bands in the infrared spectra measured at low temperature; they did not observe O—H bands characteristic of a non-ordered solvent layer separating the solvates from the bulk solvent. Essentially, according to the assumption of Symons, in a protic medium the anions are surrounded by a solvate sheath (bound by hydrogen bonds) consisting of 4–6 solvent molecules, these serving as acceptors towards the molecules of the following solvent sheath. The latter are then linked in the customary way to the bulk of the solvent.

Particularly valuable information was provided by a series of such Symons-type studies of frozen methanolic solutions of tetraalkylammonium salts. It was established earlier that solvation of the cation is negligible in solutions of these salts. On this basis, it was to be expected that the new O—H bands appearing as a consequence of cooling can be ascribed to solvates of the anion. With the exception of the OH band of methanol bound to the fluoride ion, these bands occur on the high-frequency side of the bulk methanol band. This indicates that the hydrogen bonds between the solvent and dissolved anion may be weaker than the hydrogen bonds binding the solvent molecules together. When comparing the stabilities of the hydrogen bonds holding together the solvent–solvent and the solvent–solute associations, it must not be forgotten that almost always more than one molecule of methanol is bound to the anion, while two methanol molecules are joined by a single hydrogen bond. The formation of the latter increases the base strength of the oxygen, and hence favours the binding of the hydrogen of a subsequent methanol molecule. Thus, the oxygen of the methanol may participate in at most two hydrogen bonds whereas, depending on their coordination numbers, the anions bind four or possibly six solvent molecules in their solvent sheaths. The increase in the number of hydrogen bonds bound to a given central atom is accompanied by a decrease in the stability of the hydrogen bond. Dilution of the methanol with an apolar solvent, for instance, by leading to the partial cleavage of the above-mentioned double hydrogen bonds, increases the strength of the remaining bonds, and this is manifested in the O—H vibrations.

Accordingly, the reason why the anion dependence of the OH vibrations of the methanol molecules linked by hydrogen bonds to the various anions in methanolic solution is so small (barely larger than the experimental error) is that the hydrogen-bond strength decrease caused by the decrease in the electron density of the anions (in the sequence $Cl^- > Br^- > I^-$) is partly compensated for by the hydrogen-bond strength increasing effect of the decrease in the coordination number (in the above sequence of the anions). The overall picture is that the chloride ion binds the methanol molecules in its solvate with higher coordination number, with roughly the same strength as does the iodide ion in its solvate with lower coordination number. The fact that the difference between the O—H bands characteristic of the chloride and the iodide solvates is scarcely larger than the experimental error can be explained in this way.

It follows that the effect of the basicities of various anions on the strength of the hydrogen bond can be established unambiguously only if the analogy of the coordination spheres is ensured. The simplest means for this is to examine the interaction between the tetraalkylammonium salt and methanol in an inert solvent. The reactants may be brought together in this way in such proportions that only 1:1 associates are formed. The O—H frequencies of these clearly demonstrate the decrease in the strength of the hydrogen bond in the sequence $Cl^- > Br^- > I^- > NO_3^- > ClO_4^-$. The O—H frequency difference between the individual ions is as high as 40–60 wavenumbers [St 77].

Working in an inert solvent and varying the salt to methanol ratio, it was possible to ensure the formation of various species: those containing only one or two methanol molecules as primary solvate sheath, and analogous solvates also containing methanol in the second sphere, linked to the inner sphere by hydrogen bonds. Surprisingly, the study of these types of solvates showed that the strength of the binding of the methanol in the inner sphere (the stability of the hydrogen bond linking it to the anion) will be higher if this methanol is further linked through its oxygen atom and by hydrogen bond to methanol in the second sphere. The secondary solvate sheath therefore enhances the stability of the primary solvate sheath. At the same time, these examinations confirmed the finding that an increase in the coordination number, that is, the binding of a further methanol molecule in the inner sphere, is accompanied by a decrease in the stability of the hydrogen bonds involved.

A considerable contribution to the interpretation of the fine structure of IR spectra recorded at low temperature was made by an investigation of the dependence of the O—H stretching bands on the electrolyte concentration. It was found by Strauss and Symons [St 78] that, in addition to the two bands in the O—H vibration range in dilute solutions (the narrower of which can be ascribed to the solvate sheath of the anion, and the broader one jointly to the free solvent and the solvate sheath of the cation), new bands appear in concentrated (1–2 M) solutions. For example, the IR spectrum measured at −140°C had, in a methanolic solution of tetrabutylammonium chloride (where solvation of the cation can be neglected), six absorption bands (partly overlapping) in the wavenumber range 3100–3500 (see Fig. 5.4).

The starting point for assignment of the bands in the spectrum was the following solvate model:

```
          R       R       R
          |       |       |
Cl⁻---H—O---H—O---H—O---
          S1      S2      S3
```

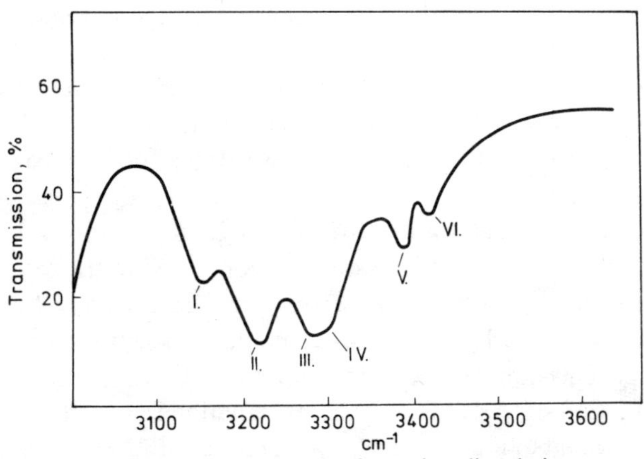

Fig. 5.4. Infrared spectrum of a methanolic solution of tetrabutylammonium chloride, recorded at −140 °C [St 78]

In dilute solution the oxygen atoms of the methanol molecules hydrogen-bonded to the chloride ion are linked by hydrogen bonds to the methanol molecules of the next solvate sphere. Accordingly, there is a continuous transition between the main bulk of the solution and the inner solvate sphere. With an increase in the electrolyte concentration of the solution, increasing amounts of solvent are required for formation of the inner solvate spheres. Accordingly, the lengths of the $S_1 \ldots S_2 \ldots S_3 \ldots$ chains progressively decrease. In a solution of appropriate concentration, these chains terminate with a methanol molecule that is connected to the chain by only a single hydrogen bond. It was concluded that one or other of the two bands occurring at the highest wavenumbers (denoted by V and VI in Fig. 5.4) can be assigned to the O—H group of the chain-terminating methanol.

The intensities of the bands at 3150 cm^{-1} (I) and 3220 cm^{-1} (II), characteristic of the most stable hydrogen bonds, increase with increasing electrolyte concentration. This can be explained by ion-pair formation in the system as a result of the increase in the salt concentration. Association of the tetrabutylammonium cation to the chloride ion decreases the number of sites available for binding of methanol molecules (the cation shields the anion). Thus, whereas the chloride ion has a coordination number of four in dilute methanolic solution, ion-pair formation reduces this number to two or three with respect to methanol. In agreement with the fact that a decrease in the number of hydrogen bonds to a given central atom causes an increase in the strength of the hydrogen bond (a decrease in its polarity), which is manifested in the shift of the O—H band in the direction of lower wavenumbers, band I (assigned to the inner coordination sphere of two methanol molecules) and band II (assigned to the inner coordination sphere consisting of three methanol molecules) appear at lower frequencies than the O—H bands characteristic of the full solvate sheath. Bands III and IV in the spectrum seem to be analogous with these latter ones. The broad band characteristic of the bulk solvent did not occur at all in such concentrated solutions.

In the assignment of the above bands, Strauss and Symons [St 78] also relied on the effects brought about in the spectrum of the methanolic solution by the addition of a small amount of water. It was found that even 0.1 M of water causes the O—H band ascribed to the chain-terminating methanol to disappear. It might be concluded from this that each water molecule bridges together two chains in such a way that the hydrogen atoms of the water are connected to the oxygen atom of a chain-terminating methanol, which possesses a free electron pair. However, band I, assigned to the strongest hydrogen bond, similarly disappeared, while band II became stronger. This double effect might point to the entry of the water molecule into the coordination sphere, although this explanation is in contradiction with the fact that in methanolic solution the hydrogen bonding ability of the water molecule is less than that of methanol. Greatest probability was finally attributed to the formation of the following cyclic solvate:

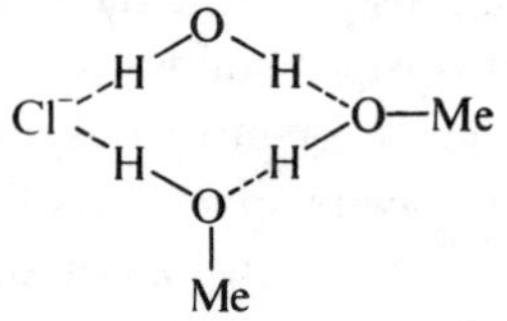

O
H   H
Cl⁻   O—Me
H   H
O
|
Me

According to this model, the ring closure due to the coordination of water leads simultaneously to the disappearance of the chain-terminating methanol and to an increase in the coordination number of the chloride, which, by reducing the strength of the hydrogen bond of the methanol in the inner sphere, may be the cause of the disappearance of band I.

Although these band assignments are qualitative and fairly uncertain, the model of the solution structure deduced from them seems to be convincing.

An analogous study of alkali metal halides led to even more complicated results. In these systems the effect of solvation of the cation cannot be neglected either. For example, of the three partially overlapping bands occurring in the IR spectrum of lithium iodide in amorphous frozen methanol, one varied as a function of the anion concentration, another as a function of the cation concentration and the third as a function of the concentrations of both ions. This last, surprisingly narrow, band can be assigned to the solvent-shared lithium iodide ion pair. Hence, it is understandable that its intensity increases strongly with increasing salt concentration, following the increase in the concentration of the ion pair [St 77].

The experimental procedure of Symons is not only of use for the investigation of solutions prepared with methanol and with other protic non-aqueous solvents, but has recently also proved applicable to the examination of hydration processes predominating in water [St 78].

Infrared spectra in the O—H and N—H stretching regions could be used also to measure the interaction between halide ions (Cl^-, Br^-, I^- as tetra-alkylammonium salts) and water or N-methyl acetamide (NMA) in tetrachloromethane. Formation constants for the monosolvates (at 22.5°C) have been estimated by Symons *et al.* [Sy 80]. The frequency shifts were approximately proportional to the free energy of hydrogen-bond formation. Formation constants for the monoaquo complexes and the NMA complexes were similar to those of the corresponding methanol solvates; however, the sensitivity of the N—H frequency was less than that of the O—H frequency.

Nuclear magnetic resonance (NMR) spectroscopy

Of the more recent structural examination methods, nuclear magnetic resonance (NMR) spectroscopy is particularly helpful in the study of solvation and related effects [Go 81, He 71a, La 81, Po 79].

The primary sources of information are the chemical shifts of the magnetic nuclei examined and the corresponding coupling constants. Information on the system may also be given by the spin–lattice and spin–spin relaxation times.

The value of the chemical shift depends on the chemical environment; on the structure of the electron sheath surrounding the magnetic nucleus and on the neighbouring atomic nuclei. Changes are brought about in the electronic structures and chemical environments of the magnetic nuclei, and hence in the values of the NMR chemical shifts, by the interactions occurring in protonic solvents, hydrogen bonding between the solvent molecules and/or between the solvent and the solute, coordination of the donor atoms, etc. The proton resonance spectrum can thus give the most direct information on the various hydrogen bonding interactions, or on the coordination of donor atoms close to groups containing protons, etc. NMR studies on donor solvents containing a magnetic nucleus as the donor atom, or on solvate complexes containing a magnetic nucleus as the central atom, can be just as valuable.

The coupling constants reflect the interactions between the nuclei. The distance between the nuclei participating in the coupling is a function of the chemical environment: for example, proton–proton coupling generally acts through only two or three bonds, whereas if an unsaturated system is located between the protons there is a possibility of long-range coupling. In addition, coupling also depends on the geometry of the system. Hence, changes may also appear in the coupling constants as a result of changes in the solvation conditions.

In the NMR investigation of solutions, the effects of paramagnetic substances in solution on the proton resonance peaks of the solvent may be of particular significance. Certain europium(III) and praseodymium(III) chelates, for example, behave as Lewis acids towards donor solvents such as alcohols, ethers, esters and ketones. As a result of this acid–base interaction the proton resonance peaks undergo shifts which favour, among others, the resolution of the PMR spectrum.

The quantitative analytical evaluation of NMR spectra is made possible by the fact that the areas under the NMR absorption bands relating to equivalent nuclei are proportional to the numbers of the corresponding atomic nuclei. Thus, in the case of kinetically inert solvates, for instance, if the analytical composition of the solution is known, the solvation number of the dissolved ions can be calculated from the ratio of the areas under the NMR bands assigned to the solvent molecules bound in the solvate sheath and to the free solvent. In the optimal case, even the complicated solvation conditions arising in solvent mixtures may be interpreted by this means. For ions of high charge and low radius, e.g., Al^{3+}, Ga^{3+}, Be^{2+} and Mg^{2+}, the lifetime of the solvent molecules bound in the first coordination sphere is fairly long, generally being more than 10^{-3}–10^{-4} s [Di 39]. Under favourable conditions, therefore, in solutions in mixed solvents the occurrence of solvates of different compositions is indicated by separate resonance lines [Li 71, Fr 72].

The formation of less inert solvates can be investigated on the basis of the dependence of the chemical shift of the solvent on the electrolyte concentration [Hi 71].

Even this brief introduction shows that the effect of solvation on the electronic structure can be followed from two approaches by this method: by observing the changes due to solvation in the NMR signal of some magnetic nucleus of the solvent molecule [Na 77], or in the signal of the dissolved substance [Ba 81, Di 79, Ue 79].

Proton magnetic resonance (PMR) studies of lithium, ammonium and sodium salts in dimethyl sulphoxide [Ma 68] resulted in the determination of the number of solvent molecules coordinated to the cation. These studies indicated that all three cations coordinate two dimethyl sulphoxide molecules. It must be noted that with increase in the size of the dissolved ion the chemical shift of the solvent becomes less and less sensitive to the effect of the cation [Bu 68].

Proton resonance measurements can also give information on the structure-building or structure-breaking effects of various ions on the solvent [Fr 65, Ha 67a, Kr 67a].

A proton resonance spectroscopic study of methanolic solutions of magnesium, zinc and aluminium perchlorates well reflected the effects of the cations on the solvent [Bu 69]. In the case of salts with different anions, however, the different hydrogen-bonding abilities of the anions could also be detected in the proton resonance spectrum. This is clearly seen, for example, from the data from PMR measurements on dimethylformamide solutions of halides [Mo 68].

In the optimal case, even the compositions of mixed solvates formed in solvent mixtures can be determined by means of proton resonance measurements. As an example, measurements at $-75\,^{\circ}\mathrm{C}$ on a solution of magnesium perchlorate in a 10:1 methanol–water solvent mixture revealed that statistically 5 methanol molecules and only 0.7 water molecules were coordinated to the magnesium [Am 66].

With the employment of an appropriate model system, proton resonance spectroscopy has proved applicable also for following the transformation of a contact ion pair into a solvent-separated ion pair. Naseer Ahmad and Day [Na 77] used the system sodium tetraethylaluminate–benzene–donor solvent for this purpose. The spin–spin interaction between the ^{1}H and the ^{27}Al in the tetraethylaluminate served as the source of information; this is reflected by the resonance spectrum of the methylene protons.

The Day system has the advantage that the benzene used as the parent solvent, with a low dielectric constant, dissolves both sodium tetraethylaluminate and the donor solvents, without entering into coordination interaction with the components. Therefore, the proton resonance spectrum of the tetraethylaluminate is governed by the interaction between the sodium and the tetraethylaluminate, this interaction depending on the solvating effect of the donor solvent.

In the solvent-separated ion pair the tetraethylaluminate has regular T_d symmetry. Accordingly, as a result of the spin–spin interaction with the neighbouring methylene groups and the aluminium nucleus, the methylene protons give a 9-line spectrum. The tetraethylaluminate in the contact ion pair with the sodium is strongly distorted. Hence, the ^{27}Al–^{1}H spin interaction does not show up in its proton resonance spectrum. The residual ^{1}H–^{1}H spin interaction causes the appearance of a four-line pattern.

When various donor solvents were employed in the Day model system, and the ratio of donor solvent to sodium tetraethylaluminate was varied, the above method could be used to determine when contact or solvent-separated ion pairs were formed. The low relative permittivity ensured by the benzene parent solvent acts against the total dissociation of the ion pair; this may therefore be neglected.

Naseer Ahmad and Day [Na 77] found a good correlation between the Gutmann donicities of the solvents and their ability to transform contact ion pairs to solvent-separated ion pairs. The increase in the donicity of the solvent favours the above transformation. Hexamethylphosphoramide, with a donicity of 33.1, causes the formation of the solvent-separated ion pair at a donor solvent to $Na[AlEt_4]$ ratio of 4 : 1, whereas in the case of tetrahydrofuran, with a donicity of 20.0, the ratio is 18 : 1. In diethyl ether the solvent-separated ion pair did not appear at all.

The process is perceptibly controlled by the solvation of the sodium ion. With an increase in the donicity and concentration of the donor solvent, the solvation of the sodium ion increases. The higher the donicity of the solvent, the lower is the donor solvent to $Na[AlEt_4]$ ratio at which the solvation of the sodium ion attains a value such that it is bound to the tetraethylaluminate ion only *via* the solvate sphere (coordinated solvent), and hence a solvent-separated ion pair is formed.

The examinations thus yield direct information on the solvation of the sodium ion. For example, it has been established as a result of such investigations that pyridine is a weaker donor towards the sodium ion than would be expected on the basis of the Gutmann donicity.

It emerges from the foregoing that in most of the investigations the conclusions were drawn not from the NMR coupling constant of the donor atom, but from the results of proton resonance measurements, recorded in effect as secondary data. Although the results obtained in this way are of value, more direct information may be expected from the NMR examination of the donor atom. For instance, from the ^{14}N NMR hyperfine coupling constant of acetonitrile, conclusions could be drawn about the electronic structures of acetonitrile solvates of transition metal ions, and even on the kinetic data of pseudo-first-order solvent-exchange reactions. In an analogous manner, the ^{17}O coupling constants of aquo complexes were used to draw conclusions of the structures of copper [Sm 66], nickel and thallium [Ev 53, Po 56] hydrates.

The solvent induced shift (SIS) in ^{13}C NMR spectroscopy was also used in the study of structural, stereochemical and conformational problems of solvation [Ue 80].

The rates of ligand-exchange reactions of solvent molecules bound in the first coordination sphere of nickel(II) ions depend on the natures of the other ligands attached to the metal [Hu 71]. By means of ^{14}N NMR measurements in systems containing nitrogen donor atoms, Lincoln and West [Li 74] studied how various chelate-forming ligands affect the lability of acetonitrile solvent molecules coordinated to nickel. In the case of complexes containing several acetonitrile molecules, a correlation was found between the position (*cis* or *trans*) of the coordinated solvent with reference to the chelate-forming ligand and its kinetic lability.

Clearer information can similarly be expected from the NMR parameters of the solvated ion than from the proton resonance data relating to the solvent, which are in effect indirect data. In the following, a few examinations of this type will be presented to indicate how NMR spectroscopy can be used for the study of non-aqueous solutions and solvation phenomena in them.

Maciel *et al.* [Ma 66] used the chemical shift of ^{7}Li to study solutions of lithium salts in eleven organic solvents. Symons [Sy 67] used the chemical shift of ^{23}Na to investigate solutions of sodium iodide in fourteen donor solvents. The data show a correlation with the Lewis base strength of solvent. Sodium resonance measurements in acetonitrile solution clearly point to a π-interaction between the dissolved salt and the C≡N bond of the solvent [Bl 68]. Sodium NMR has been used to study the solvation of the sodium ion in solutions of sodium methoxide in methanol and in mixtures of methanol and dimethyl sulphoxide [Ba 81].

^{109}Ag NMR studies of organic silver complexes in DMSO (combined with a ^{13}C NMR investigation of the system) indicated the presence of a variety of complex species in the solution [He 79].

The chemical shifts of ^{27}Al determined in solutions of aluminium chloride in various organic solvents showed the solvation strengths of the solvents to decrease in the following order [Ha 69]:

$$H_2O > C_2H_5OH, \quad C_3H_7OH > C_2H_5NCS > C_6H_5NC > CH_3CN.$$

Evaluation of NMR measurements on ^{27}Al is facilitated by the fact that the spectrum is comparatively simple; the line broadening caused by quadrupole relaxation is extremely small even in solvent mixtures. This permitted the use of this method to distinguish between the different solvates formed in solvent mixtures. The coordination number of the aluminium ion is known to be 6 in both dimethylformamide and dimethyl sulphoxide solution [Mo 67, Mo 68, Th 66c]. In a mixture of these two solvents, however, the solvating effect of DMSO is more marked than that of DMF [Gu 74]. The solvation conditions are also influenced by dilution with an inert solvent. Thus, for instance, if a solution containing DMSO and DMF as solvents is diluted with nitromethane, which can be regarded as inert from the aspect of solvation, then the earlier difference between the solvating powers of the two donor solvents no longer appears. This experiment also reflects the effect of the solvent–solvent interaction on the solvation of a solute.

NMR spectroscopy has proved applicable to the study of the exchange rate between the bulk solvent and the solvent molecules coordinated to paramagnetic ions. The examination is based on the measurement of the NMR line broadening

[La 73, Li 77, Me 80b]. The combination of stopped-flow Fourier-transform NMR and conventional NMR line broadening studies made the determination of more accurate rate data and ΔH^* and ΔS^* values possible [Me 80b].

In the course of solvent exchange investigations, it was found that the rate of solvent exchange depends not only on the nature of the solvent, but also on the coordination site of the dissolved ion to which the solvent molecule is bound, as well as on the nature of other ligands in the coordination sphere [Wa 76].

Studies dealing with the copper ion and copper complexes, for example, have demonstrated that coordinated DMSO is exchanged considerably more slowly than coordinated water. For a given solvent, the rate of exchange of the *axially* bound solvent is higher than in the case of the *equatorial* attachment [Sw 62, Vi 71]. In mixed complexes containing DMSO and ethylene diamine, the exchange rate of the DMSO is higher than that of any of the DMSO ligands in solvate complex containing only DMSO.

If the complex formation takes place stepwise, the exchange rates of the solvent bound in the different species may also differ. Thus, if the solvent exchange rates of DMSO solvates of Ni^{2+}, $Ni(en)^{2+}$ and $Ni(en)_2^{2+}$ are investigated in DMSO solution, the exchange rate is found to increase in the above sequence, which can be attributed to the *trans* effect of the individual ligands. As a consequence of the coordination of the nitrogen donor atoms of ethylenediamine, the DMSO molecules bound *via* oxygen in *trans* positions to the nitrogen atoms become more readily exchangeable [Wa 76].

The effect of the electrolyte on the solution structure is not confined to the solvation of the ions, but may also play a role in altering the structure of the solvent itself. Hence, these two effects appear together in the NMR shift. This is well reflected by studies on the system $AgNO_3$–water–acetonitrile [Sc 66, Sc 76]. The chemical shift of water in this system yields relatively little information on its selective solvation effect, since the addition of acetonitrile causes the hydrogen-bonded structure of the water to break down; this causes a shift in the NMR spectrum in the opposite direction to that due to the coordination of water to silver. More meaningful information on the composition of the solvate is provided by the chemical shift of the acetonitrile. In agreement with the results of potentiometric titrations, this shows that two acetonitrile molecules are coordinated in the first coordination sphere of the silver.

The $AgNO_3$–H_2O–DMSO system behaves in a different manner to the corresponding acetonitrile system [Cl 73]. In this system the effect of the DMSO on the water structure appears so predominantly in the NMR spectrum that the line shift characteristic of the DMSO solvate of silver shows up only at high DMSO concentrations. Hence, the measurement cannot be utilized to draw conclusions on the composition of this solvate.

The composition of the DMSO solvate can be interpreted more unambiguously in the $AgClO_4$–CH_3OH–DMSO system [Ro 76]. From the dependence of the chemical shift on the $AgClO_4$ concentration and on the composition of the mixed solvent, it could be established that the silver ion coordinates four DMSO molecules at most.

From the above and similar investigations [Mo 71, Be 73], it could be seen that, in various binary solvent mixtures containing water, acetonitrile, methanol and dimethyl sulphoxide, the solvating effects of the individual components are generally proportional to the effects displayed by the pure solvents. An exception is the H_2O–DMSO mixture, in which the silver ion occurs only in the forms of its pure aquo complex and its pure DMSO complex.

In solutions of nickel(II) and aluminium(III) in the above mixture [Fr 68, Th 70], formation of the aquo complex was favoured; this can be explained in that the addition of DMSO, by causing a certain degree of breakdown of the hydrogen-bonded water structure, makes the water molecules more suitable for coordination than in the case of pure water.

Among the results of NMR studies of solvation which have aroused great interest are those aimed at interpreting the effect of the solvent on the chemical shifts of various NMR nuclei based on model calculations [La 66, Em 71, Bu 60, We 71, etc.].

According to Buckingham [Bu 60], the effect of the solvent on the chemical shift of the proton can be expressed by the linear combination of four terms: two magnetic terms, one of which takes into account the anisotropy of the bulk susceptibility of the solvent, and two electrostatic terms, one of which accounts for the Van der Waals forces and the other the reaction field of the solvent. Detailed models were elaborated for the electrostatic terms. It turned out that the reaction field term depends in a complex manner on the relative permittivity and refractive index of the solvent [Bu 60, Em 71]. Weiner and Malinovski [We 71], who attempted to ensure reaction conditions such that the chemical shift is governed by the electrostatic term alone (primarily by using a suitable reference substance), demonstrated that in a series of polar solvents the chemical shift of the proton cannot be described by the previously proposed electrostatic term. Assuming that the cause of the difference is that the reaction field term cannot be estimated on the basis of the bulk susceptibility, they suggested the introduction of a fifth term taking into account the effect of the permanent dipole.

Examinations have also been carried out to interpret the effect of the solvent on the chemical shifts of compounds of ^{13}C and ^{19}F [Be 65, Em 66, Ta 63]. It appears that the solvent dependence of the ^{13}C chemical shift [Be 65] cannot be described with the equation of Buckingham [Bu 69]. Emsley and Phillips [Em 66] suggested that in fluorine-substituted aromatic compounds the effect of the reaction field is directly connected with the polarity of the solvent. However, the results indicate that this correlation is a fairly uncertain one [Fa 76].

On the basis of a study of the solvent dependence of the chemical shift of ^{19}F, Taft *el al.* [Ta 63] suggested that the solvent effect can be attributed to donor–acceptor chemical interactions. The first really convincing confirmation of this concept was provided by the ^{23}Na NMR studies by Erlich, Popov *et al.* [Er 70, Er 71, Gr 73, Po 79]. In this work a linear correlation was demonstrated between the Gutmann donicities of the solvents and the chemical shift values for solutions of sodium perchlorate and sodium tetraphenylborate in various organic solvents. This clearly showed that the variation of the chemical shift of the ^{23}Na in the systems investigated was caused predominantly by the donor–acceptor interaction between the donor solvent and the sodium ion.

The analogy of the above concept for acceptor solvents led to the construction of the most recent solvent acceptor strength scale. Mayer *et al.* [Ma 75, 79b] examined the variation of the ^{31}P chemical shift of triethylphosphine oxide in various acceptor solvents. Assuming that the cause of the variation is exclusively, or at least predominantly, the donor–acceptor interaction (hydrogen bond formation, or coordination) between the oxygen donor atom of triethylphosphine oxide and the electron-pair acceptor (e.g., hydrogen bond forming) solvent, they regarded the sequence of chemical shifts as the sequence of acceptor strengths of the solvents. Appropriate normalization of these led to the Gutmann "acceptor number" concept (see also Chapter 4).

The model of Krygowski and Fawcett [Kr 75], developed for the description of the solvent effect and taking into account exclusively the Lewis acid–base properties of the solvent, also appears suitable for the description of the solvent dependence of the chemical shift. For example, this model reflects well the results of the ^{23}Na resonance studies by Erlich *et al.* referred to earlier [Er 70, Er 71, Gr 73]. In addition to the interaction between the sodium ion and the solvent, it also points to the dependence of the chemical shift on the concentration as a result of ion-pair formation. However, the authors themselves [Fa 76] reported that the model was unsuitable for the description of other NMR data reflecting the solvent effect.

Happer [Ha 82] has correlated the ^{13}C NMR chemical shift of the vinyl group of styrene and some of its *meta*- and *para*-substituted derivatives in seven solvents by means of the Hammett equation.

A simplified reaction field theory was used by Abboud and Taft [Ab 79] for the description of the solvent dependence of NMR shift data. Their concept proved to be useful for the generalization of solvent properties in polar solvents only.

To sum up, it may be stated that there is still no correlation of general validity that is suitable for a quantitative description of the solvent dependence of NMR chemical shifts. Nevertheless, the experimental data reflecting such an effect may be of fundamental assistance for a closer understanding of the systems investigated.

The different variants of nuclear magnetic relaxation examinations have recently been employed more extensively for the study of solvation processes. Particularly worthy of mention is the work directed to the investigation of preferential solvation in solvent mixtures.

From the matter discussed so far, but especially from that to be surveyed in Chapter 8, it is evident that in solvent mixtures the affinities of the component solvents for the dissolved ions or molecules may differ greatly. Consequently, the ratios of the solvent molecules bound in the solvate sheaths of the dissolved ions may be different from the ratio of the components in the solvent mixture, (e.g., ref. [De 81, Ka 80]). This well-known effect is preferential solvation, and numerous procedures have been used for its detection. It really is necessary to have a range of procedures available, as the applicability of a method of examination depends on the various chemical and physical properties of the solvent mixture. Hence, a method which gives an excellent reflection of this effect in one system may be entirely useless in another system.

Nuclear magnetic relaxation examinations can follow interactions of two kinds: those between nuclei of the dissolved ions or molecules and the nucleus of the atom of the solvent (generally the proton) involved in the particular interaction [Gi 76, Ho 77, Ho 78]; and interactions of the solvent molecules with one another [Ca 78a, Ca 78b]. All of these interactions are reflected in the relaxation rate.

Because of the novelty of such examinations, one example of these indirect procedures deserves special mention. In systems where there is no magnetic or other interaction between the spins of the dissolved ions and the solvent molecules, nuclear magnetic relaxation may be used to follow the solvate effect and, within this, preferential solvation. In such cases, the nuclear magnetic interaction of the solvent molecules with one another may be the factor determining the relaxation.

As a consequence of the preferential solvation of the dissolved ions in a solvent mixture, one or other solvent may become enriched in the solvate sphere of a given ion, and this changes the relaxation conditions reflecting the solvent–solvent interaction. Equations describing these have been elaborated by Capparelli *et al.* [Ca 78a]. Their method was used to draw conclusions on the changes in the compositions of the solvate sheaths in solutions of magnesium perchlorate, potassium iodide and rubidium iodide in water–methanol solvent mixtures, purely from the influence of the interaction of the water molecules alone on the rate of intermolecular relaxation. It was found that all three cations coordinate water in a higher proportion than corresponds to the composition of the solvent mixture, i.e., they are preferentially hydrated.

Even from this brief account, it is obvious that NMR spectroscopy is applicable to the solution of many different types of problems in the study of non-aqueous systems.

Magnetic susceptibility examinations

One of the most promising areas of application of magnetochemistry is the study of transition metal complexes, in both the solid and dissolved states. Accordingly, magnetic susceptibility measurements may provide valuable information on the structures of solutions containing transition metal ions or complexes, on the symmetry conditions of transition metal ion solvates, etc. From magnetic measurements it is possible to establish the electronic structure of the central atom of the transition metal complex, its oxidation state and, in certain cases, its symmetry conditions.

In complexes, metal ions are generally separated from one another by ligands that are magnetically inactive (the solvent may have a similar role in solutions containing transition metal ions). Hence, magnetic interaction between the metal atoms, which might give rise to ferromagnetism or antiferromagnetism, is hindered. In such systems, therefore, these forms of magnetism are of subordinate importance. It is all the more important to know the paramagnetic properties; as a consequence of the isolating effect of the ligands (or solvent), they yield direct information about the electronic structure of the unfilled *d* shell of the central atom.

From the experimentally determined paramagnetic susceptibility, one can calculate the magnetic moments of the transition metal ions in the system. If the magnetic moment is known, it is simple to calculate the number of unpaired electrons. This throws light on the oxidation number of the metal ion being examined or the central atom of the complex, and also on its low- or high-spin electronic structure. Further, if the number of unpaired electrons in the complex is known, the value of the orbital momentum contribution can be calculated; in certain cases this permits conclusions to be drawn about the symmetry of the complex (solvate) (see, e.g., ref. [Bu 73]).

Many literature data [Ba 53, Cl 55, Ma 65, Sa 60] indicate that the magnetic properties of complexes in solution may be different from those in the solid state; for example, a number of nickel(II) complexes that are diamagnetic in the solid state become paramagnetic in solution. The first such effect was described by French *et al.* [Fr 42]. Since then, many authors have dealt with this phenomenon. The diamagnetic → paramagnetic transformation has been ascribed to various causes. Basolo [Ba 53] and Holm [Ho 63] suggest that nickel(II) complexes, which are square planar in the solid state, assume tetrahedral symmetry in solution. This has the result that the electronic structure, which is low-spin in the solid state, becomes high-spin in solution, thereby giving rise to a magnetic moment characteristic of two unpaired electrons. Maki [Ma 58a] and Ballhausen [Ba 59] consider that nickel(II) complexes, square planar in the solid state, coordinate solvent molecules in basic solvents, and thus assume distorted octahedral (square bipyramidal) symmetry. This causes the low-spin electronic structure of the nickel(II) atom in the solid complex to be transformed into the high-spin state. Cotton [Co 60] explains the occurrence of paramagnetism by the formation of a complex with octahedral symmetry even in those solvents where coordination of the solvent is excluded. His reasoning is that polymerization *via* the coupling of a number of square planar molecules can result in an octahedral coordination sphere around the nickel(II) atom. In the case of the nickel(II)-acetylacetone complex, he used spectrophotometric examinations to demonstrate the monomer–dimer equilibrium producing paramagnetism. It also appears possible to utilize this polymerization mechanism to interpret the magnetic behaviour of the nickel(II)–salicylaldimide system investigated by Sacconi [Sa 60]. On the basis of his [Fa 65, Fa 66] and literature data [An 63, Br 63, Cu 65, Go 63, Ny 64], Farago came to the conclusion that the paramagnetic nickel(II) complexes must possess octahedral or distorted octahedral symmetry. Such complexes are formed by the coordination of solvent molecules or of anions in the solution or, in certain cases, by dimerization.

It is known from the investigations of Brown [Br 63] and Nyberg [Ny 64] that the metal–nitrogen bond distance in the diamagnetic nickel(II) complex with square planar symmetry is shorter than in the corresponding paramagnetic complex. In the complexes of diamine ligands substituted on the nitrogen atom, therefore, the steric effect of the substituents favours formation of the paramagnetic complex [Fa 66, Go 63]. Substituents bound to the carbon atoms of diamines, on the other hand, promote the formation of the square planar complex by shielding the *axial* coordination sites of the central atom [Co 60, Fa 66]. The diamagnetic–paramagnetic equilibrium, therefore, depends not only on the solvent and on the anion present in the solution, but also on the structure of the chelating ligand.

By means of the NMR examination of nickel(II) complexes of Schiff's bases of salicylaldehyde with N,N-substituted ethylenediamines, Thwaites *et al.* [Th 66a, Th 66b] demonstrated that, in addition to the diamagnetic square planar and the paramagnetic octahedral complexes, other paramagnetic species containing a nickel(II) atom with a coordination number of five are also formed in chloroform solution. No difference between the two types of paramagnetic complex can be established by means of magnetic moment measurements, and only the NMR spectra provide a possibility for distinguishing between them.

Information of various kinds can be obtained on complexes, including solvates, from the dependence of the magnetic susceptibility on the temperature. The change in magnetic susceptibility caused by the temperature change of a solution frequently points to different reactions occurring at the different temperatures. For example, studies by Meek [Me 60] showed that the affinity of the chloride ion for the nickel ion increased when the temperature of a DMSO solution of nickel chloride was raised. Accordingly, with increase in temperature the chloride ion, previously situated in the outer sphere, replaces one of the DMSO molecules in the inner sphere [Wa 76].

Nuclear quadrupole resonance (NQR) spectroscopy

The energy levels of atomic nuclei possessing a quadrupole moment ($I \geqq 1$) are known to split in an inhomogeneous electric field. The magnitude of this splitting, which depends on the electric field gradient induced by the inhomogeneous electric field, can be determined with the aid of NQR spectroscopy.

If the electric field gradient (i.e., the inhomogeneous electric field) causing the splitting of the nuclear levels originates from the electron shell surrounding the given nucleus, then the NQR parameter giving the magnitude of the splitting can be used to draw conclusions on the structure of the electron shell, and primarily its symmetry, and hence on the symmetry of the molecule.

This method has proved applicable to the investigation of molecular complexes of metal halides with donor molecules [Kr 68, Ma 69, Ma 70a, Ma 73, To 69]. NQR examinations yield information on the effect of complex formation on the electronic structure of the halide, and hence indirectly on the locations of the halide ions and the donor molecules in the complex. NQR data have similarly thrown light on the structures of halogen complexes of organic amines [Bo 69, Br 72].

More recently, the method has also found use in the study of solvation processes. The advantages of using NQR spectroscopy in this field are well illustrated by a series of examinations by Petrosyan *et al.* [Pe 73a], dealing with non-aqueous solutions of organotin complexes. From NQR studies of the solvate complexes of methyltin halides with various solvent molecules, these authors drew conclusions not only on the structures and symmetries of the complexes, but also on the sequence of donor strengths of the solvents. In the case of solvates formed with trimethyltin bromide as the reference acceptor, it could be assumed that the shift in the NQR resonance frequency of bromine is governed predominantly by the change due to the solvation of the tin central atom, i.e., by the alteration of the strength (covalency) of the tin–halogen bond. NQR examinations of the resonance frequency of bromine, in agreement with the data of infrared spectroscopic [Be 63b, Ma64b] and X-ray structural [Hu 63] studies of analogous systems, have shown that coordination of the solvent to the tin atom increases the ionic character of the Sn—Br bond in the $(CH_3)_3SnBr$ molecule, which results in a decrease in the resonance frequency of the bromine.

The greater the interaction between the solvent molecule and the tin atom (i.e., the stronger the donor character of the solvent), the lower is the value to which the NQR frequency of the bromine will shift. On this basis, the following sequence of increasing donor strengths of solvents was established: diethyl ether < acetone < tetrahydrofuran < dimethylformamide < hexamethylphosphorotriamide < pyridine < dimethyl sulphoxide. This sequence differs from the donor strength sequence constructed on the basis of the Gutmann donicities of the solvents (see Chapter 4). A likely explanation is that the covalency of the Sn—Br bond is affected by the solvation not only of the tin, but also of the bromine atom, and the latter process is not negligible in this system. On the other hand, the difference in the reference acceptors [here $(CH_3)_3SnBr$, but $SbCl_5$ in the Gutmann series] may also be the cause of the difference. Even minor differences in the reference acceptor can give rise to large changes in the NQR parameters of the solvate. This can be seen from the fact that, in the case of dimethyltin halides, which are very similar to trimethyltin halides, the correlation between the NQR data and the donor properties of the solvent is much more complicated than for the latter system. Thus, in examinations with $(CH_3)_2SnBr_2$ as the reference acceptor, the previous donor strength sequence was not obtained: instead, the results only reflected a division of the solvents into two groups: weak donors (dioxane, tetrahydrofuran, acetone) and strong donors (pyridine, dimethylformamide, dimethyl sulphoxide).

In donor solvents, tin tetrahalides are able to coordinate two donor molecules. By means of analogous examinations, even the *cis* and *trans* forms of the mixed complexes formed could be differentiated [Pe 73a]. From the NQR frequency of ^{81}Br in tin tetrabromide, it has been established that *trans* mixed complexes are formed in diethyl ether, dioxane, pyridine and dimethyl sulphoxide, and *cis* mixed complexes in hexamethylphosphortriamide.

Many new results are to be expected in this field from the further application of NQR spectroscopy.

Electron paramagnetic resonance (EPR) spectroscopy

Electron paramagnetic resonance (EPR) or electron spin resonance (ESR) spectroscopy is suitable for the study of systems containing unpaired electrons, i.e., free radicals, transition metal ions, lanthanide metal ions, transuranium ions and complexes. Systems containing transition metal ions or lanthanide metal ions are particularly important in the study of solution structure.

The parameters of EPR spectroscopy are the g factor (spectroscopic splitting factor), the hyperfine coupling constants, line intensities, line widths and relaxation times. Changes in the first two of these parameters provide the most information in structural studies on solutions.

The value of the g factor for the free electron is 2.002319. An experimentally determined value of the g factor different from this reflects the influence of the chemical environment. Particularly in transition metal ions and complexes, the value of the g factor is considerably affected by the influence of the chemical environment on the angular orbital momentum of the electron.

If there is an interaction between the unpaired electron and the magnetic nuclei, this causes further splitting of the resonance line. This hyperfine splitting is characterized by the hyperfine coupling constant. An atomic nucleus with a nuclear spin of I causes the electron resonance line to split into $2I+1$ lines. The resulting new lines have equal intensities. Thus, on the action of the ^{1}H, ^{19}F, ^{13}C, ^{15}N or ^{31}P nuclei, for instance, which have a nuclear spin $I=1/2$, the EPR spectrum will be split into two lines of equal intensity; on the action of the ^{14}N atomic nucleus, with a nuclear spin $I=1$, the spectrum is split into three equal lines; etc. The distance between the lines gives the value of the hyperfine coupling constant, which is a measure of the probability of residence of the electron in the environment of the given nucleus.

The situation is more complicated if the unpaired electron is in simultaneous interaction with several magnetic nuclei. These exert their effects independently of one another. Hence, if the electron is in interaction with one atomic nucleus with a nuclear spin of I_1 and with another nucleus with a nuclear spin of I_2, for example, the number of lines appearing in the spectrum will be $(2I_1+1)(2I_2+1)$. The distances between the lines originating from the same nucleus give the coupling constants relating to the individual nuclei, which in turn give the probabilities of residence of the electron in the environments of the various atomic nuclei. Accordingly, the distribution of the unpaired electron within the molecule can be determined.

The area under the EPR curve (the intensities of the bands) is proportional to the number of unpaired electrons. Owing to the high sensitivity of the method, it may be used to determine the number of unpaired spins even in very dilute solutions and hence, for example, the oxidation numbers and low- or high-spin electronic structures of transitional metal ions.

Even this brief account reveals that EPR spectroscopy can be employed with advantage in investigating the effects of the immediate environment of paramagnetic transition metal ions (and thus their solvate sheaths) on the electronic structures of these ions. The parameters of the EPR spectrum shed light on the locations of the unpaired electrons of the transition metal ion in the available d orbitals. Since EPR parameters are affected by the immediate environment of the paramagnetic ion, i.e., by the donor atoms (even if these are diamagnetic), the EPR spectrum may yield information on the symmetry conditions of the diamagnetic neighbours and on the nature of the bonds existing between the paramagnetic transition metal central atom and the donor atoms [Ab 50, Ab 51, Ab 55].

EPR spectroscopy was employed by Rockenbauer *et al.* [Ro 72] to study the equilibrium stability constants of low-spin cobalt(II) mixed complexes. They determined the equilibrium constants of formation of the mixed complexes of the bis(dimethylglyoximato)cobalt(II) parent complex with pyridine ligands, in methanol. It was shown that in a methanolic solution of the parent complex, two methanol molecules are coordinated along the z axis, and these methanol molecules can be replaced stepwise with pyridine.

Very important results of these investigations were the establishment of the temperature dependence of the stepwise stability constants, and the determination of the changes occurring in this kinetically labile system in the course of the preparation of a rapidly frozen solution. At room temperature, about 90% of the total complex concentration is present as the monopyridine mixed complex in the solution, whereas in the rapidly frozen solid the dipyridine mixed complex is the predominant ionic species. These studies emphasize that information obtained from rapidly frozen solutions must be employed with caution for characterizing the original solution system of room temperature.

The Rockenbauer method also appears to be applicable to the equilibrium investigation of other low-spin complex systems with d^7 electronic structure.

From the fine structure of the EPR spectra of copper(II)–dimethylglyoxime in the solid state and in solution, conclusions could be drawn about the solvation of the complex [Fa 70, Ti 63]. In the complex prepared in the solid state, the central copper(II) atom is bound to four nitrogen atoms, and also to one of the oxime oxygen atoms of a neighbouring complex molecule. In the solid state, therefore, the complex is dimeric and the coordination number of the central atom is 5. This finding is in agreement with the results of the X-ray structural studies by Frasson [Fr 59, Fr 60]. In aqueous solution, on the other hand, apart from the nitrogen atoms of the dimethylglyoxime, the copper(II) atom is bonded to the oxygen atoms of two solvent molecules, and its coordination number is therefore 6. In solution the fifth and sixth coordination sites are thus occupied by the solvent. Hence, these examinations confirmed the assumption made by Dyrssen [Dy 61] with regard to the solvation of the copper(II) complex, based on solvent-extraction studies.

EPR investigations have also proved convenient for following other solvation effects, solvent substitution, etc. For example, McClung *et al.* [Mc 66] have employed this method to demonstrate that when the tetrachlorodimethoxy complex of molybdenum(V) is dissolved in dimethyl sulphoxide, the chloride ligand is partly replaced with the solvent. By adding an excess of chloride, however, the solvent molecules can be expelled. The investigations also proved that the complex has a *trans* structure of D_{4h} symmetry.

It was shown by Symons *et al.* [Br 76] that when solutions of silver nitrate in methyl cyanide are exposed to ^{60}Co γ-rays at 77 K, Ag^0 and Ag^{II} centres are formed, in addition to various solvent radicals, and $NO_3^{2+} + NO_2$ from the anions. The EPR results were remarkable in that there was no evidence of superhyperfine coupling to ^{14}N, which had been expected from MeCN ligands, especially for the Ag^{II} centre. The explanation of the authors was that these centres were formed in clusters, the ligands being NO_3^- rather than MeCN. When the nitrate is replaced by the perchlorate both centres (Ag^0 and Ag^{II}) displayed well defined hyperfine coupling to four equivalent ^{14}N nuclei showing that $Ag(MeCN)_4^+$ ions had gained and lost electrons [Al 80]. The tendency for ClO_4^- ions to co-ordinate to silver is namely very much less than that for NO_3^- ions and hence that clustering is also less important in such solutions. Addition of water or methanol had little effect on the EPR spectra up to ≈ 0.5 mole fraction (MF) of protic solvent, showing that Ag^+ is solvated preferentially by methyl cyanide. However, in the 0.45–0.90 MF region (D_2O) a new species exhibiting hyperfine coupling to only one ^{14}N nucleus was detected. This is thought to be formed from $Ag^+(H_2O)_3(MeCN)$ ions.

EPR coupling constants of semiquinone radicals recorded in different solvents showed good correlations with solvent polarity scales [Ho 82]. In some of the systems even solvation equilibrium constants could be calculated.

It is obvious from even this short review that the possibilities afforded by EPR spectroscopy in the investigation of solvation are far from having been exhausted.

Mössbauer spectroscopy in frozen solutions

It is only in recent years that recoilless γ-ray resonance spectroscopy (Mössbauer spectroscopy) [Mö 58] has been used for the study of solvation processes [Bu 72, Bu 73, Vé 70a, b, Vé 78, Vé 79]. The parameters obtainable from the Mössbauer spectrum (primarily the *isomer shift**, δ, which is proportional to the electron density at the site of the atomic nucleus, and the *quadrupole splitting*, ΔE, which is suitable for the study of the symmetry of the electric field surrounding the nucleus) yield valuable information on the electronic configuration of the atom examined, and on the changes occurring in this configuration as a result of the action of various chemical effects such a solvation. (In some cases *magnetic hyperfine splitting* can also give important chemical information.)

Coordination of solvent molecules changes the electronic configurations and symmetries of the dissolved ions and solvent molecules, the dissociation conditions in the solution, etc. These changes largely depend on the properties of the solvent (with an increase in the donicity of the solvent, for instance, the covalency of the solvate complexes increases; an increase in the relative permittivity favours heterolytic dissociation in the solution, etc.).

Mössbauer spectroscopy therefore appears in principle to be suitable for following changes brought about by solvation. However, the practical application is considerably limited by the fact that the Mössbauer effect (recoilless γ-ray resonance absorption) appears only in solid substances, where the Mössbauer nucleus is so firmly bound in the crystal lattice that recoilless γ-ray emission and absorption are ensured.

With a view to eliminating the above limitation, a number of research groups ([Bu 72, Bu 73, Dé 65a, b, Ru 71, Vé 78, Vé 79] and references cited therein) have dealt with the Mössbauer study of frozen solutions. This methodology permitted

*An increase in the electron density at the site of the nucleus is shown by a decrease in the isomer shift with iron and antimony, and by an increase with tin [Bu 73, Vé 79].

the investigation of molecules or ions made independent, to a certain extent, of intermolecular interactions. The study of substances that are stable only in solution, but cannot be prepared in the solid state, may also become possible by this means. Several authors have attempted to use this method to follow the effect of a solute on the structure of the frozen solution.

A problem of long standing, however, was to establish the degree to which the Mössbauer parameters obtained with a frozen solution give information regarding the structure of the original liquid solution. Accordingly, various research groups [Ca 71, Dé 65a, b, Na 70b, Pe 69, Ru 71] studied the effects of the freezing conditions (rate of freezing, etc.), the storage conditions of the resulting solid, its possible pre-treatment, the chemical composition of the solution, etc., on the structure of the frozen solution. The effects examined included the changes occurring in the compositions and concentrations of the ionic species in solution as a result of freezing and the structural properties of the solid that are different from those of the liquid. With the above aim, investigations were carried out on the effect of the freezing rate on the Mössbauer parameters. The changes in the system in the course of ageing and thermal treatment of frozen solutions were followed by conductivity measurements, NMR spectroscopy and DTA examinations.

The results of these complex investigation methods revealed that the rapid freezing of solutions (at a freezing rate higher than 15 °C/s) results in a glass-like amorphous solid, in which the immediate environment of the dissolved ions corresponds to the state in the original liquid solution [Vé 70a]. Naturally, one can hope to draw realistic conclusions about the system in the original solution from the Mössbauer spectrum of the frozen solution only if the system is such that, when the solution is suddenly frozen, the equilibria in the solution are also "frozen in", and thus the original compositions of the immediate environments (inner ligand spheres) of the dissolved ions and molecules in the solution are retained during the transition into the solid state. The validity of this assumption is not influenced by the circumstance that ligand exchange may possibly continue at a different rate in the solid phase. (Of course, the possibility is excluded that there is a phase effect of the ligand exchange, i.e., that the ratio of the exchange rates of the individual solvate sheath components is different in the liquid and solid phases.)

It follows that the Mössbauer parameters of frozen solutions are suitable for establishing the structure of the original liquid solution primarily in kinetically inert systems, since in these the rates of the changes (association reactions, polymerization, etc.) occurring as a result of the cooling are so low that the compositions and concentrations of the dissolved ions hardly vary during the time necessary for rapid freezing.

In kinetically labile systems, the equilibria prevailing in solution are shifted during the cooling. Hence, the Mössbauer parameters will reflect the picture characteristic of the temperature at which the solution solidifies. However, since (in most of the simpler systems) these processes cause changes mainly in the concentrations of the stepwise-formed complexes in solution, but do not result in the the formation of new, different complexes, these investigations have also proved applicable to the study of solvation effects.

In the following a survey of some of the more important results achieved in this field with the aid of Mössbauer spectroscopy will be given.

The Mössbauer study of solids prepared by rapid freezing of solutions of organotin complexes in various solvents has provided valuable information relating to the solvate complexes existing in these solutions and the solvating powers of the solvents [Go 65, Go 70, Kh 65, Pe 73b, Ro 68].

Goldanskii *et al.* [Go 65] added an increasing amount of the donor solvent to methylene chloride solutions of *dialkyltin dihalide* complexes, and studied the dependences of the Mössbauer isomer shift and quadrupole splitting on the nature and amount of the donor solvent. They found that the isomer shifts were constant within the limits of experimental error, whereas the quadrupole splitting values increased with increase in the solvating power of the donor solvent. They explained the latter effect by an increase in the ionic character of the tin–halide bond as a result of coordination of the solvent to the tin central atom. Consequently, the difference between the Sn—C and the Sn—halide bonds increases, which causes an increase in the electric field gradient around the tin nucleus, and hence in the quadrupole splitting. According to these considerations, the magnitude of the quadrupole splitting can be regarded as proportional to the donor strength of the solvent. They therefore proposed the introduction of a coefficient, calculable from the quadrupole splitting, to characterize the donor strengths of solvents.

In reality, however, the quadrupole splitting in the given systems is governed by numerous other factors also. Examples of these are the variation of the C—Sn—C bond angle and steric effects. Thus, it is not surprising that the donor strength sequence constructed on the basis of the Mössbauer quadrupole splitting values is not in complete agreement with sequences established by other methods, e.g., from nuclear quadrupole resonance [Pe 73a], or the Gutmann donicities.

A small change in the composition of the reference acceptor may also cause drastic changes in the Mössbauer quadrupole splitting values. The Mössbauer study of frozen solutions of *methyltin trihalide* reference substances revealed, e.g., a completely different picture from that for the analogous dihalides. In these systems the isomer shift values decreased as a result of coordination of the solvent, which indicates a decrease in the electron density at the site of the tin nucleus. The explanation is to be sought either in an increase in the polarity of the tin–halide bond, or in the larger shielding effect caused by filling of the 5*d* orbitals of tin. Both explanations can be correlated with the sequence of donor strength changes of the coordinated solvents. Hence, a decreasing sequence of the isomer shift values may indicate an increasing sequence of the donor strengths of the solvents. It is further of interest that the solvent dependence of the quadrupole splitting of the trihalide complex is exactly the opposite of that observed for the dihalide. An increase in the solvating power of the solvent is accompanied by a decrease in quadrupole splitting in these systems. This contradiction could be explained as follows. NQR examinations [Pe 73a] have shown that alkyltin trihalides contain two types of tin–halide bond, with different ionic characters; the Sn—X bond *trans* to the alkyl group differs from the Sn—X bond *trans* to halide (where X = halide). The difference between the extents of the ionicity of the two types of tin–halide bond is 12.5% for chlorides and 7.5% for bromides. As a consequence of the coordination of donor solvents, the difference between the two types of tin–halide bond decreases [Ma 70], which causes a decrease in the electric field gradient, and hence in the quadrupole splitting.

When *trimethyltin monohalides* were used as reference acceptors, the behaviour observed was again completely different from that of the two previous systems [Pe 73b]. In these systems, neither the isomer shift nor quadrupole splitting exhibited a solvent dependence large enough to permit interpretation of the solvent effect.

It must be noted that, in the mono- and dialkyltin halides, one cause of the change in the quadrupole splitting as a result of solvation is that the coordination number of tin, which is four in the solid compound, changes to five in solution on coordination of the solvent. Accordingly, in systems such as $(C_6H_5)_3SnF$, $(C_2H_5)_3SnF$ and $(C_2H_5)_3SnCl$, in which tin has a coordination number of five even in the solid state (these compounds are polynuclear formations containing halide bridges), the quadrupole splitting in the frozen solution will be of approximately the same magnitude as it was in the solid sample. Hence, these systems are not so sensitive to changes in the solvating powers of solvents. On the other hand, a comparison of the quadrupole splittings of the solid sample and the frozen solution prepared from it provides indirect information as to whether the sample in the solid state was a monomer. It should be noted that for compounds where the data from these two types of measurements are indicative of the polymeric nature of the solid sample, the magnitude of the Mössbauer effect (the Mössbauer-Lamb factor) is also suggestive of a polymeric structure.

On the basis of Mössbauer studies on frozen solutions of iron(II) salts in various non-aqueous solvents, Burger *et al.* [Bu 70] drew conclusions about the structures of the iron(II) solvates. High-spin iron(II) salts were selected as their first model system, as the Mössbauer parameters here, and primarily the high quadrupole splitting, are sensitive indicators of even small changes in the electronic configuration of the iron.

On dissolution of iron(II) salts in various solvents, the electronic structure and hence the Mössbauer spectrum of the iron are determined by various effects. On these, the fundamental two are:

(1) coordination of the solvent molecules to the iron;

(2) the extent of the association of the iron and its anion.

From conductivity measurements it may be assumed that in dilute methanolic and formamide solutions of iron(II) chloride all of the coordination sites of the iron are occupied by solvent molecules, whereas in ethylene glycol, dimethylformamide and pyridine the chloride ion also is to be found in the inner coordination sphere. If it is assumed that the equilibria prevailing in the solution are also frozen in during the rapid freezing of the solutions, and that new complex ions are not formed during

Table 5.2. Mössbauer isomer shifts (δ) and quadrupole splittings (ΔE) of frozen iron(II) chloride and iron(II) sulphate solutions measured at liquid nitrogen temperature [Bu 70]
Radiation source: ^{57}Co in stainless steel

Solute	Solvent	ΔE, mm/s	δ, mm/s
$FeCl_2$	Methanol	3.50	1.41
$FeSO_4$	Methanol	3.44	1.37
$FeCl_2$	Ethylene glycol	3.20	1.20
$FeSO_4$	Ethylene glycol	3.06	1.48
$FeCl_2 . 4H_2O$	Methanol	3.24	1.48
$FeCl_2$	Methanol + water	3.33	1.40
$FeCl_2$	Water	3.20	1.51
$FeCl_2$	Pyridine	3.52	1.36
$FeCl_2$	Formamide	1.89	1.43
$FeCl_2$	Dimethylformamide	1.38	0.54
$FeCl_2 . 4H_2O$	Formamide	2.71	1.41
$FeCl_2 . 4H_2O$	Dimethylformamide	2.67	1.25

this period, the Mössbauer spectra of frozen solutions of iron(II) chloride in methanol and in formamide should give a picture characteristic of the solvate parent complex; for solutions in pyridine, ethylene glycol and dimethylformamide the picture obtained should be characteristic of mixed complexes of iron(II), containing not only solvent molecules but also chloride ion in the inner coordination sphere of the metal. The above assumptions were confirmed by the fact that in frozen methanol iron(II) sulphate gave a Mössbauer spectrum coinciding with that of iron(II) chloride, while in frozen ethylene glycol the Mössbauer parameters of the two salts were different (Table 5.2). In methanolic solution, therefore, the two iron(II) salts form iron(II) solvate complexes of identical composition, whereas in ethylene glycol solution the composition of the coordination sphere of the iron(II) also depends on the anion.

A comparison of the Mössbauer parameters of the methanol solvate with those of the aquo complex revealed that the electric field gradient at the site of the iron nucleus is greater in the former than in the latter.

The isomer shifts of iron(II) solvates that also contain chloride ion in the inner coordination sphere was found to be smaller than those of the solvate complexes with homogeneous coordination spheres, indicating the nephelauxetic effect of the coordinated chloride. By means of this effect, the chloride in the first coordination sphere decreases the *d*-electron density on the iron. Thus, with the decrease in the shielding effect of the *d*-electrons, the electron density at the nucleus increases. This shows up in a decrease in the isomer shift values.

In the case of the high-spin iron(II) compounds, by increasing the symmetry of the electron shell around the iron, the nephelauxetic effect generally decreases the ΔE values also. The symmetry-decreasing effect of the chloride ion in the solvate sheath, however, may act in the opposite direction, thereby resulting in an increase in ΔE. Depending on which of these two opposite-acting forces is the more dominating, the value of the quadrupole splitting may be smaller or larger than that for the analogous solvate which does not contain chloride. In an ethylene glycol solution of iron(II) chloride $\Delta E=3.20$, and in an ethylene glycol solution of iron(II) sulphate $\Delta E=3.06$, i.e., the action of the nephelauxetic effect in decreasing the quadrupole splitting is not manifested. In contrast, the isomer shift values ($FeCl_2$, $\delta=1.20$; $FeSO_4$, $\delta=1.48$) clearly demonstrate this effect. In a frozen solution in dimethylformamide, however, the iron(II) quadrupole splitting ($\Delta E=2.67$) and isomer shift ($\delta=1.25$) both show the nephelauxetic effect of the chloride.

If solvent mixtures consisting of two solvents are used and the ratio of these components is varied, it is possible to demonstrate the differences between the Mössbauer spectra of the two parent solvate complexes and the various mixed solvates.

This is shown clearly by the Mössbauer spectra of frozen solutions of anhydrous iron(II) chloride in methanol–formamide solvent mixtures of various compositions [Vé 69]; each of these spectra consists of the superposition of at least two spectra displaying quadrupole splitting (Fig. 5.5). It may be concluded that the iron in these solutions is present in at least two different solvate sheaths. The spectral part with the smaller quadrupole splitting exhibits values lying close to the ΔE measured in pure formamide solution (cf., Table 5.3). Thus, this is probably a line pair originating from a solvate sheath consisting completely or predominantly of formamide molecules. The larger ΔE can be related to a solvate sheath built up from methanol molecules, for this is close to the ΔE value obtained in pure methanolic solution.

It must be noted that the larger and smaller ΔE values cannot be identified completely with the quadrupole splitting values of ΔE = 3.58 ± 0.05 mm/s and E = 1.89 ± 0.05 mm/s relating to the pure methanolic and the pure formamide solution, respectively. (The isomer shift values in the mixtures and in the pure solvents are identical, within the limits of error.) The differences between the quadrupole splitting values measured in the pure solvents and in their mixtures may arise partly from the fact that the two types of solvate sheath formed do not consist only of methanol or only of formamide molecules: mixed solvate sheaths may also be formed. In addition, a change to this extent in the quadrupole splitting may also be caused by the different structures of the frozen solvent mixtures compared with the structure of pure frozen methanol or formamide; such a difference may result in minor differences in the symmetry of the charge distribution around the iron nucleus.

Accordingly, the Mössbauer spectra do not allow an unambiguous decision as to whether mixed or pure solvate sheaths are formed in methanol–formamide mixtures. It may be stated with certainty only that at least two types of solvate are present.

Figure 5.5 also shows the dependence of the spectra on the composition of the solvent mixture. With variation of the concentrations of the components, clear changes occur in the intensities of the spectral patterns assigned to them.

Similar investigations were made by Vértes *et al.* [Vé 73a] with frozen solutions of iron(II) chloride in other solvent mixtures. Some of the important results are described below.

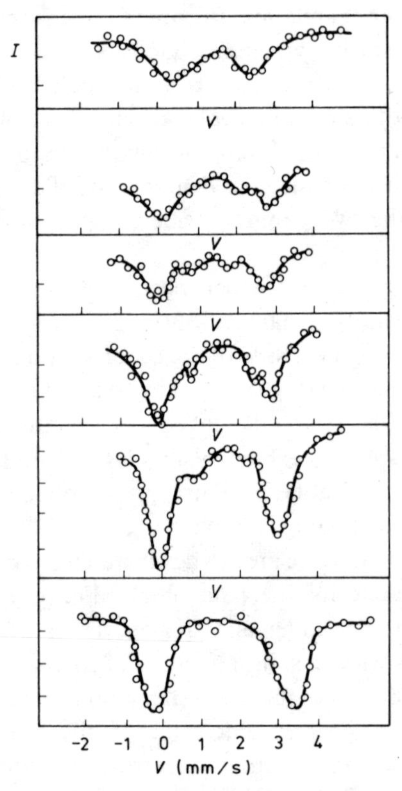

Fig. 5.5. Mössbauer spectra of rapidly frozen solutions of iron(II) chloride in methanol–formamide solvent mixtures [Vé 69]

Table 5.3. Mössbauer parameters of iron(II) chloride, measured in methanol–formamide solvent mixtures at liquid nitrogen temperature [Vé 69]
Radiation source: ^{57}Co in stainless steel

Composition of the mixture, %(w/w)		Mössbauer parameters	
Methanol	Formamide	ΔE, mm/s*	δ, mm/s*
0.00	100.00	1.89	1.43
20.04	79.96	2.80 1.45	1.50
37.60	62.40	2.86 1.36	1.41
59.00	41.00	2.92 1.60	1.40
79.50	20.50	3.10 1.45	1.47
100.00	0.00	3.58	1.34

* The accuracy of the measurements was ±0.05 mm/s.

Table 5.4. Mössbauer parameters of anhydrous iron(II) chloride in frozen solutions at liquid nitrogen temperature, I [Vé 73a]
Radiation source: ^{57}Co in stainless steel

Methanol–dimethylformamide molar ratio	Isomer shift, δ mm/s	Quadrupole splitting, ΔE mm/s
1.00:0	1.35	3.45
3.85:1	1.32	3.08
1.92:1	1.32	3.03
0.96:1	1.31	3.00
0.64:1	1.31	2.93
0.49:1	1.25	2.63
0.00:1	1.21	2.12

In methanol–dimethylformamide mixtures, it was not possible to break down the Mössbauer spectrum into several components; in this system, it was only the broadening of the lines which indicated the presence of different types of solvate. The Mössbauer parameters of some mixtures with different compositions are listed in Table 5.4.

Even these few data show that there is a sudden change in the composition of the solvate sheath at a methanol to dimethylformamide ratio of 0.64 : 1. In systems with a lower dimethylformamide content, methanol clearly predominates in the inner ligand sphere.

The parameters of the Mössbauer spectra recorded in pyridine–hexamethylphosphoramide agreed, within the limits of experimental error, with the data for solutions containing only pyridine as solvent, even in systems containing a large excess of hexamethylphosphoramide (e.g., in a solution with an HMPA content of 90%, v/v).

Mössbauer spectra recorded in hexamethylphosphoramide–water mixtures had two line pairs superimposed (Table 5.5). The Mössbauer parameters of these agreed, within the given limits of experimental error, with the parameters of the Mössbauer spectra recorded in pure water or in pure hexamethylphosphoramide.

Table 5.5. Mössbauer parameters of anhydrous iron(II) chloride in frozen solutions at liquid nitrogen temperature, II [Vé 73a]
Radiation source: ^{57}Co in stainless steel

HMPA–water molar ratio	Isomer shift, δ mm/s	Quadrupole splitting, ΔE mm/s
1.0:0	1.16	2.77
8.0:1	1.16	2.68
	1.40	3.24
1.7:1	1.20	2.72
	1.46	3.24
0.5:1	1.19	2.72
	1.48	3.20
0.0:1	1.51	3.20

This result indicates that after attaining equilibrium, in HMPA–water mixtures more than 90% of parent solvates containing either only water or only HMPA as ligands; hence the stability constants of the mixed solvates are small compared with those of the parent solvates.

As regards the quantitative ratios of the individual components, it could be stated that the amounts of the two parent solvates are roughly identical in a mixture containing the two solvents in a volumetric ratio of HMPA : H_2O = 17 : 1 (molar ratio 1.74 : 1); in mixtures containing more water than this, the aquo complex will predominate. This result demonstrates that in solvent mixtures, the competition for solvate formation is not necessarily decided by the Gutmann donicities; in the present system, the coordination of water, which has the lower donicity, but a high relative permittivity and smaller spatial requirement, takes place to the greater extent. It should be noted in this connection that, in solvent mixtures, the cluster-type structure characteristic of pure water breaks down in proportion to the decrease in the water content, and the donor properties of the smaller associates, and particularly of monomeric water, are better than those of water molecules bound in a large cluster.

The effects of various organic solvents on the Mössbauer parameters are well illustrated by the studies by Vértes and Burger [Vé 72] on rapidly frozen non-aqueous solutions of antimony pentachloride and tin(IV) halides (Table 5.6).

The three data relating to antimony pentachloride appear to reflect a simple connection between the isomer shift of the antimony and the donicities of the solvents.

Table 5.6. Mössbauer parameters of $SbCl_5$ and tin tetrahalides measured in rapidly frozen solutions, and donicities (DN) and relative permittivities (ε) of the solvents [Vé 72]
Radiation sources: $Ca^{121}Sn(Sb)O_3$ and $Ba^{119}SnO_3$, respectively

Reference acceptor	Solvent	Isomer shift, δ, mm/s	Line width, ΔE, mm/s	DN_{SbCl_5}	ε
$SbCl_5$	Dimethylformamide	-1.9 ± 0.2	3.0 ± 0.25	26.6	36.1
	Tributyl phosphate	-2.2 ± 0.2	3.0 ± 0.25	23.7	6.8
	Acetonitrile	-2.6 ± 0.2	3.0 ± 0.25	14.1	38.0
$SnCl_4$	Dimethyl sulphoxide	0.32 ± 0.08	1.18	29.8	45.0
	Dimethylformamide	0.29 ± 0.08	1.11	26.6	36.1
	Tributyl phosphate	0.24 ± 0.08	1.50	23.7	6.8
	Acetonitrile	0.64 ± 0.08	1.19	14.1	38.0
	No solvent	0.38 ± 0.08	1.02		
SnI_4	Dimethyl sulphoxide	0.48 ± 0.08	1.37	29.8	45.0
	Dimethylformamide	0.51 ± 0.08	1.81	26.6	36.1
	Ethanol	0.40 ± 0.08	1.25	30.4	24.3
	Tributyl phosphate	1.50 ± 0.08	1.60	23.7	6.8
	Carbon tetrachloride	1.60 ± 0.10	1.25	0	2.2
	No solvent	1.47 ± 0.05	0.98		

An increase in the donicity of the solvent results in an increase in the isomer shift of the antimony, which is indicative of a decrease in the electron density at the site of the antimony nucleus. (An increase in the donicity of the coordinated solvent molecules increases the electron density in the *d* and *p* orbitals shielding the *s* electrons, which causes a decrease in the electron density at the nucleus, and hence an increase in the isomer shift.)

In the compounds SnI_4 and $SnCl_4$ there is no electric field gradient at the position of the nucleus, and the value of the quadrupole splitting is therefore zero. The greater line widths and the shapes of the lines in the Mössbauer spectra of the frozen solutions, however, indicate that some effect causing the broadening of the Mössbauer lines does occur [(the half-width relating to β-tin (white tin) for the radiation source ($BaSnO_3$) is $\Gamma = 0.9$ mm/s)]. This line broadening can be explained by the fact that, as a consequence of the stepwise dissociation of tin(IV) halides in the concentration range employed for the measurements, different solvates are present together in solution; the superposition of the Mössbauer lines corresponding to these give rise to the comparatively large line widths given in Table 5.6.

The Mössbauer parameters of the tin halides show that in these systems the correlation between the isomer shifts and the donicity values is by no means as clear as for antimony pentachloride. The isomer shifts of tin tetrachloride agree, within experimental error, in frozen dimethyl sulphoxide, dimethylformamide and tributyl phosphate solutions, and those of tin tetraiodide agree in dimethyl sulphoxide, dimethylformamide and ethanol solutions. The Gutmann donicities of these solvents lie in the range 20–30. For both tin compounds, a further decrease in the donicity of the solvent causes an increase in the isomer shift, which indicates an increase in the electron density at the tin nucleus. This effect appeared in acetonitrile solution with tin tetrachloride, and in tributyl phosphate and carbon tetrachloride solutions with tin tetraiodide. (It must be emphasized that the last two solvents have low relative permittivities, whereas that of acetonitrile is high.)

The effects of the halides on the electronic structure of tin in the frozen solutions are well illustrated by the fact that the isomer shift values measured in solutions of the two tin halides in the same solvent exhibited differences in each case larger than the experimental error. This shows that at least in one of the tin compounds, halide belongs to the coordination sphere of the tin even in a solvent such as dimethyl sulphoxide, which has a high relative permittivity and a high donicity.

The data reported in Table 5.6 reveal an indisputable correlation between the donicities of the solvents and the isomer shift values. Appreciable effects are also exerted on the isomer shift values, however, by a number of other factors influencing the electronic structure of the reference acceptor, such as dissociation reactions affected by the relative permittivity of the solvent.

The elimination of these secondary effects and hence a clearer indication of the correlation between the Mössbauer parameters and the donicity of the solvent can be expected from the investigation of systems in which the relative permittivity is kept constant or nearly so. For this reason, Vértes and Burger [Vé 72] continued their Mössbauer spectroscopic studies of solvation by comparing the Mössbauer parameters measured in mixtures of a given inert solvent with the solvents under examination.

Table 5.7 shows the Mössbauer parameters of frozen solutions containing 0.02 mole of SnI_4 per kilogram of CCl_4 and 1 mole donor solvent per kilogram of CCl_4. The relative permittivity of CCl_4 is small ($\varepsilon = 2.2$), and thus the extent of dissociation in the solutions is slight. Accordingly, a considerable proportion of the tin is present in the solution in the form of a solvate with the composition $SnI_4.D_2$ (where D is the donor solvent molecule). The Mössbauer isomer shift values measured in such solutions indicate a clear correlation between the donicity and the isomer shift.

If the ratio of the inert solvent to the donor solvent is varied, under optimal conditions the Mössbauer lines of the solvated and the non-solvated tin tetrahalide may appear side by side in the Mössbauer spectrum (Fig. 5.6). After suitable computer resolution of the complex spectrum, the concentrations of the two types of

Table 5.7. Relationship between the isomer shift (δ) and the donicity (DN_{SbCl_5}) in solutions of tin(IV) iodide in CCl_4–donor solvent mixtures [Vé 72]. Radiation source: $Ba^{119}SnO_3$

Donor solvent	δ ($\pm$0.08) mm/s	DN_{SbCl_5}
Hexamethylphosphoramide	0.56	38.8
Dimethylformamide	0.65	26.6
Ethyl acetate	0.96	17.1
Acetone	0.95	17.0
Acetonitrile	1.04	14.1
Nitrobenzene	1.40	4.4
Carbon tetrachloride	1.60	0.0

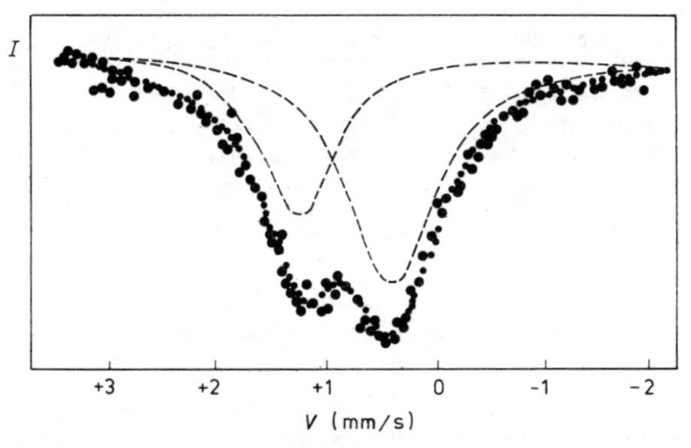

Fig. 5.6. Mössbauer spectrum of a rapidly frozen solution of $SnBr_4$ in tetrahydrofuran–carbon tetrachloride solvent mixture [Vé 76]

species can be calculated from the areas under the parts of the curve assigned to them. From an appropriate number of experimental data of satisfactory accuracy, the equilibrium constant of solvation of the complex can be obtained. By means of such studies, it was possible to determine the equilibrium constants of solvation of tin tetrabromine in acetic anhydride, acetonitrile, acetone and tetrahydrofuran [Vé 76]. The equilibrium constants are listed in Table 5.8, together with the values of the donicity and the relative permittivity of the solvent. Even these few data demonstrate that the stability of the solvate formed increases not only with the increase in the donicity of the solvent, but also with the decrease in the relative permittivity, indeed, it is highly probable that other factors (e.g., spatial requirements, polarizability) also affect the solvation.

The complexity of the solvent effect in solvent mixtures is illustrated well by Mössbauer studies of tin tetraiodide in carbon tetrachloride–dimethylformamide mixtures [Vé 73b]. The data in Table 5.9 show that the value of the isomer shift of tin(IV) iodide dissolved in carbon tetrachloride is larger than that of crystalline tin(IV) iodide, which indicates a higher electron density at the tin nucleus. This demonstrates that the 5*p* (and possibly the 5*d*) electron orbitals, which shield the *s* electrons, are filled to a greater extent in tin (IV) iodide in the crystalline state than in carbon tetrachloride solution; this permits the conclusion that tin(IV) iodide dissolved in carbon tetrachloride has a tetrahedral structure, and the SnI_4–CCl_4 interaction (overlapping of the electron orbitals) is smaller than the interaction between the SnI_4 molecules in the crystalline substance.

Table 5.8. Equilibrium constants of solvation of tin tetrabromide [Vé 76].

$$K = \frac{[SnBr_4D_2]}{[SnBr_4][D]^2}$$

Solvent	DN_{SbCl_5}	ε	K
Acetic anhydride	10.5	20.7	5
Acetonitrile	14.1	38.0	25
Acetone	17.0	20.7	39
Tetrahydrofuran	20.0	7.6	93

Table 5.9. Mössbauer isomer shifts of tin(IV) iodide in CCL_4–DMF solvent mixtures [Vé 73b]. Radiation source: $Ba^{119}SnO_3$

Composition of solution (mole/kg CCl_4)		Isomer shift, δ, mm/s	Intensities of lines, as percentage of area
$^{119}SnI_4$	DMF		
0.020	0.00	1.64 ± 0.03	
0.022	0.04	0.69 ± 0.03*	
0.022	1.10	0.63 ± 0.03	97 ± 1
		$\delta < 0$	3 ± 1
0.024	3.12	0.70 ± 0.03	94 ± 2
		$\delta < 0$	6 ± 2
0.023	4.62	0.59 ± 0.03	87 ± 2
		$\delta < 0$	13 ± 2
0.024	6.93	0.55 ± 0.03	95 ± 2
		$\delta < 0$	5 ± 2
0.022	pure DMF	0.56 ± 0.03	92 ± 2
		$\delta < 0$	8 ± 2
0.023	4.62	0.26 ± 0.03	69 ± 2
saturated with NaI		$\delta < 0$	31 ± 2
solid SnI_4	—	1.47 ± 0.03	

* A precipitate formed in the solution. The value given refers to the isolated precipitate.

If dimethylformamide is added to a carbon tetrachloride solution of tin(IV) iodide, a precipitate of composition $SnI_4(DMF)_2$ is obtained, the isomer shift of which is considerably smaller than that of the non-solvated SnI_4 molecule. This shows that the coordination of DMF results in a large decrease in the electron density at the tin nucleus, owing to the higher occupation of the 5*p* and 5*d* orbitals shielding the *s* orbitals.

The isomer shift values measured in carbon tetrachloride–DMF mixtures display a slight tendency to decrease as the DMF concentration increases. This can be interpreted in terms of exchange occurring between DMF molecules and the iodide ions in the coordination sphere of the tin.

As a consequence of the exchange, both the 5*s* and the 5*p* orbitals may be occupied to a lower extent. However, whereas the effect of the change in the 5*s* orbital is manifested directly in the isomer shift, alteration in the 5*p* orbitals has only an indirect and weaker influence. Hence, the resultant of the two opposing effects will be a decrease in the electron density at the nucleus. Another explanation could be based on the assumption that the coordination of DMF will result in increased shielding of the 5*s* orbitals of tin by the *p* and *d* orbitals.

In addition to the high-intensity line in the spectra of the mixtures, a line of lower intensity can also be observed at around $\delta = -0.2$ mm/s, which is indicative of a very low electron density at the tin nucleus. This might be explained by the strongly ionic nature of the tin atom, i.e., by the dissociation of SnI_4. However, earlier conductometric measurements showed that, although SnI_4 does undergo extensive dissociation, it is far from complete under the given conditions. It seems more probable that complexes with higher iodide content are formed in the solution. This view is supported by the experimental result that on addition of iodide ion the intensity of the line at $\delta = -0.2$ mm/s significantly increased. This definitely suggests the formation of ions containing more than four iodide atoms (SnI_5^- or SnI_6^{2-}). In agreement with the earlier equilibrium investigations by Mayer and Gutmann [Ma 70b], the formation of such ions in a solution containing no excess of iodide ions can be conceived according to the following auto-complex-formation processes:

$$2\,SnI_4D_2 \rightleftharpoons SnI_3D_3^+ + SnI_5D^-$$

or

$$2\,SnI_4D_2 \rightleftharpoons SnI_2D_4^{3+} + SnI_6^{2-}$$

We shall return to an interpretation of the phenomenon in the section dealing with the coordination chemistry of solvent mixtures (see Chapter 8).

The effect of solvation on the Mössbauer parameters was investigated for iron(III)–dithiocarbonate complexes by De Vries *et al.* [De 71]. They compared the Mössbauer parameters of mixed complexes of the composition $Fe(dtc)_2X$ (where $X = Cl^-$, SCN^-) in the solid state and in frozen solutions prepared with various solvents. These complexes are known to contain iron with a coordination number of five in the solid state [Ha 66, Ho 66]. Furlani [Fu 68] assumed that in solution the coordination number of the iron will be six. De Vries *et al.* [De 71] demonstrated that the quadrupole splitting values of the complexes [$Fe(dtc)_2Cl$] and [$Fe(dtc)_2SCN$] in solution were lower than in the solid state, and agreed roughly with the ΔE values of the parent complex [$Fe(dtc)_3$], which barely differ in the solid state and in solution. The magnetic susceptibilities of the mixed complexes and the parent complex proved to be almost identical in solution [Ma 64a, Wh 64].

All of these data show that in frozen solutions the immediate environment of the central atom of the mixed complexes has almost the same symmetry as the tris parent complex. That is, the pentacoordinated mixed complex of square pyramidal symmetry in the solid state will presumably assume in solution a distorted octahedral, hexacoordinated structure as a result of the coordination of a solvent molecule.

The assumption concerning the coordination number of iron in solution, was later confirmed by Angelos Malliaris and Niarchos [An 76] on the basis of an analysis of the temperature dependence of the Mössbauer parameters and the paramagnetic hyperfine structure of the spectra of the solvate complexes $Fe(dtc)_2BrX$ formed with tetrahydrofuran (X).

In solutions of metal salts in non-aqueous solvents (particularly in systems with low permittivities), it is frequently necessary to take into account the formation of polynuclear species. Differentiation of the monomeric and homopolynuclear formations in solution is a difficult task in most cases. This is well reflected by investigations of various non-aqueous solutions of iron(III) chloride, for instance, which led to contradictory results in the above respect (*cf.*, e.g., [We 62, Fa 68, Ca 62, Gu 70, Ar 65]). Study of the paramagnetic spin relaxation by Mössbauer spectroscopy is an excellent means for the differentiation of monomeric and polynuclear high-spin iron(III) species [Vé 78].

If the paramagnetic spin relaxation time is longer than the reciprocal frequency of the Larmor precession of the magnetic moment of the atomic nucleus, and also longer than the average lifetime of the nuclear excited state of the Mössbauer atom, then magnetic hyperfine splitting will appear in the Mössbauer spectrum. In measurements at constant temperature in rapidly frozen, glass-like solutions, however, the paramagnetic spin relaxation time depends only on the distance of the iron(III) centres from one another. In dilute frozen solutions containing the mononuclear species, this distance is sufficiently great for the system to satisfy the above conditions and for a magnetic hyperfine structure to appear in the spectrum (Fig. 5.7). The formation of polynuclear (even dimeric) species causes the iron(III) centres to approach one another, which results in the disappearance of the magnetic hyperfine structure because of the acceleration of the spin-spin relaxation (Fig. 5.8).

On the basis of the Mössbauer spectra of solutions of iron(III) chloride in seven solvents, Vértes *et al.* [Vé 78] unambiguously established which solvents favour

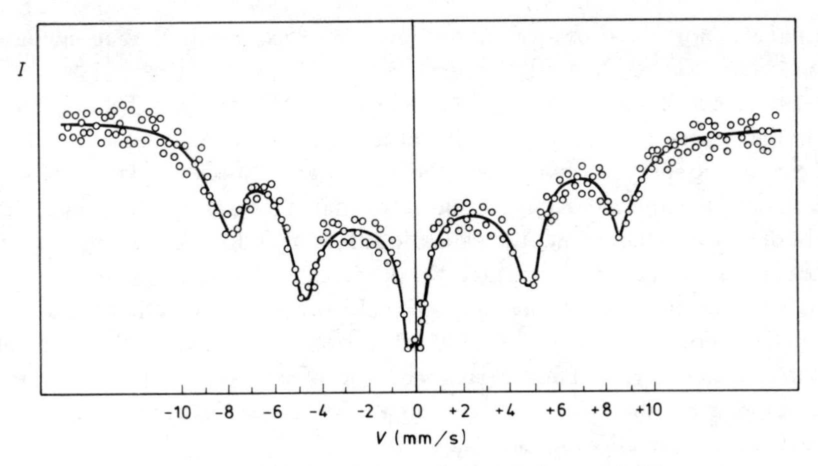

Fig. 5.7. Mössbauer spectrum of a rapidly frozen solution of iron(III) chloride in dimethylformamide [Vé 78]

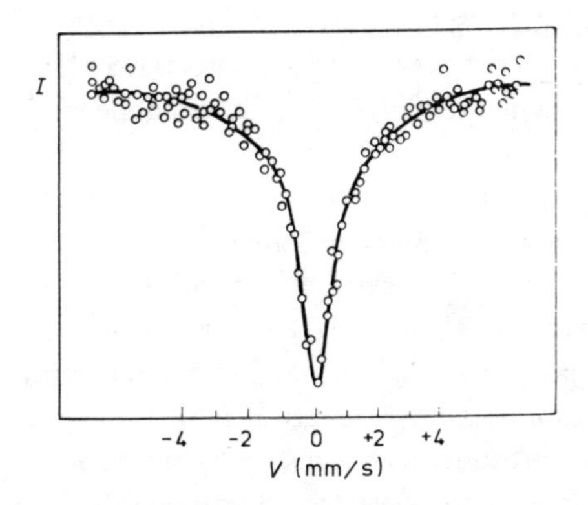

Fig. 5.8. Mössbauer spectrum of a rapidly frozen solution of iron(III) chloride in benzene [Vé 78]

the formation of monomeric species (ethanol, DMF), and which favour that of dimeric or polymeric species (benzene, nitrobenzene, acetone, acetonitrile, pyridine).

In ethanol only the formation of monomeric and in benzene and nitrobenzene only that of dimeric species was observed. In the other solvents, solvates of both kinds formed, but in DMF the monomeric species predominated, whereas in acetone, acetonitrile and pyridine the dimeric or oligomeric species predominated.

From these results one can conclude that for the formation of the monomeric solvates a solvent of high donicity and relative permittivity is necessary (e.g., ethanol). High relative permittivity (nitrobenzene) or high donicity (pyridine) alone is insufficient for the transformation of the dissolved $FeCl_3$ into mononuclear solvate.

Measurement of the Mössbauer spectra in an external magnetic field has shown that in the dimeric solvates the iron(III) atoms are antiferromagnetically coupled.

The magnetic splitting observed for the mononuclear species makes possible the calculation of the internal magnetic field at the iron nucleus. From this, conclusions may be drawn as to the extent of coordination of the ligand, since bonding electrons, by means of their shielding effect, decrease the magnetic field induced at the nucleus by the $3d^5$ electrons. The internal magnetic field measured in an ethanolic solution of iron(III) chloride ($H_{5/2}=513\pm10$ kOe*) proved to be smaller than the value measured in dimethylformamide ($H_{5/2}=540\pm5$ kOe*). This indicates that more chloride ions are bound in the inner coordination sphere of the iron(III) in ethanolic solution than in dimethylformamide.

From the quadrupole splitting values of the Mössbauer spectra of polynuclear (dimeric, possibly oligomeric) species, conclusions may be drawn about the symmetry conditions, and hence about the effect of solvation on these.

The Mössbauer parameters of iron(III) chloride dissolved in benzene ($\delta=0$, $\Delta E=0.36$) are almost the same as those of anhydrous iron(III) chloride measured in the solid state. This shows that the interaction between the benzene molecule and the iron atom is very weak; the charge symmetry and the *s*-electron density on the iron nucleus do not change in the course of dissolution. Accordingly, in an iron(III) chloride solution in benzene dimers or polymers are to be present which are unsolvated or solvated to only an extremely small extent.

In the Mössbauer spectrum of an iron(III) chloride solution in nitrobenezene, the quadrupole splitting is larger ($\Delta E=0.70$) than in benzene. In this solvent also, within the limit of error, the isomer shift is identical with the δ value for solid iron(III) chloride, indicating that, presumably because of its low donicity, the coordination of nitrobenzene does not change the *s*-electron density on the iron nucleus. The larger ΔE value, however, is evidence of an interaction between the solvent and iron(III). It can be seen that the nitrobenzene molecule, which has a high dipole moment, distorts the electron cloud around the iron nucleus.

On the basis of the conductivity of the solution, Gutmann and Wegleitner [Gu 70] assumed the occurrence of auto-complex formation in nitrobenzene. The results of Mössbauer studies suggest that this assumption is probably incorrect; the following equilibrium, proposed by de Main [De 59], is presumably established in the solution:

$$Fe_2Cl_6 + 2\,S \rightleftharpoons Fe_2Cl_5S^+ + Cl.S^-$$

In solutions of iron(III) chloride in acetonitrile or acetone, the quadrupole interaction was indicated only by the large line width. Of all the solvents examined,

* 1 Oe = 79.6 A/m

it is in these two that the electron sheath of the iron(III) nucleus in the solvate formed has the highest symmetry.

A strikingly large quadrupole splitting is observed in pyridine solution ($\Delta E = 1.72 \pm 0.05$ mm/s), which points to the considerable asymmetry of the electron sheath of the iron(III) in the solvate (Fig. 5.9). The isomer shift, $\delta = 0.15 \pm 0.05$ mm/s, which is larger than those measured in the previously mentioned solvents, also reflects the stronger interaction between the iron and pyridine.

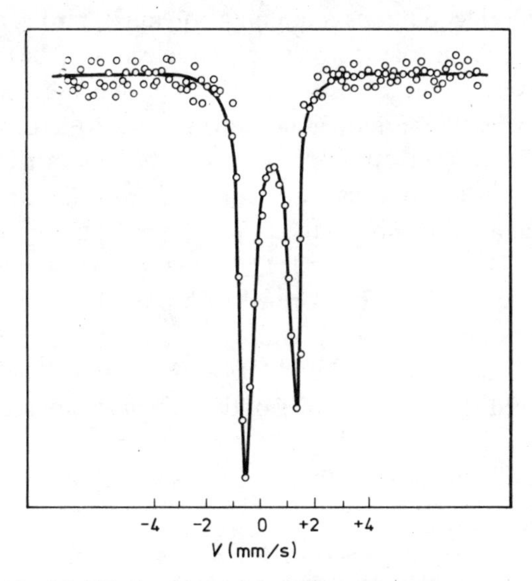

Fig. 5.9. Mössbauer spectrum of a rapidly frozen solution of iron(III) chloride in pyridine [Vé 78]

The large difference between the magnitudes of the quadrupole splittings observed in nitrobenzene and in pyridine solutions permitted the separation of two line pairs in the Mössbauer spectra of iron(III) chloride in a mixture of these two solvents and thereby the study of the formation of the pyridine solvate in the solvent mixture.

When solutions of iron(III) chloride in nitrobenzene are mixed with pyridine, the line pair characteristic of the iron(III)–pyridine complex appears; the intensity of the line pair increases with increase in the pyridine concentration, whereas the intensity of the line pair observed in pure nitrobenzene decreases proportionally. After computer resolution of the spectra, a knowledge of the ratios of the intensities of the lines and the total iron(III) and pyridine concentrations of the solutions permits calculation of the composition and formation equilibrium constant of the complex. In contrast with the early work of Weinland [We 22], assuming a composition $[Fe_2Cl_2py_8]^{4+}$, these calculations pointed to the composition

$[Fe_2Cl_4py_2]^{2+}$. The symmetry of the iron(III) in the complex can be regarded as that of a distorted tetrahedron. The assumed structural formula is

$$\left[\begin{array}{ccccc} py & & Cl & & Cl \\ & Fe & & Fe & \\ Cl & & Cl & & py \end{array}\right]^{2+}$$

probably forming an ion pair with chloride ions in the solution.

If the iron(III) chloride solutions containing polynuclear iron(III) are diluted with dimethylformamide, the magnetic structure characteristic of the monomeric species appears progressively in the Mössbauer spectrum. On the assumption that the Debye–Waller factors of the individual solvates are to a good approximation identical, the concentration ratios of the various monomeric and polynuclear iron(III) complexes in solution can be estimated from the areas belonging the Mössbauer spectral lines relating to these complexes. Thus, one can determine the parameter $S = A_D/A_{DMF}$, where A_D and A_{DMF} are the areas of the Mössbauer lines assigned to the solvates of iron(III) formed with the solvent in question and with dimethylformamide, respectively. S values relating to the concentrations 0.02 mole · dm^{-3} $FeCl_3$ and 0.2 mole · dm^{-3} DMF are listed in Table 5.10.

It may be observed that the stabilities of the solvates compared with that of the DMF solvate increase as the donicity of the solvent increases. The stability of the iron(III)–pyridine solvate exceeds those of the solvate complexes formed in the other solvents by more than two orders of magnitude. This is a consequence not only of the high donicity of pyridine, but also of its low relative permittivity, which counteracts heterolytic dissociation.

Table 5.10. Stabilities (S) of solvates of iron(III) chloride referred to the DMF solvate, and the donicities (DN) and relative permittivity (ε) of the solvents [Vé 78]

Solvent	S	DN	ε
DMF	1	26.6	36.1
Benzene	0.14	0.1	2.3
Nitrobenzene	0.18	4.4	34.8
Acetonitrile	0.21	14.1	38.0
Acetone	0.24	17	20.7
Pyridine	50	33.1	12.3

It can be seen from this brief survey that Mössbauer spectroscopy usefully supplements the information on solvation and solvate complexes obtained by other methods. The various equilibrium study methods establish the compositions of the species formed in solution; the Mössbauer spectrum generally yields only indirect information on this. However, Mössbauer spectroscopy reveals the symmetries and electronic structures of the species.

It also emerges that when the solvation equilibria give rise to Mössbauer atom-containing ions whose electronic configurations are sufficiently different to allow differentiation in the spectrum and the equilibria are "frozen in" in the course of the rapid freezing of the solution, Mössbauer spectroscopy can further be used for equilibrium measurements in the classical sense. The Bjerrum formation functions can be calculated from the dependences of the intensities of the Mössbauer lines, assigned to the individual ionic species, on the ligand concentration (possibly the solvent concentration) [Vé 76]. In most cases the equilibrium constants determined in this way refer to the temperature of solidification of the solution. When it is taken into consideration that the recording of each individual spectrum is fairly time consuming, and that the recording of a large number of spectra is necessary, it is clear that examinations of this type are profitable only if, for some reason, the customary, simpler procedures of studying equilibria cannot be employed.

Recently a new procedure — Capillary Mössbauer Spectroscopy (CMS) — has been elaborated by Burger *et al.* [Bu 82, 83] which made possible the Mössbauer study of liquid samples trapped in the pores of "*thirsty*" glass of 4 nm pore size. In the case of solutes with closed coordination sphere and suitable size (tris/2,2′-dipyridil/iron(II) and hexaaqua iron(II) complexes in aqueous solution) the Mössbauer spectra of the solution in the capillaries of the glass carrier corresponded to the species in the original liquid solution. In tin tetrahalide systems (liquid $SnCl_4$ and solutions of SnI_4) interaction with the glass surface could also be observed. CMS may open a new field also in the study of solvation phenomena.

X-ray diffraction studies

Amorphous and liquid materials and solutions can also be examined by X-ray diffraction, and hence this is another procedure suitable for the study of solvation processes and solvates.

In systems containing coordination compounds of metals, such as solvates, solution X-ray studies are possible because the environments of the identical atoms of the solute agree up to the second or third neighbour. However, this small degree of order, comprising only a few neighbours, is not repeated. The small microstructures are connected "irregularly" to one another, and this connection varies with time. Consequently, a clear diffraction picture similar to those of crystalline substances cannot be obtained on the X-ray diffraction photograph, yet it may be expected that some interference will occur reflecting the ordered structures of the solution and the dissolved molecules.

It is apparent from the foregoing that these investigations can provide less, and less clear, information on the structure than X-ray structural examinations of single crystals or even powders. However, since the compositions and structures of complexes isolated from solution in the solid state are not necessarily the same as they were in solution (in fact they do differ in most cases), the X-ray study of solutions is of great importance.

The evaluation of the results of such studies may be facilitated considerably if the compositions and stability conditions of the species in solution is also determined by means of equilibrium measurements. This is shown well by the investigations of Österberg and Sjöberg [Ös 68], who utilized equilibrium data, based on electromotive force measurements, to establish the compositions of the complexes formed between glycyl-L-histidylglycine and the copper (II) ion in solution; they then prepared some complexes in the solid state, and compared the results of X-ray structural studies on the solid single crystals with the results of solution X-ray structural studies and with potentiometric equilibrium measurement data. In this way they obtained a reliable picture of the system in solution.

There is another means of combining solution X-ray studies with classical complex equilibrium investigations in order to determine the structure of the species in solution. When the equilibrium constants are known, solutions are prepared with compositions in which one of the stepwise-formed complexes is the predominant ionic species. Hence, with the aid of solution X-ray studies, the structures of these complexes can be established.

In the above manner, in dimethyl sulphoxide and dimethylformamide solutions of mercury (II) iodide–sodium iodide systems of various compositions, Gaizer and Johansson [Ga 69] determined the Hg—I bond lengths and also the I—I distances for the dissolved complexes HgI_4^{2-}, HgI_3^- and HgI_2, and from these data conclusions were drawn about the symmetries of the species.

In a solution of the complex HgI_4^{2-}, the Hg—I bond length is 0.28 nm, which, in agreement with the X-ray data on the various solid iodomercurates [Ha 55, Fe 66], corresponds to a regular tetrahedral complex ion.

In a solution of the complex ion HgI_3^-, the Hg—I distance is 0.273 nm and the I—I distance is 0.453 nm, which corresponds to a slightly flattened tetrahedral complex, the fourth apex of which is occupied by a solvent molecule. In the solid $[(CH_3)_3S]HgI_3$ crystal, the HgI_3^- moiety has planar triangular symmetry [Fe 66].

In solution, just as in the vapour [Ak 59] and crystalline [Je 67] states, the complex HgI_2 has an almost linear structure; the Hg—I distance is 0.26 nm.

A study was also made of the system mercury (II) iodide–cadmium (II) iodide in the above solvents, and it was found that, in contrast to the results of spectrophotometric and conductivity measurements [Ga 67], the formation of polynuclear complexes could not be detected.

In the following, some examples will be presented of how the interaction between the solvent and solute is reflected by the results of solution X-ray studies.

X-ray structural investigations on alcoholic solutions of a number of transition metal salts [We 69], such as iron(II) chloride, iron(II) bromide and cobalt(II) bromide, revealed that in the case of the two iron(II) salts the transition metal ion is surrounded by four alcohol molecules, in accordance with tetrahedral symmetry. In an alcoholic solution of cobalt(II) bromide, on the other hand, cobalt solvates containing two bromide ions and two solvent molecules could be detected. In more concentrated iron(II) chloride solutions, a dimeric complex of composition Fe_2Cl_6 was found.

An investigation of formamide solutions of potassium iodide [De 68] demonstrated the dependence of the solvation number on the concentration. In dilute solutions, four formamide molecules were coordinated through the oxygen of the carbonyl group to the potassium ion, whereas in concentrated, nearly saturated solutions the number of coordinated formamide molecules was only two.

Similarly to solution X-ray studies, electron and neutron diffraction methods are also applicable to investigations of liquid structures, and hence solution structures. Whereas X-ray [Na 70a, Na 71, Na 72] and neutron diffraction [Pa 71, Po 72] measurements have achieved fairly widespread application, primarily in structural examinations of water and aqueous solutions, experiments on such a use of electron diffraction began only comparatively recently [Ká 74a, Ká 74b]. The first results were obtained in the structural examination of water. The methods developed for these purposes appear suitable (even though to a limited extent) for subsequent solution structural investigations.

X-ray photoelectron spectroscopy (XPS, ESCA) in rapidly frozen solutions

X-ray excitation photoelectron spectroscopy renders possible the measurement of the binding energy of electrons in the internal orbitals of every atom with the exception of hydrogen [Si 67]. Since the electron binding energies (ionization energies) depend on the electronic configuration and electron density of the atom examined and on the chemical environment affecting these, ESCA may be suitable for the study of all effects that cause a change in the electronic structure, provided that the change in the electron binding energies is large enough to be detected by the method [Bu 74a, b, c, Bu 75b, Jø 71, Jø 72]. Investigations by Burger *et al.* [Bu 75b] have shown that even such "weak interactions" as hydrogen bond formation and outer-sphere coordination cause changes which can be followed by ESCA.

Utilization of the method for the study of solvation processes and solvate complexes was for many years prevented (and is still restricted) by the fact that the sample is kept under high vacuum during the recording of the photoelectron spectra. This led to great difficulties in the ESCA study of volatile substances, including liquids solutions. Recently, Siegbahn and Siegbahn [Si 73] devised a cell in which the ESCA spectrum of a liquid can be recorded by the transmission of a jet of the liquid. The range of application of the method is very limited, however (liquids with low tensions, e.g., formamide, can be examined), and the reproducibility of the data is much lower than with other techniques; this excludes the study of finer effects.

Burger and Fluck [Bu 74a] were the first to employ ESCA for the study of solvation. They solved the problem outlined above by preparing a rapidly frozen solid from the solution to be examined (in an analogous way as in the Mössbauer studies of solvation), and measured the electron binding energies in the orbitals of the corresponding atoms of the solute in this solid.

The method was later further refined [Bu 77]. On the basis of this proposal, Miksche *et al.* [Mi 77] constructed a simple apparatus attached to the ESCA instrument.

The introduction of the ESCA study of rapidly frozen solutions also led to the elimination of one of the major possibilities of error in this method [Bu 78]. It is known that in the ESCA study of electrically insulating samples the emission of the photoelectrons causes the occurrence of a positive charge on the surface of the sample. This charge diminishes the kinetic energy of the photoelectrons. Since the electron binding (ionization) energies are calculated from these experimentally determined kinetic energies, the positive charge developing on the surface can cause a considerable error, making the results uncertain.

Elimination of this error is hampered by the fact that the extent of the charging of the surface, and hence the magnitude of the error, depends on so many varied factors that the reproducibility and the possibility of making corrections for the error are fairly uncertain. It is also known, however, that the above effect is identical for every atom and every orbital of a uniform sample. Accordingly, the elimination of, or at least a decrease in, this error is possible by employing some appropriate internal standard.

For this purpose, various research groups have proposed the application of different internal standards. For example, a thin gold layer was sublimed on to the surface of the solid sample; this lets through the photoelectrons and serves as a reference. Some workers used as reference the C 1*s* data for the Scotch tape adhesive serving to keep the solid sample on the sample holder; others used the C 1*s* data for the hydrocarbons of the traces of pump oil precipitated on to the surface of the sample.

Each of the solutions is based on the assumption that the charging effect appearing on the reference substance is the same as that for the sample under study. However, the validity of this assumption is more or less limited. The charging effect appearing on the internal standard will perfectly coincide with that on the sample only if these two substances are situated in a common phase. This condition is met if the ESCA electron binding energy value for some orbital of the "bulk" solvent in the frozen solution is used as the reference [Bu 78].

Recent investigations by Burger [Bu 78] have shown on the basis of the ESCA study of several model systems in rapidly frozen (quenched) solutions that one of the electron lines of the bulk solvent in each system can be used as internal standard for charge correction. In carbon tetrachloride or tetrachloroethylene solutions the Cl $2p_{1/2,\,3/2}$ lines, in dimethyl sulphoxide the S $2p_{1/2,\,3/2}$ lines and in aqueous solutions the O $1s$ lines were shown to be suitable internal standards. The reproducibility of the ESCA binding energies referred to these standards has been found to be ± 0.1 eV. By this method the ESCA data of all substances soluble in the same solvent can be brought to a common scale.

Owing to the high reproducibility of the corresponding energy differences obtained in this way, the effect of weak interactions causing small changes in the electron binding energies can also be studied with success.

As the model system in their ESCA studies of solvation, Burger and Fluck [Bu 74a] employed various solutions of antimony pentachloride, which had been used as a reference acceptor previously by Gutmann in his investigations of the donicities of solvents. In addition to the information provided by the Gutmann calorimetric method, the Mössbauer studies by Vértes and Burger [Vé 72] in rapidly frozen (quenched) solutions also yielded many new data on these systems. These Mössbauer studies revealed that the parameters measured in the rapidly frozen solutions objectively reflected the conditions in the original liquid solution.

In the ESCA investigations, 1.2-dichloroethane, which has a low relative permittivity ($\varepsilon = 10.1$), was used as the inert solvent. Solutions of antimony pentachloride in this were prepared, and the donor solvent was added in small amounts, so that one molecule of this should be bound at the sixth coordination site of the $SbCl_5$, but the relative permittivity of the system should not be increased appreciably. In this way it could be ensured that the $SbCl_5$ did not dissociate even partially, and hence the change in the electron binding energies of the antimony on addition of the donor solvent could be attributed only to the coordination of this solvent (to the solvation of $SbCl_5$).

In order to be able to carry out the measurement in the shortest possible time and to reduce disturbing effects originating from surface changes due to evaporation of the sample, measurements were made only of the electron binding energy of one orbital (the comparatively high-intensity $3d_{5/2}$ orbital) for the antimony, and the electron binding energy of the $2p_{1/2,\,3/2}$ orbital of the chlorine of the dichloroethane solvent as the reference. With the aid of this reference atom, it was possible to eliminate the shifts in the absolute values of the electron binding energies caused by the charge developing on the electrically insulating samples (such as frozen $SbCl_5$ solutions) as a result of the emission of a photoelectron.

The electron binding energies of the $3d_{5/2}$ orbital of antimony, determined in various $SbCl_5$ solutions and referred to the chlorine in the solvent dichloroethane, are listed in Table 5.11, together with the Gutmann donicity values of the donor solvents. Figure 5.10 depicts these electron binding energy differences as a function of the donicity of the donor solvents. It can be seen that the relation between the donor strengths of the solvents (the ΔH values for solvation of $SbCl_5$) and the electron binding energy differences is linear: with an increase in the donicity of the donor molecule, the electron binding energy on the antimony increases.

The above correlation can be explained as follows. A greater effect is exerted on the ionization energy (the electron binding energy) of the antimony by the five more

Table 5.11. Electron binding energies of the $3d_{5/2}$ orbital of antimony, measured in rapidly frozen $SbCl_5$ solutions in dichloroethane containing various donor solvents [Bu 74a]

Donor solvent	DN_{SbCl_5}	ΔI_{Sb-Cl}
Hexamethylphosphoramide	39.8	323.25
Dimethyl sulphoxide	29.8	322.95
Dimethylformamide	36.1	322.80
Trimethyl phosphate	23.0	322.75
Diethyl ether	19.2	322.55
Acetonitrile	14.1	322.45
No donor solvent	—	321.90

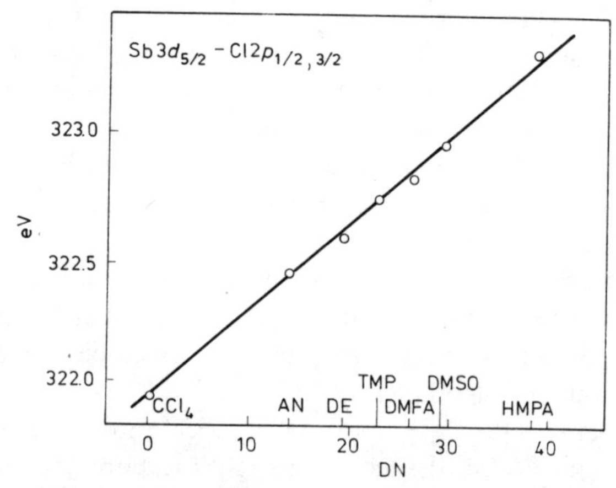

Fig. 5.10. ESCA data on solvates of $SbCl_5$ with various donor solvents, determined in rapidly frozen dichloroethane solution (the Sb $3d_{5/2}$ electron binding energies are referred to the Cl $2p_{1,2\;3/2}$ electron binding energies of the solvent), as a function of the donicities of the solvents [Bu 74a]

covalent Sb–Cl bonds than by the one coordinate bond between the antimony and the donor atom of the solvent. Coordination of this donor atom to the free coordination site of $SbCl_5$, however, decreases the covalency (bond strength) of the Sb—Cl bonds. (This can be seen, for instance, if a comparison is made of the Sb—Cl bond lengths in an analogous system [Ta 58], e.g., in $SbCl_5$ and in $SbCl_6^-$, or of the force constants of these bonds, calculated from the infrared spectrum [Be 63a, Kr 67b].) Thus with increasing interaction between the antimony and the coordinated solvent the covalency of the Sb—Cl bond decreases. This favours the solvation of chloride. Both effects, the increased ionicity of the Sb—Cl bond and the increased solvation of its chloride, cause the electron density on the central antimony atom to decrease with increasing solvation and hence the electron binding energy of its *d* orbitals to increase. This is the explanation for the correlation displayed by the data in Table 5.11 and Figure 5.10.

These studies also show that the changes in the electronic structure of the solute caused by the solvation can be followed by various methods. Under appropriately selected experimental conditions, where the measured parameter is not influenced by side reactions, the results obtained in the course of an investigation of the electron sheath of the acceptor may exhibit a close correlation with the reaction heats of coordination of the solvents (i.e., with the Gutmann donicities).

Positron annihilation and positronium chemistry*

In the past few decades a new method for the examination of matter has developed from the study of positron annihilation. This method utilizes the interaction of matter and the antielectron, i.e., the positron, one of the elementary particles of antimatter most readily accessible under terrestrial conditions. Positron annihilation reacts sensitively to changes occurring in the physical and (as a consequence of the formation of positronium atoms, which behave as chemical elements) the chemical properties of the medium.

The results and developments attained in this field are indicated by the several hundred publications, including a number of reviews and monographs, which have appeared [De 53, Fe 56, Go 68a, Go 71a, Gr 64, Wa 60], and by the numerous international conferences dealing with the results achieved in this new research field [Po 67, Pr 71, Pr 75, Pr 79]. The annihilation of positrons is dealt with by researchers in numerous branches of science, such as nuclear physics, solid-state physics, chemistry and biology, and useful information has been obtained. The method has recently also been employed in the field of solution studies.

* This section was compiled by Prof. Attila Vértes and Dr. Béla Lévay.

In the course of annihilation resulting from the collision of a positron and an electron, one, two or three γ-photons may be formed, and annihilation may also occur without the formation of a γ-quantum.

The probabilities of annihilation with the formation of no or one γ-photon are extremely low. The occurrence of 2γ- or 3γ-annihilation as a result of the interaction of a slow positron and an electron depends on whether these two particles meet with parallel or antiparallel spins. At the moment of their meeting, the electron–positron pair come into a triplet or a singlet state in the case of parallel spins or antiparallel spins, respectively, when their total moments are 1 or 0. Because of the law of parity conservation, only an even number (e.g., 2) of γ-quanta can be formed from the singlet state, and only an odd number (e.g., 3) from the triplet state. The theoretically calculated ratio of the probabilities of 3γ- and 2γ-annihilations is 1:372.

Free collision of the two particles (free annihilation) is not the only means whereby annihilation may occur. This may also be preceded by a bound state formed between the positron and neutral molecules or ions.

The third, and from a chemical aspect the most important, possibility and means of annihilation is that which is preceded by the formation of a positronium atom.

A positronium (Ps) is a particle with a structure reminiscent of that of the hydrogen atom, but it contains a positron instead of a proton, and is therefore 920 times lighter. Thus, it is the simplest chemical element, and might be regarded as an isotope of hydrogen.

The spin value of both the electron and the positron is 1/2. (The absolute value of the magnetic moment of the positron agrees with that of the electron, i.e., 1 Bohr magneton, but its sign is positive and hence its direction is parallel with the spin.) Accordingly, depending on whether the spins of the electron and the positron in the positronium display antiparallel or parallel orientations, the resultant spin value of the Ps may be 0 or 1 ($h/2\pi$). The name of the former is singlet or *para*-Ps ($^1S_0$:^{S}Ps), and that of the latter is triplet or *ortho*-Ps ($^3S_1$:^{T}Ps). The former decomposes by 2γ- and the latter by 3γ-annihilation.

The mean lifetime of the free *para*-Ps *in vacuo* is $\tau_S^0 = 1.25 \times 10^{-10}$ s, and that of *ortho*-Ps is 1.4×10^{-7} s. These lifetimes appear to be very short, but if they are compared with the shortest time intervals experimentally measurable, or, for example, with the period of intramolecular vibration of atoms (10^{-14} s), it can be seen that they are sufficiently long for the positronium atom to take part in chemical reactions or to enter into other interactions with the particles of the medium, and to allow the following of these processes in time. This is particularly the case as regards *ortho*-positronium.

All of these processes are influenced by the material and chemical environments of the positronium, and hence it is clear that the study of positron annihilation may lead to information relating to the chemical structure of the material interacting with the positron radiation. In general, only the fate of the *ortho*-positronium can be followed experimentally, and in the following, therefore, the factors affecting this will be discussed.

The possible mechanisms of interaction of the positronium atoms are as follows:

(1) The most general interaction, which occurs in all materials, is "pick-off" interaction. The essence of this is that positron of the positronium atom in the triplet state is annihilated not by its "own" electron, but undergoes 2γ-annihilation with some antiparallel-spin electron of a molecule encountered during collision with the molecules of the medium. As a result of the interaction, the lifetime of the *ortho*-positronium is shortened, although, because of the shielding effect of its own electron, not to the extent that it would have been if it had been destroyed in free annihilation. Pick-off interaction occurs to a significant extent particularly in the condensed phase.

(2) Another important type of positronium interactions is *ortho–para* conversion. This can take place if the medium includes paramagnetic species containing unpaired electrons. On collision with such a species, the direction of one of the parallel spins in the *ortho*-positronium is reversed, and at the same time the direction of the spin of the unpaired electron in the colliding molecule also reverses. In accordance with its lifetime, the resulting *para*-positronium is then rapidly annihilated. Hence, this effect also leads to a decrease in the lifetime of the positronium.

Free radicals, or the cations of transition metals, containing an unpaired electron or electrons, generally come into consideration as such converters. However, no close correlation can be found between the strengths of the converters and the number of unpaired electrons in them.

(3) Positronium reactions of a chemical nature form the third, and from the chemical aspect the most important, group of interactions. (It must be noted that certain types of the *ortho–para* conversion reactions are also of a chemical nature, e.g., the free radical reactions.) The main types of chemical reactions of positronium are illustrated by the following examples:

(*a*) addition reactions:

$$Ps+(CF_3)_2NO \rightarrow (CF_3)_2NO.Ps \rightarrow 2\gamma$$

(*b*) substitution reactions:

$$Ps+Cl_2 \rightarrow Cl+PsCl \rightarrow 2\gamma$$

(*c*) oxidation reactions:

$$Ps+Fe^{3+} \rightarrow Fe^{2+}+e^+ \rightarrow 2\gamma$$

(*d*) reduction reactions:

$$Ps+e^- \rightarrow Ps^- \rightarrow 2\gamma$$

The chemical reactions also lead to a decrease in the lifetime of the *ortho*-positronium. In the positronium compounds formed in processes (*a*) and (*b*), the positron resides in a location with a higher electron density than when migrating from molecule to molecule, and this increases the probability of annihilation. The positron formed in reaction (*c*) will, with high probability, decompose by rapid free annihilation. The Ps^- formed in process (*d*) is a much less stable formation than the Ps atom, its binding energy being merely 0.3 eV. Since the new electron may be coupled to the positron with either parallel or antiparallel spin, the average annihilation rate of the Ps^- will be practically the same as that of free annihilation.

The processes that reduce the lifetime of positronium are collectively referred to as quenching of the positronium.

One of the reasons why the study of the chemical reactions of positronium is extremely interesting is that, in a certain sense, we may speak here of the extreme limit of microchemistry, when the reaction times of the individual reacting atoms can be directly followed and quantitatively measured. In these processes the positronium atom at the same time also features as a labelled atom, giving an indication of its own fate by means of the annihilation.

The most interesting information from the chemical point of view may be obtained by measurement of the distribution of the lifetimes of the positrons or positronium atoms. In addition to the lifetime, this method also gives a measure of the relative probability of *ortho*-positronium formation as a separable parameter. Hence, from the lifetime distribution, information is obtained not only on the rates of the chemical reactions and other interactions of the positronium, but also on the probability of positronium formation, e.g., the extents of the possible inhibitory processes. The only deficiency of the method is that it does not provide a possibility for simple differentiation between *ortho–para* conversion processes and the chemical reactions, since both processes cause a decrease in the lifetime.

In the course of the positronium lifetime measurements, radioactive isotopes of positron decay serve as sources, e.g., sodium-22, copper-64. At the moment of the emission of the positron from this source a γ-photon is also released. (In the case of, e.g., sodium-22 its energy is 1.28 MeV.) This γ-photon serves as the start signal in the coincidence equipment used. The γ-photon produced by the 2γ-annihilation process to be studied (0.51 MeV) is the stop signal. The magnitude of the time measured between the start and stop signals (the positronium lifetime) is in the range $10^{-10}-10^{-7}$ s. To get a lifetime curve of adequate statistics, the apparatus repeats the time measurement about 10^6-10^8 times. For the details of the experimental technique see, e.g., refs. [De 53, Fe 56, Go 71a].

Pick-off annihilation and the molecular and solution structures

It has been seen that pick-off annihilation is a general means of annihilation, occurring in all types of material and media. Since the positron of the positronium is annihilated here by the electrons of the medium, it is to be expected that a connection exists between the rate of annihilation and the structure of the molecules in the medium, as well as the interactions between them, and thus information indicative of the liquid structure may also be obtained.

By measuring the rates of pick-off annihilation in nearly 200 organic liquids, Gray *et al.* [Gr 68] found a correlation between the lifetime of the triplet positronium and the structures of the molecules. In the case of groups of compounds of the same type, e.g., n-alkanes, normal primary alcohols or 1-chloro-n-alkanes, the quenching cross-section of the positronium varied linearly with the number of carbon atoms. It was possible to establish the separate quenching cross-section contributions of the individual groups building up the molecules; these depended not only on the nature of the atoms comprising the corresponding groups, but also on the bond types. Correlations were also found with other molecular properties, e.g., the electron polarizations of the molecules.

The dependence of pick-off annihilation on the various parameters was also studied in solutions and liquid mixtures.

In aqueous solutions of hydroxy acids it was observed that the pick-off annihilation was predominantly a function of the molar density, being nearly additive [Ta 69].

In contrast, when methanol–water and dioxane–water mixtures were examined and the literature data for other alcohol–water mixtures were analyzed, it could be concluded that there is a significant deviation from additivity in alcohol–water mixtures [Le 72]. A simple correlation can be deduced between the rate of pick-off annihilation and the molar volume, with the assumption that the cross-section of annihilation is a linear function of the molar fraction in the mixture, i.e., it is additive:

$$\lambda_{AB}V_{AB}=\lambda_A V_A-(\lambda_A V_A-\lambda_B V_B)X_B$$

where λ is the rate constant of the pick-off annihilation, V is the molar volume and X is the molar fraction; the subscripts A and B refer to the individual pure components and the subscript AB to the mixture.

In the event of additivity, therefore, $\lambda_{AB}V_{AB}$ varies linearly with the molar fraction; this was so for the dioxane–water mixtures, but for the methanol–water and other alcohol–water mixtures the results revealed a considerable negative deviation.

Further investigations disclosed a fundamental correlation between the rate constant (λ_p) of the pick-off annihilation observed in liquids and the surface tension (γ) of the liquids:

$$\lambda_p = \kappa \gamma^\alpha$$

where the values of the constants κ and α (if λ_p was measured in units of ns^{-1} and γ in units of dyne cm^{-1}) were 0.061 and 0.50, respectively. This empirical correlation proved to hold from the liquid noble gases (with very low surface tensions), over a wide range of organic liquids, up to water (with a high surface tension) [Ta 72].

It was further demonstrated that the above correlation is also valid for liquid mixtures [Lé 73a]. The decomposition constants measured in the methanol–water and dioxane–water systems, for example, lay on a common curve if they were plotted as a function of the surface tension of the mixtures.

When the correlation with the parachor (which bears an additive connection with the molecular structure) was utilized, it was found [Lé 73a] that the conditions are better described by the equation

$$\sqrt{\lambda_{AB}}\, V_{AB} = \sqrt{\lambda_A}\, V_A - (\sqrt{\lambda_A}\, V_A - \sqrt{\lambda_B}\, V_B) X_B$$

than by the previous equation of additivity. The values measured for the methanol–water system are also in agreement with this equation.

Correlations could further be deduced from the dependence of the rate of pick-off annihilation on temperature and viscosity [Lé 73a]: the square root of the lifetime of the pick-off annihilation was found to vary linearly with the absolute temperature and with the reciprocal of the viscosity. Hence, the temperature and viscosity dependences of the pick-off annihilation could be described well even in associating liquids, where other theories were inapplicable.

A theoretical explanation of the close correlation between pick-off annihilation and the surface tension is given by the "bubble model". According to this model, a small "bubble" develops around the Ps atom, owing to the repulsion potentials arising between the molecules of the medium and the neutral Ps atom formed in liquids. The surface tension acting on the surface of the bubble tends to decrease the surface area, and thus the resulting radius of the bubble will depend on the equilibrium of the opposing forces.

For a quantitative discussion of the model, the bubble is regarded as a spherically symmetric potential well in the interior of which the Ps atom moves in a potential-free space. The parameters of the bubble may be determined by quantum-mechanical methods on the basis of the experimentally measured lifetime and angular distribution data [Lé 73b].

The model was first elaborated to explain the anomalously high lifetime observed in liquid helium, which has a large electron density compared with the gaseous state [Fe 57, Ro 67]. It was later extended to molecular liquids with low surface tensions [Bu 71], and it was subsequently demonstrated [Lé 76] that it can also be applied to aqueous solutions of inorganic substances where surface tensions are high.

An interesting result was provided by the investigation of aqueous solutions of surface-active agents of high molecular weight, which, even at a low concentration, bring about a large decrease in surface tension. In these solutions, the change in the surface tension did not lead to a change in the pick-off lifetime [Lé 74]. This fact provides indirect support for the bubble model, as it is not to be expected that these large molecules, which diffuse slowly and which are present in very low concentration, would take part in the development of the microscopic bubbles around the Ps atom.

Coordination chemical aspects of positron annihilation

Positron annihilation studies are suitable for the sensitive detection of delocalized unpaired electrons and spin density, and by this means for following changes in the coordination spheres of complexes in solution, e.g. in solvate complexes. In these studies the Ps atom plays the role of a free radical, and its reactions are comparable to the analogous reactions of other free radicals.

The reactions of the Co^{2+} ion (containing unpaired electrons) with positronium and various free radicals (e.g., diphenylpicrylhydrazyl or 2,2,6,6-tetramethyl-4-oxopiperidin-1-oxyl) were investigated as a function of the composition of, for example, propanol–water mixtures [Go 68b]. Angular correlation measurements indicated that the Co^{2+} ion interacts with positronium in an *ortho–para* conversion reaction, the rate of which was determined on the basis of the lifetime spectra.

The variation observed in the reaction of positronium with cobalt as a function of the solvent composition, and the essentially different natures of this with chloride and perchlorate as the anion, can be explained by the changes occurring in the coordination sphere of the Co^{2+} ion.

As confirmed by spectrophotometric measurements, in a pure propanolic solution of $CoCl_2$ the coordination sphere of Co^{2+} has tetrahedral symmetry, and chloride ions are also involved in the coordination. With $Co(ClO_4)_2$, the coordination sphere is composed of molecules of the solvent only, in octahedral symmetry. In this case, the rate of conversion with positronium is much lower, which points to the important role of the coordinated chloride ions in this reaction. Further evidence of the participation of the chloride ions in the conversion reaction is that whereas the rate constant varies to only a slight extent with increasing water content in the case of $Co(ClO_4)_2$, it decreases rapidly in the case of $CoCl_2$. The explanation is that water displaces the chloride ions from the coordination sphere. In solutions with high water contents, when the coordination sphere consists of water molecules, the conditions become independent of the nature of the anion.

The important role of the chloride ions in the conversion reactions can be explained as follows. The coordination (e.g. solvate) sphere prevents the

Ps from entering into direct reaction with the unpaired electrons of the Co^{2+}. With respect to the *ortho–para* conversion, therefore, only the electron density delocalized at various sites of the coordination sphere, owing to the unpaired electrons, can come into consideration. Although this electron density is merely 0.02 on the chloride ion, this value is nevertheless more than an order of magnitude larger than the electron density on the carbon or hydrogen atoms of the solvent. The strong effect of chloride ions proves the extremely high sensitivity of positronium to changes in the density of the unpaired electron. Although the density of the unpaired electrons on the oxygen of the hydroxyl group of the solvent (alcohol and water), is comparable to that of the chloride ion (ca. 0.01), these electrons are much less accessible for the Ps because of the steric hindrance of the hydrogen bonds and the hydrocarbon chains of the alcohol molecules; their effect is exerted only when the coordination sphere consists only of water molecules. The experimental results also show that the water molecules first expel from the coordination sphere the chloride ions, which are the most effective from the point of view of the conversion. With increase in the water concentration, this give rise to a decrease in the rate constant in the first section of the curve. Then, as a result of a further increase in the water concentration, when water starts to displace also the alcohol molecules from the coordination sphere, the value of the rate constant again begins to increase.

A similar effect of chloride ions was also observed in an oxidation-type reaction of positronium with $Fe(ClO_4)_3$ in aqueous solutions at various pH values [Go 72]. As the pH was increased from 0 to above 2, the rate constant of the Ps reaction markedly decreased, but on the addition of chloride ions it became higher again.

These results are in close agreement with those from Mössbauer spectroscopic measurements in a similar system [Ko 70]. At pH=0 the iron is present in the form of $[Fe(H_2O)_6]^{3+}$ ions, whereas at pH=2 prevailing form is the dimeric complex ion:

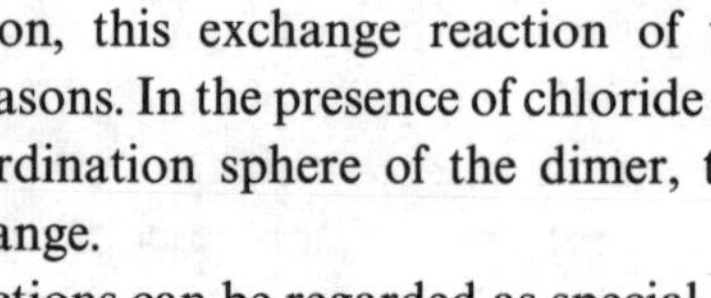

$$\left[(H_2O)_4Fe \begin{matrix} \overset{H}{O} \\ \underset{H}{O} \end{matrix} Fe(H_2O)_4 \right]^{4+}$$

While hexaquoiron(III) ions participate at a very high rate in the Fe^{3+}–Fe^{2+} electron-exchange reaction, this exchange reaction of the dimeric complex is hindered for structural reasons. In the presence of chloride ions, however, which are incorporated in the coordination sphere of the dimer, the dimer also becomes capable of electron exchange.

Electron-exchange reactions can be regarded as special cases of redox reactions, and in the given case, therefore, the parallel between the electron-exchange reaction of the iron ions and their redox reaction with the Ps atoms is obvious.

Another interesting observation achieved during the study of certain reactions of Ps was the effect of solvation due to a charge-transfer interaction on the annihilation.

In a study of the reaction of Ps and iodine in various solvents [Ta 70], it was found that the rate constant of the reaction varied linearly with the reciprocal of the viscosity of the solvent. The reaction is thus a diffusion-controlled fast reaction, similarly to the reaction between Ps and dissolved paramagnetic (and therefore strongly quenching) oxygen [Po 54].

A noteworthy conclusion in the publication cited is that the explanation of the deviation from a linear correlation may be the interaction between the iodine and the individual solvents, i.e., the formation of charge-transfer complexes. This would alter the diffusion coefficients, and additionally, positronium may react at different rates with free iodine and with iodine in the complex form.

These assumptions were later proved directly, on the example of the formation of the pyridine–iodine charge-transfer complex [Lé 72b]. It was shown that the rate of the reaction of Ps with iodine bound by pyridine in a comparatively stable complex was only half of the value expected on the basis of the theory of diffusion-controlled reactions, taking into account the changes due to complex formation in the diffusion coefficients.

This vividly confirms that one of the causes of the fast reaction between the Ps and iodine is the electron-acceptor property of the iodine. Hence, the explanation of the halving of the reaction rate is that the interaction between the electron acceptor iodine and the strong electron donor pyridine hinders the reaction between the iodine and the Ps. On the other hand, if the electron donor pyridine is not present, then complex formation of a donor–acceptor nature between the Ps and the iodine may be counted upon.

In contrast, however, it must be considered that the rate of the reaction involving iodine bound in the complex by pyridine is "only" halved, and is still very high (at 20 °C in cyclohexane, $k = 2.9 \times 10^{10}$ l mole^{-1}.s^{-1}). This points to the possibility of another reaction between the Ps and iodine, proceeding in parallel with the previous one. This other reaction may consist of dissociation of the iodine molecule and formation of the compound PsI as an intermediate. This reaction path is not influenced substantially by the bond formed between the pyridine and iodine.

Thus, with the complex formation being utilized as a model, the following scheme could be suggested for the mechanism of the reaction of Ps and iodine:

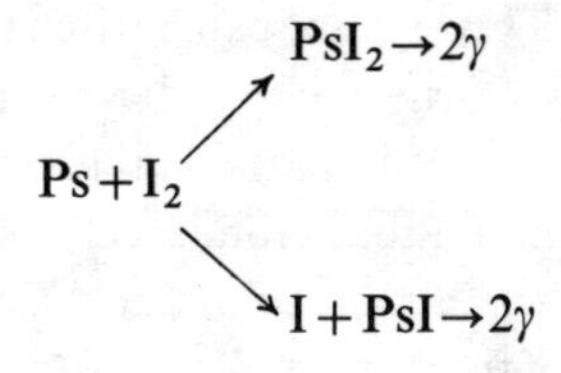

In addition, an effect resulting in inhibition of positronium formation was also observed during the development of the strongly polar pyridine–iodine complex. It is probable that a reaction competing with positronium formation occurs between the positrons and the negative charge localized on the molecule, and this causes the inhibitory effect.

From the different quenching rates of free iodine and iodine bound in the complex, the equilibrium constant of complex formation can also be determined. It may be assumed that the overall quenching rate constant (λ_{ox}) is the additive resultant of the quenching rate constants of the free iodine and of the iodine bound in the complex ($\lambda_{ox,\,I}$ and $\lambda_{ox,\,K}$, respectively). Therefore,

$$\lambda_{ox} = \lambda_{ox,\,I} - (\lambda_{ox,\,I} - \lambda_{ox,\,K}) X_K$$

where X_K is the proportion of the total iodine bound in the complex used for the calculation of the equilibrium constant.

Even from the few investigations surveyed here, it is clear that the method of examination of positron annihilation can frequently provide basic or supplementary information in the study of the chemical structures of solutions.

Ultrasonic velocity investigations

The velocity of elastic ultrasonic waves in solution is strongly influenced by solute–solvent and solute–solute interactions which are determined by the chemical structure of the solute and solvent molecules. Still, acoustical methods have made only minor contributions to the detailed description of solute–solvent interactions. Ultrasonic velocity measurements are mostly limited to obtaining "hydration numbers" of molecules in aqueous solution [Br 75]. The successful application of acoustical methods to physico-chemical investigation of solutions became possible after development of adequate theoretical approaches and methods for precise ultrasonic velocity measurements in small volumes of liquids [Sa 77, Bu 79].

Sarvazyan *et al.* [Sa 80, Bu 80] presented the results of ultrasonic velocity studies of various nucleosides and nucleotides in aqueous solutions. Their parameter derived from experimental sound velocity values is the relative change in sound velocity per unit concentration. This A value can be used to estimate the dependence of hydration of the molecule on its structure and charge distribution. The changes in A which occur upon protonation can similarly be interpreted. The most significant changes in hydration occur at individual atomic groups (site hydration) and only involve a hydration layer 1–1.5 water molecules thick. The results could be used for the estimation of the solvation of these molecules and gave information of the effect of solvation on its structure and charge distribution. Similar investigations lead to the determination of other solute–solute interactions [He 80]. This procedure used mainly till now for the study of aqueous solutions seem to be applicable in the study of solvation in non-aqueous systems too.

Electron spin echo modulation studies

The geometrical structures of solvated electrons in several organic and aqueous glasses formed by freezing have been elucidated with the help of electron spin echo modulation patterns [Ke 79]. A detailed picture of the water, methanol, ethanol and methyltetrahydrofuran (MTHF) molecular arrangement around excess electrons in these glassy matrices has been obtained.

Solvated electrons appear to be characteristic of small anions with regard to their equilibrium solvation shell structures. For example, the OH bond orientation of water around solvated electrons can be compared with the same water orientation around fluoride ion in hydrated crystals of KF [Be 64]. Electron spin echo modulation studies have been suggested [Ic 80] therefore for the study of the solvation shell structure of anions. Tetracyanoethylene anion ($TCNE^-$) was found to be a suitable model. The investigations were performed in the frozen solutions with CD_3OH, CH_3—OD, CD_3CO and $(CD_3)_2SO$ at 4.2 K temperature. The results indicated that $TCNE^-$ is solvated by four methanol molecules with the methanol molecular dipole oriented toward the anion; the distances from the anion to the hydroxyl and methyl deuterons are 0.59 and 0.38 nm, respectively. $TCNE^-$ is solvated by two molecules of acetone or dimethyl sulphoxide; it is suggested that these solvating molecules are above and below the $TCNE^-$ plane with their molecular dipoles oriented toward the C═C bond in $TCNE^-$.

References

Ab 50 Abragam, A.: Phys. Rev., **79,** 534 (1950).

Ab 51 Abragam, A., Price, M. H. L.: Proc. Roy. Soc., **A205,** 135 (1951); **A206,** 164, 173 (1951).

Ab 55 Abragam, A., Horowitz, J., Price, M. H. L.: Proc. Roy. Soc., **A230,** 169 (1955).

Ab 79 Abboud, J–L. M., Taft, R. W.: J. Phys. Chem., **83,** 412 (1979).

Ad 64 Adamski, P., Kryszewsky, M.: J. Chim. Phys., **1964,** 708.

Ad 68 Addison, C. C., Amos, D. W., Sutton, D.: J. Chem. Soc., **1968,** 2285.

Ad 71 Adams, D. M., Blandamer, M. J., Symons, M. C. R., Waddington, D.: Trans. Faraday Soc., **67,** 611 (1971).

Ah 78 Ahrland, S.: Acta Chem. Scand., **A32,** 607 (1978).

Ah 80 Ahrland, S., Bläuenstein, P., Tagesson, B., Tuhtar, D.: Acta Chem. Scand. **A34,** 265 (1980).

Ah 81 Ahrland, S., Björk, N., Persson, I.: Acta Chem. Scand. **A35,** 67 (1981).

Ak 59 Akishin, P. A., Spiridinov, V. D., Khodchenkov, A. N.: Zhur, Fíz. Khim., **33,** 20 (1959).

Al 67 Alexander, R., Ko, E. C., Mac, Y. C., Parker, A. J.: J. Am. Chem. Soc., **89,** 3703 (1967).

Al 80 Alesbury, C. K., Symons, M. C. R.: J. Chem. Soc. Faraday I, **76,** 244 (1980).

Am 66 Amis, E. S.: Solvent Effects on Reaction Rates and Mechanisms. Academic Press, New York, London, 1966.

An 63 Antsiskina, A. S., Porai-Koshits, M. A.: Dokl. Akad. Nauk SSSR, **143,** 105 (1963).

An 76 Angelos Malliaris, Niarchos, D.: Inorg. Chem., **15,** 1340 (1976).

Ar 66 Arvedson, D. F., Larsen, E. M.: Inorg. Chem., **5,** 779 (1966).

Ar 70 Arnett, E. M., Joris, L., Mitchell, E., Murty, T. S. S. R., Gorry, T. M., Schleyer, P. von R.: J. Am. Chem. Soc., **92,** 2365 (1970).

Ba 37 Badger, P. M., Bauer, S. H.: J. Chem. Phys., **5,** 839 (1937).

Ba 51 Baker, E. B.: Rev. Sci. Instr., **22,** 34 (1951).

Ba 53 Basolo, F., Matoush, W. R.: J. Am. Chem. Soc., **75,** 5663 (1953).

Ba 54 Bayliss, N. S., MacRae, E. G.: J. Phys. Chem., **58,** 1002 (1954).

Ba 59 Ballhausen, C. J., Liehr, A.: J. Am. Chem. Soc., **81,** 538 (1959).

Ba 65 Baddiel, C. B., Tait, M. J., Tait, G. J.: J. Phys. Chem., **69,** 3634 (1965).

Ba 68 Barthel, J.: Angew. Chem., **80,** 253 (1968); Angew, Chem. Int. Ed., **7,** 260 (1968).

Ba 70 Bardin, J. C.: J. Electroanal. Chem., **28,** 157 (1970).

Ba 71 Barthel, J., Behret, H., Schmithals, F.: Ber. Bunsenges. phys. Chem., **75,** 305 (1971).

Ba 76 Barthel, J.: Ionen in nichtwässerigen Lösungen. Steinkopff Verlag, Darmstadt, 1976.

Ba 77 Barthel, J., Wachter, R., Gores, H. J.: Ion-Ion and Ion-Solvent Interactions. The Chem. Soc., Faraday Division, General Discussion 64/7, Oxford, 1977.

Ba 78a Barthel, J., Schmeer, G., Strasser, F., Wachter, R.: Rev. chim. minérale, **75,** 99 (1978).

Ba 78b Barthel, J.: Chem. Ing. Tech., **50,** 259 (1978).

Ba 81 Baltzer, L., Bergman, N., Drakenberg, T.: Acta Chem. Scand. **A35,** 759 (1981).

Be 49 Benesi, H. A., Hildebrand, J. H.: J. Am. Chem. Soc., **71,** 2703 (1949).

Be 61 Bent, H. A.: Chem. Rev., **61,** 275 (1961).

Be 63a Beattie, I. R., Veboner, M.: J. Chem. Soc., **1963,** 38.

Be 63b Beattie, I. R., McQuillan, G. P.: J. Chem. Soc., **1963,** 1519.

Be 64 Beurskens, G., Jeffrey, G. A.: J. Chem. Phys., **41,** 917 (1964).

Be 65 Becconsall, J. K., Hampson, P.: Mol. Phys. **10,** 21 (1965).

Be 67 Bennett, M. J., Cotton, F. A., Weaver, D. L.: Acta Crystallogr., **23,** 788 (1967)

Be 69 Benoit, R. L., Beauchamp, A., Deneux, M.: J. Phys. Chem., **73,** 3268 (1969).

Be 70 Beronius, P., Pataki, L.: J. Am. Chem. Soc., **92,** 4518 (1970).

Be 71 Beronius, P., Pataki, L.: Acta Chem. Scand., **25,** 3705 (1971).

Be 71b Bellamy, L. J., Blandamer, M. J., Symons, M. C. R., Waddington, D.: Trans. Faraday Soc., **67,** 3435 (1971).

Be 73 Benter, G., Schneider, H.: Ber. Bunsenges. Physik. Chem., **77,** 997 (1973).

Be 79 Beronius, P.: Acta Chem. Scand., **A33,** 101 (1979).

Bj 26 Bjerrum, N.: Kgl. Danske Videnskab., **7,** 9 (1926).

Bl 68 Bloor, E. G., Kidd, R. G.: Canad. J. Chem., **46,** 3425 (1968).

Bl 70 Blandamer, M. J., Fox, M. F.: Chem. Rev., **70,** 59 (1970).

Bo 59 Bock, E.: Canad. J. Chem., **37,** 1883 (1959).

Bo 69a Bordewijk, P., Gransch, F., Böttcher, C. J. F.: J. Phys. Chem., **73,** 3255 (1969).

Bo 69b Bowmaker, G. A., Hacobian, S.: Austral. J. Chem., **22,** 2047 (1969).

Bo 72 Bosnich, B., Harrowfield, J. M.: J. Am. Chem. Soc., **94,** 989, 3425 (1972).

Bo 79 Bose, K., Kundu, K. K.: Indian J. Chem., **17A,** 122 (1979).

Br 61 Briegleb, G.: Elektronen-Donator-Akzeptor-Komplexe. Springer Verlag, Berlin, 1961.

Br 63 Brown, B. W., Lingafelter, E. C.: Acta Cryst., **16,** 753 (1963).

Br 65 Brooker, L. G. S., Craig, A. C., Heseltine, D. W., Jenkins, P. W., Lincoln, L. L.: J. Am. Chem. Soc., **87**, 2443 (1965).

Br 70a Bréant, M., Buisson, C., Porteix, M., Sue, J., Terrat, J.: J. Electroanal. Chem., **24**, 409 (1970).

Br 70b Brink, G., Falk, M.: Canad. J. Chem., **48**, 2096 (1970).

Br 71 Braunstein, J., Robbins, G. D.: J. Chem. Educ., **48**, 52 (1971).

Br 72 Brüggemann, R., Reiter, F., Voitlander, L.: Z. Naturforsch., **27a**, 1525 (1972).

Br 75 Braginskaya, F. I., Sadikhova, S. Kh.: Biofizika, **20**, 20 (1975).

Br 76 Brawn, D. R., Findlay, T. J. V., Symons, M. C. R.: J. Chem. Soc. Faraday I. **72**, 1792 (1976).

Br 80 Brittain, H. G.: J. Am. Chem. Soc., **102**, 1207 (1980).

Br 81 Brzezinski, B., Szafran, M.: Organic Magnetic Resonance **15**, 78 (1981).

Bu 59 Bufalini, J., Stern, K. H.: Science, **130**, 1249 (1959).

Bu 60 Buckingham, A. D., Schaffer, T., Schneider, W. G.: J. Chem. Phys., **32**, 1227 (1960).

Bu 61 Bufalini, J., Stern, K. H.: J. Am. Chem. Soc., **83**, 4362 (1961).

Bu 67 Butler, J. N.: Anal. Chem., **39**, 1799 (1967).

Bu 68 Burgers, J., Symons, M. C. R.: Quart. Rev. (London), **22**, 768 (1968).

Bu 69 Butler, R. N., Symons, M. C. R.: Trans. Faraday Soc., **65**, 945, 2559 (1969).

Bu 70 Burger, K., Vértes, A., Nagy-Czakó, I.: Acta Chim. Acad. Sci. Hung., **63**, 115 (1970).

Bu 71a Buchikin, A. P., Goldanskii, V. I., Shantarovich, V. P.: Zhur. Exp. Theor. Fiz., **13**, 624 (1971).

Bu 71b Buchikin, A. P., Goldanskii, V. I., Gagur, A. O.: Zhur. Exp. Theor. Fiz., **60**, 1136 (1971).

Bu 72 Burger, K.: Inorg. Chim. Acta Rev., **6**, 31 (1972).

Bu 73 Burger, K.: Coordination Chemistry: Experimental Methods. Butterworths, London, 1973.

Bu 74a Burger, K., Fluck, E.: Inorg. Nucl. Chem. Letters, **10**, 171 (1974).

Bu 74b Burger, K., Fluck, E., Várhelyi, Cs., Binder, H., Speyer, I.: Z. anorg. allg. Chem., **408**, 304 (1974).

Bu 74c Burger, K., Buvári, Á.: Inorg. Chim. Acta, **11**, 25 (1974).

Bu 75a Burger, K.: Unpublished results. See also: Andrási, E., Barcza, L., Burger, K.: Proc. Euroanal. 2nd Conf. Budapest, 1975, p. 179.

Bu 75b Burger, K., Fluck, E., Binder, H., Várhelyi, Cs.: J. Inorg. Nucl. Chem., **37**, 55 (1975).

Bu 77 Burger, K., Tschismarov, K., Ebel, H.: J. Electron Spectrosc., **10**, 461 (1977).

Bu 78 Burger, K.: J. Electron. Spectrosc. Rel. Phenom., **14**, 405 (1978).

Bu 80 Buckin, V. A., Sarvazyan, A. P., Dudchenko, E. I., Hemmes, P.: J. Phys. Chem., **84**, 696 (1980).

Bu 82 Burger, K., Vértes, A.: XXII International Conference on Coordination Chemistry, Budapest, Hungary, August 23–27, 1982, Abstract of Papers, Vol. 1. p. 77.

Bu 83 Burger, K., Vértes, A., Zay, I.: Inorg. Chim. Acta **76**, L 247 (1983).

Ca 62 Carlson, R. L.: Doctoral Dissertation, University of Illinois, Urbana, Ill., 1962.

Ca 71 Cameron, J. A., Keszthelyi, L., Nagy, G., Kacsóh, L.: Chem. Phys. Letters, **8**, 628 (1971).

Ca 78a Capparelli, A. L., Gill, D. S., Hertz, H. G., Tutsch, R., Weingärtner, H.: J. Chem. Soc. Faraday Trans. I., **74**, 1834 (1978).

Ca 78b Capparelli, A. L., Gill, D. S., Hertz, H. G., Tutsch, R.: J. Chem. Soc. Faraday Trans. I., **74**, 1849 (1978).

Ca 80 Calderwood, J. H.: Nature, **285**, 616 (1980).

Ch 73a Chantooni, M. K., Kolthoff, I. M.: J. Phys. Chem., **77**, 1 (1973).

Ch 73b Chamberlain, J., Chantry, G. W. (Eds.): High Frequency Dielectric Measurements. I.P.C. Science and Technology Press, Guilford, 1973.

Cl 55 Clark, H. C., Odell, A. L.: J. Chem. Soc., **1955**, 3431.

Cl 73 Clausen, A., El-Harakany, A. A., Schneider, H.: Ber. Bunsenges. Physik. Chem., **77**, 994 (1973).

Co 54 Conway, B. E., Bockris, J. O'M.: Modern Aspects of Electrochemistry, Vol. I. Butterworths, London, 1954, Chapter II.

Co 60a Cotton, F. A., Fackler, J. P.: J. Am. Chem. Soc., **82**, 5005 (1960).

Co 60b Cotton, F. A., Francis, R., Horrocks, W. D.: J. Phys. Chem., **64**, 1534 (1960).

Co 60c Cotton, F. A., Francis, R.: J. Am. Chem. Soc., **82,** 2986 (1960).
Co 61a Cotton, F. A., Francis, R.: J. Inorg. Nucl. Chem., **17,** 62 (1961).
Co 61b Cotton, F. A., Horrocks, W. D.: Spectrochim. Acta, **17,** 134 (1961).
Co 80 Cotton, F. A., Felthouse, T. R.: Inorg. Chem., **19,** 2347 (1980).
Cr 68 Criss, M., Lukshe, E.: J. Phys. Chem., **72,** 2966 (1968).
Cr 70 Crossley, J., Glasser, L., Smyth, C. P.: J. Chem. Phys., **52,** 6203 (1970).
Cr 71a Crossley, J.: Canad. J. Chem., **49,** 712 (1971).
Cr 71b Crossley, J.: J. Phys. Chem., **75,** 1790 (1971).
Cu 65 Curtis, N. F., Curtis, Y. M.: Inorg. Chem., **4,** 804 (1965).
Cu 67 Currier, W. F., Weber, J. H.: Inorg. Chem., **6,** 1539 (1967).
Da 68 Dannhauser, W.: J. Chem. Phys., **48,** 1918 (1968).
Da 70 Dannhauser, W., Fluekinger, A. F.: Phys. Chem. Liquids, **2,** 37 (1970).
Da 71 Daumezon, P., Heitz, R.: J. Chem. Phys., **55,** 5704 (1971).
Da 77 Day, M. C.: Pure Appl. Chem., **49,** 75 (1977).
De 53 Deutsch, M.: Progr. Nucl. Phys. **3,** 131 (1953).
De 55 Denison, J. T., Ramsay, J. B.: J. Am. Chem. Soc., **77,** 2615 (1955).
De 59 de Maine, P. A. D., Koubek, E.: J. Inorg. Nucl. Chem., **11,** 329 (1959).
De 68 DeSando, R. J., Brown, G. H.: J. Phys. Chem., **72,** 1088 (1968).
De 71 De Vries, J. I. K. F., Trooster, J. M., De Boer, E.: Inorg. Chem., **10,** 81 (1971).
De 81 Delville, A., Detellier, C., Gerstman, A., Laszló, P.: Helv. Chim. Acta **64,** 556 (1981).
Dé 65a Dézsi, I., Keszthelyi, L., Molnár, B., Pócs, L.: Phys. Letters, **18,** 28 (1965).
Dé 65b Dézsi, I., Keszthelyi, L., Pócs, L., Korecz, L.: Phys. Letters, **14,** 14 (1965).
Di 63 Dimroth, K., Reichardt, C., Siepmann, T., Bohlmann, F.: Liebigs Ann. Chem., **661,** 1 (1963).
Di 69 Diebler, H., Eigen, M., Illgenfritz, G., Maas, G., Winkler, L.: Pure Appl. Chem., **20,** 93 (1969).
Dr 61 Drago, R. S., Meek, D. W.: J. Phys. Chem., **65,** 1446 (1961).
Dr 63 Drago, R. S., Wajland, B. B., Carlson, R. L.: J. Am. Chem. Soc., **85,** 3125 (1963).
Dr 65 Drago, R. S., Purcell, K. F.: Coordinating Solvents. In: Non-Aqueous Solvent Systems (Ed. Waddington, D. C.). Academic Press, London and New York, 1965, p. 241.
Dr 69 Drago, R. S., Epley, T. D.: J. Am. Chem. Soc., **91,** 2883 (1969).
Dr 70 Drago, R. S., O'Bryan, N., Vogel, G. C.: J. Am. Chem. Soc., **92,** 3924 (1970).
Dy 61 Dyrssen, D., Hennichs, M.: Acta Chem. Scand., **15,** 47 (1961).
Dy 65 Dyrssen, D., Petkovic, D.: Acta Chem. Scand., **19,** 653 (1965).
Eb 78 Ebeling, W., Feistel, R., Kelbg, G., Sändig, R.: J. Nonequil. Thermodyn., **3,** 11 (1978).
Ed 67 Edwards, J. O., Goetsch, R. J., Strital, J. A.: Inorg. Chim. Acta, **1,** 360 (1967).
Em 66 Emsley, J. W., Phillips, L.: Mol. Phys., **11,** 437 (1966).
Em 71 Emsley, J. W., Phillips, L.: Prog. NMR Spectroscopy, **7,** 1 (1971).
Ep 67 Epley, T. D., Drago, R. S.: J. Am. Chem. Soc., **89,** 5770 (1967).
Er 70 Erlich, R. H., Poack, E., Popov, A. I.: J. Am. Chem. Soc., **92,** 4489 (1970).
Er 71 Erlich, R. H., Popov, A. I.: J. Am. Chem. Soc., **93,** 5620 (1971).
Ev 53 Evans, M. G., Nancollas, G. N.: Trans. Faraday Soc., **49,** 363 (1953).
Ev 65 Evans, L. C., Lo, G. Y. S.: Spectrochim. Acta, **21,** 1033 (1965).
Fa 59 Falkenhagen, H., Kelbg, G.: in: Modern Aspects of Electrochemistry, No. 2 (Ed.: Bockris, J. O'M.). Butterworths, London, 1959.
Fa 65 Faragó, M. E., James, J. M.: Chem. Comm., **1965,** 470.
Fa 66 Faragó, M. E., James, J. M., Trew, V. C. G.: Proc. 9th ICCC, St. Moritz-Bad, 1966, p. 225.
Fa 68 Fajer, J., Linschitz, H.: J. Inorg. Nucl. Chem., **30,** 2259 (1968).
Fa 70 Falk, K. E., Ivanova, E., Roos, B., Vängard, T.: Inorg. Chem., **9,** 556 (1970).
Fa 71 Falkenhagen, H., Ebeling, W.: in: Ionic Interaction (Ed.: Pertucci, S.). Academic Press, Vol. I, p. 1, New York, 1971; Falkenhagen, H., Ebeling, W., Kraeft, W. D.: ibid., p. 61.

Fa 76 Fawcett, W. R., Krygowski, T. M.: Canad. J. Chem., **54,** 3283 (1976).
Fe 56 Ferrell, R. A.: Rev. Mod. Phys., **28,** 299 (1956).
Fe 57 Ferrell, R. A.: Phys. Rev., **108,** 167 (1957).
Fe 66 Fenn, R. H.: Acta Cryst., **20,** 2024 (1966).
Fe 73 Fernandez-Prini, R.: in: Physical Chemistry of Organic Solvent Systems (Eds.: Covington, A. K., Dickinson, T.). Plenum Press, London, 1973.
Fr 42 French, H. S., Magee, M. Z., Sheffield, E.: J. Am. Chem. Soc., **64,** 1924 (1942).
Fr 57 Frank, H. S., Wen, W. Y.: Discuss. Faraday Soc., **24,** 133 (1957).
Fr 59 Frasson, E., Panattoni, C., Zannetti, R.: Acta Cryst., **12,** 1027 (1959).
Fr 60 Frasson, E., Panattoni, C.: Acta Cryst., **13,** 893 (1960).
Fr 65 Franconi, C., Dejak, C., Conti, F.: In: Nuclear Magnetic Resonance in Chemistry (Ed.: Presces, B.). Academic Press, New York, 1965, p. 363.
Fr 68a Frankel, L. S., Stengle, T. R., Langford, C. H.: Canad. J. Chem., **46,** 3183 (1968).
Fr 68b French, M. J., Wood, J. L.: J. Chem. Phys., **49,** 2358 (1968).
Fr 72 Fratiello, A.: Prog. Inorg. Chem., Vol. 17 (Ed.: Edwards, J. O.). Interscience, New York, 1972, pp. 57–92.
Fu 58 Fuoss, R. M.: J. Am. Chem. Soc., **80,** 5059 (1958).
Fu 59 Fuoss, R. M., Accascina, F.: Electrolytic Conductance. Interscience, New York, 1959.
Fu 62 Fuoss, R. M., Onsager, L.: J. Phys. Chem., **66,** 1722 (1962); **67,** 621, 628 (1963); **68,** 1 (1964).
Fu 65 Fuoss, R. M., Onsager, L., Skinner, J. F.: J. Phys. Chem., **69,** 2581 (1965).
Fu 68 Furlani, C.: Coord. Chem. Rev., **3,** 141 (1968).
Fü 24 Fürth, R.: Physik. Z., **25,** 676 (1924).
Ga 50 Gast, T.: Z. angew. Phys., **2,** 41 (1950).
Ga 65 Garg, S. K., Smyth, C. P.: J. Phys. Chem., **69,** 1294 (1965).
Ga 67a Gaizer, F., Beck, M. T.: J. Inorg. Nucl. Chem., **29,** 21 (1967).
Ga 67b Garg, S. K., Smyth, C. P.: J. Chem. Phys., **46,** 373 (1967).
Ga 69 Gaizer, F., Johansson, G.: Magy. Kém. Foly., **75,** 553 (1969).
Ga 79 Gaizer, F.: Coord. Chem. Rev., **27,** 195 (1979).
Gi 60 Gilkerson, W. R., Stamm, R. E.: J. Am. Chem. Soc., **82,** 5295 (1960).
Gi 76 Gill, D. S., Hertz, H. G., Tutsch, R.: J. Chem. Soc. Faraday Trans., I **72,** 1559 (1976).
Gl 64 Glueckauf, E.: Trans. Faraday, Soc., **60,** 1637 (1964).
Gl 72 Glasser, J., Crossley, J., Smyth, C. P.: J. Chem. Phys., **57,** 3977 (1972).
Go 63 Goodgame, D. M. L., Venanzi, L. M.: J. Chem. Soc., **1963,** 616, 5909.
Go 65 Goldanskii, V. I., Okhlobystin, O. Yu., Rochev, V. Ya., Khrapov, V. V.: J. Organometal. Chem., **4,** 160 (1965).
Go 68a Goldanskii, V. I.: Atomic Energy Review, Vol. VI. IAEA, Vienna, 1968, p. 1.
Go 68b Goldanskii, V. I.: Fizicheskaya Khimia Positrona. Tad Nauk, Moscow, 1968.
Go 70 Goldanskii, V. I., Rochev, V. Ya., Khrapov, V. V., Kravtsov, D. N., Rokhlina, E. M.: Dokl. Akad. Nauk SSSR, **191,** 134 (1970).
Go 71a Goldanskii, V. I., Firsov, V. G.: Ann. Rev. Phys. Chem., **22,** 209 (1971).
Go 71b Gollogly, J. R., Hawkins, C. J., Beattie, J. K.: Inorg. Chem., **10,** 317 (1971).
Go 72 Goldanskii, V. I., Lévay, B., Shantarovich, V. P., Ranogajec-Komor, M., Vértes, A.: Radiochem. Radioanal. Letters, **12,** 289 (1972).
Go 81 Golding, R. M., Pascual, R. O., Suvanprakorn, C.: in: Advances in Solution Chemistry. Plenum Press, New York, London, 1981, p. 129.
Gr 56 Griffith, J. S.: J. Inorg. Nucl. Chem., **2,** 1229 (1956).
Gr 60a Griffith, J. S.: Molec. Phys., **3,** 477 (1960).
Gr 60b Griffiths, T. R., Symons, M. C. R.: Mol. Phys., **3,** 90 (1960).
Gr 64 Green, J., Lee, J.: Positronium Chemistry. Academic Press, New York, London, 1964.

Gr 68 Gray, F., Cook, C. F., Sturm, G. P.: J. Chem. Phys., **48,** 1145 (1968).
Gr 70 Griffiths, T. R., Wijayanayake, R. H.: Trans. Faraday Soc., **66,** 1563 (1970).
Gr 73 Greenberg, M. S., Bochner, R. L., Popov, A. I.: J. Phys. Chem., **77,** 2449 (1973).
Gr 79 Griffiths, G., Thornton, D. A.: J. Mol. Struct., **52,** 39 (1979).
Gu 60 Guggenheim, E. A.: Trans. Faraday Soc., **56,** 1159 (1960).
Gu 65 Gulik-Krzywicki, T., Kecki, Z.: Rocz. Chem., **39,** 1281 (1965).
Gu 70 Gutmann, V., Wegleitner, K. H.: Monatshefte für Chemie, **101,** 1532 (1970).
Gu 73 Guryanova, E. N., Goldstein, I. P., Romm, I. P.: Donorno-Akceptornaya Svyaz, Khimiya, Moscow, 1973.
Gu 74 Gudlin, D., Schneider, H.: J. Magn. Res., **16,** 362 (1974).
Ha 22 Hantzsch, A.: Ber. Deutsch. Chem. Ges., **55,** 953 (1922).
Ha 54a Ham, J.: J. Am. Chem. Soc., **76,** 3881 (1954).
Ha 54b Hall, J. L., Gibbson, J. A., Critchfield, F. E., Phillips, H. O., Seibert, C. B.: Anal. Chem., **26,** 835 (1954).
Ha 55 Hahn, H., Frank, G., Klinger, W.: Z. anorg. allg. Chem., **279,** 271 (1955).
Ha 58a Harned, H. S., Owen, B. B.: The Physical Chemistry of Electrolytic Solutions. Reinhard, New York, 1958.
Ha 58b Hasted, J. B., Roderick, G. W.: J. Chem. Phys., **29,** 17 (1958).
Ha 65 Hawkins, C. J., Larsen, E.: Acta Chem. Scand., **19,** 1969 (1965).
Ha 66 Hamilton, W. C., Spratley, R.: Acta Cryst. Sect. A, **21,** 142 (1966).
Ha 67a Hammaker, R. M., Clegg, R. M.: J. Mol. Spectr., **22,** 109 (1967).
Ha 67b Haster, R. E., Plane, R. A.: Spectrochim. Acta, **23A,** 2289 (1967).
Ha 69 Haragucki, H., Fujiware, S. J.: J. Phys. Chem., **73,** 3467 (1969).
Ha 72 Hadži, D.: Chimia, **26,** 7 (1972).
Ha 76 Hawkins, C. J., Peachey, R. M.: Austral. J. Chem., **29,** 33 (1976).
Ha 77 Hawkins, C. J., Lawrence, G. A., Peachey, R. M.: Austral. J. Chem., **30,** 2115 (1977).
Ha 79 Haberfield, P., Lux, M. S., Jasser, I., Rosen, D.: J. Am. Chem. Soc., **101,** 645 (1979).
Ha 82 Happer, D. A. R.: Aust. J. Chem. **35,** 21 (1982).
He 71 Heitz, R., Daumezon, P.: J. Chim. Phys., **68,** 1 (1971).
He 71a Hertz, H. G.: Magnetische Kernresonanzuntersuchungen zur Struktur von Elektrolytlösungen: Theorie der Elektrolyte (Ed.: von Falkenhagen, H.). S. Hirzel Verlag, Stuttgart, 1971.
He 79 Henrichs, P. M., Sheard, S., Ackerman, J. J. H., Maciel, G. E.: J. Am. Chem. Soc., **101,** 3222 (1979).
He 80 Hemmes, P., Mayevski, A. A., Buckin, V. A., Sarvazyan, A. P., J. Phys. Chem., **84,** 699 (1980).
Hi 69 Hill, N. E., Vaughan, W. E., Price, A. H., Davies, M.: Dielectric Properties and Molecular Behaviour. Van Nostrand, London, New York, Toronto, 1969.
Hi 71 Hindman, J. C.: Chem. Phys., **44,** 4583 (1966).
Ho 63 Holm, R. H., Swaminathan, K.: Inorg. Chem., **2,** 181 (1963).
Ho 66 Hoskins, B. F., Martin, L. R., White, A. H.: Nature (London), **211,** 627 (1966).
Ho 69 Hohn, E. G., Olander, J. A., Day, M. C.: J. Phys. Chem., **73,** 3880 (1969).
Ho 72 Horne, R. A.: Water and Aqueous Solutions. Wiley-Interscience, New York, 1972.
Ho 77 Holz, M., Weingärtner, H., Hertz, H. G.: J. Chem. Soc. Trans. Faraday I, **73,** 71 (1977).
Ho 78 Holz, M.: J. Chem. Soc. Faraday Trans, I, **74,** 644 (1978).
Ho 82 Holton, D. M., Murphy, D.: J. Chem. Soc. Faraday I, **78,** 1223 (1982).
Hö 69 Höhn, E. G., Olander, J. A., Day, M. C.: J. Phys. Chem., **73,** 3880 (1969).
Hu 54 Huber, O.: Z. angew. Phys., **6,** 9 (1954).
Hu 63 Hulme, R.: J. Chem. Soc., **1963,** 1524.
Hu 71 Hunt, J. P.: Coord. Chem. Rev., **7,** 1 (1971).
Hy 62 Hyne, J. B., Levy, R. M.: Canad. J. Chem., **40,** 692 (1962).

Ic 80 Ichikawa, T., Kevan, L.: J. Chem. Phys., **72,** 2995 (1980).

Im 65 Imanov, L. M., Abbasov, Ya. M.: Zh. Fiz. Khim., **39,** 3044 (1965).

Ja 72 Jadzyn, J., Malecki, J.: Acta Phys. Polon., **A41,** 599 (1972).

Ja 77a Jackson, S. E., Strauss, I. M., Symons, M. C. R.: J. Chem. Soc. Chem. Commun., **174** (1977).

Ja 77b Jakusek, E., Sobczyk, L.: in: Dielectric and Related Molecular Processes (Ed.: Davies, M.). Vol. 3, Chem. Soc., London, 1977.

Ja 79 Jackson, M. D., Gilkerson, W. R.: J. Am. Chem. Soc., **101,** 328 (1979).

Je 48 Jeffard, N. G.: Opt. Soc., **38,** 35 (1948).

Je 67 Jeffrey, G. A., Vlasse, M.: Inorg. Chem., **6,** 396 (1967).

Jo 66 Joesten, M. D., Drago, R. S.: J. Am. Chem. Soc., **88,** 1617 (1966).

Jo 69 Johari, G. P., Smyth, C. P.: J. Am. Chem. Soc., **91,** 6215 (1969).

Jø 62 Jørgensen, C. K.: Absorption Spectra and Chemical Bonding in Complexes. Pergamon Press, Oxford, 1962.

Jø 71 Jørgensen, C. K.: Chimia, **25,** 213 (1971).

Jø 72 Jørgensen, C. K., Berthou, H.: Photo-electron Spectra Induced by X-rays of above 600 Nonmetallic Compounds Containing 77 Elements. Danske Vid. Selks. Matfys. Medd. (Coppenhagen), **38,** 15 (1972).

Ju 71a Junghähnel, G., Götz, G.: Z. Chem., **11,** 354 (1971).

Ju 71b Justice, J. C.: Electrochim. Acta, **16,** 701 (1971).

Ju 73 Junghähnel, G., Regenstein, W.: Z. Chem., **13,** 264 (1973).

Ju 75a Justice, M. C., Justice, J. C.: Colloque International du C.N.R.S., **246,** 241 (1975).

Ju 75b Justice, J. C.: J. Phys. Chem., **79,** 454 (1975).

Ju 77 Justice, M. C., Justice, J. C.: Ion-Ion and Ion-Solvent Interactions. The Chem. Soc., Faraday Division, General Discussion 64/5, Oxford, 1977.

Ká 74a Kálmán, E., Lengyel, S., Pálinkás, G., Haklik, L., Eke, A.: Water as Liquid and Solvent. Proc. International Symposium, Marburg, Physik Verlag, Weinheim, 1974.

Ká 74b Kálmán, E., Lengyel, S., Haklik, L., Eke, A.: J. Appl. Cryst., **7,** 442 (1974).

Ka 80 Kalidas, C., Sivaprasad, P.: Indian J. Chem., **18,** 532 (1980).

Ke 62 Kecki, Z.: Spectrochim. Acta, **18,** 1164 (1962).

Ke 64 Kecki, Z., Witanowski, J.: Rocz. Chem., **38,** 691 (1964).

Ke 68 Kebarle, P.: Advan. Chem. Ser., **72,** 24 (1968).

Ke 70 Kecki, Z., Wojtczak, J.: Roch. Chem., **44,** 846 (1970).

Ke 73 Kecki, Z.: Advances in Molecular Relaxation Processes, **5,** 137 (1973).

Ke 79a Kebarle, P., Dawidson, W. R., Sunner, J., Meza-Höjer, S.: Pure Appl. Chem., **51,** 63 (1979).

Ke 79b Kevan, L., in: Time Domain Electron Spin Resonance, (Ed.: L. Kevan, R. N. Schwartz), Wiley-Interscience, New York, 1979, Chap. 8.

Kh 65 Khrapov, V. V.: Candidate's dissertation. Chemical Physical Institute of the Academy of Sciences of the Soviet Union. Moscow, 1965.

Ko 65a Kosower, E. M., Klinedinst Jr., P. E.: J. Am. Chem. Soc., **78,** 3493 (1956).

Ko 65b Kosower, E. M., Burbach, J. C.: J. Am. Chem. Soc., **78,** 5838 (1956).

Ko 70 Komor, M., Vértes, A., Dézsi, I., Ruff, I.: Acta Chim. Acad. Sci. Hung., **66,** 285 (1970).

Ko 72 Kohler, F.: The Liquid State. Verlag Chemie, Berlin, 1972.

Kö 75 Kőrös, E.: Molecular Complexes. (In Hungarian), Akadémiai Kiadó, Budapest, 1975.

Kr 56 Kraus, C. A.: J. Phys. Chem., **60,** 129 (1956).

Kr 67a Krumgal, B. S., Mischenko, K. P., Ionin, B. I.: Russ. J. Phys. Chem., **41,** 1045 (1967).

Kr 67b Kryman, G. N., Pirsonius, C. V. F. T., Pirsonius, M. C.: Z. physik. Chem., **43,** 213 (1967).

Kr 68 Kravchenko, E. A., Maksutin, Yu. K., Guryanova, E. N., Semin, G. K.: Izv. Akad. Nauk SSSR Ser. Khim., **6,** 1271 (1968).

Kr 73 Krishna Bhandary, K., Manohar, W.: Acta Crystallogr., B, **29,** 1093 (1973).
Kr 75 Krygowski, T. M., Fawcett, W. R.: J. Chem. Soc., **97,** 2143 (1975).
Ku 67 Kuriyama, K., Iwata, T., Moriyama, M., Ishikawa, M., Minato, H., Takeda, K.: J. Chem. Soc., *C*, 420 (1967).
La 66 Laszlo, P., Prog. NMR Spectroscopy, **3,** 231 (1966).
La 73 La Mar, G. N., Horrocks Jr., D. W., Holm, R. H. (Eds.): NMR of Paramagnetic Molecules: Principles and Applications. Academic Press, New York, 1973.
La 81 Laszlo, P., Cornelia, A., Delville, A.: in: Advances in Solution Chemistry. Plenum Press, New York, London, 1981, p. 175.
Le 66 Leurs, D. C., Iwamoto, R. I., Klinberg, J.: Inorg. Chem., **5,** 201 (1966).
Le 70 Le Demézet, M., Madec, C., L'Her, M.: Bull. Soc. Chim. France, **1970,** 365.
Le 73 Legrand, M.: in: Fundamental Aspects and Recent Developments in Optical Rotatory Dispersion and Circular Dichroism (Eds.: Ciardelli, F., Salvadori, P.). Heyden, London, 1973, pp. 285–306.
Le 74 Lestrade, J. C., Badiali, J. P., Cachet, H.: in: Dielectric and Related Processes (Ed.: Davies, M.) The Chemical Society, London, 1974.
Lé 72 Lévay, B., Hautojärvi, P.: Radiochem. Radioanal. Lett., **10,** 309 (1972).
Lé 73a Lévay, B., Vértes, A., Hautojärvi, P.: J. Phys. Chem., **77,** 2229 (1973).
Lé 73b Lévay, B., Vértes, A.: Radiochem. Radioanal. Lett., **14,** 227 (1973).
Lé 74 Lévay, B., Vértes, A.: J. Phys. Chem., **78,** 2526 (1974).
Lé 76 Lévay, B., Vértes, A.: J. Phys. Chem., **80,** 37 (1976).
Li 71 Lincoln, S. F.: Coord. Chem. Rev., **6,** 309 (1971).
Li 74 Lincoln, S. F., West, R. J.: J. Am. Chem. Soc., **96,** 400 (1974).
Li 75 Liszi, J.: Acta Chim. Acad. Sci. Hung., **87,** 215 (1975).
Li 77 Lincoln, S. F.: Progress in Reaction Kinetics **9,** 1 (1977).
Lu 58 Lund, H.: Acta Chem. Scand., **12,** 298 (1958).
Ma 58a Maki, G.: J. Chem. Phys., **28,** 651 (1958); **29,** 1129 (1958).
Ma 58b Mandel, M., Jenard, A.: Bull. Soc. Chim. Belg., **67,** 575 (1958).
Ma 62 Mason, S. F.: Mol. Phys., **5,** 343 (1962).
Ma 64b Matwiyoff, N. A., Drago, R. S.: Inorg. Chem., **3,** 337 (1964).
Ma 65 Majumdar, A. K., Bhattacharyya, B. C.: J. Inorg. Nucl. Chem., **27,** 143 (1965).
Ma 66 Maciel, G. E., Hancock, J. K., Lafferty, L. F., Muller, P. A., Musker, W. K.: Inorg. Chem., **5,** 554 (1966).
Ma 67a Maxey, B. W., Popov, A. I.: J. Am. Chem. Soc., **89,** 2230 (1967).
Ma 67b Martin, R. L., White, A. H.: Inorg. Chem., **6,** 712 (1967).
Ma 68 Maxey, B. W., Popov, A. I.: J. Am. Chem. Soc., **90,** 4470 (1968).
Ma 69 Maksyutin, Yu. K., Baushkina, B. A., Guryanova, Ye. N., Semin, G. K.: Theor. Chim. Acta, **14,** 48 (1969).
Ma 70a Maksyutin, Yu. K., Guryanova, Ye. N., Semin, G. K.: Usp. Khim., **39,** 727 (1970).
Ma 70b Mayer, M., Gutmann, V.: Mh. Chem., **101,** 997 (1970).
Ma 73 Maksyutin, Yu. K., Guryanova, E. N., Kravchenko, E. A., Semin, G. K.: J. Chem. Soc. Dalton, **13,** 429 (1973).
Ma 75 Mayer, U., Gutmann, V., Gerger, W.: Monatshefte Chem., **106,** 1235 (1975).
Ma 79a Matsubara, T., Efrima, S., Metiw, H. I., Ford, P. C.: J. Chem. Soc. Faraday Trans. II., **75,** 390 (1979).
Ma 79b Mayer, U.: Pure Appl. Chem., **51,** 1697 (1979).
Mc 62 McDuffic, G. E., Litovitz, T. A.: J. Chem. Phys., **37,** 1699 (1962).
Mc 63 McDuffic, G. E., Litovitz, T. A.: J. Chem. Phys., **39,** 729 (1963).

Mc 66 McClung, D. A., Dalton, L. R., Brubaker, C. H.: Proc. 9th ICCC, St. Moritz-Bad, 1966, p. 406.

Mc 70 McKinney, W. J., Popov, A. I.: J. Phys. Chem., **74,** 535 (1970).

Me 60 Meek, D. W., Straub, D. K., Drago, R. S.: J. Am. Chem. Soc., **82,** 6013 (1960).

Me 80a Mejean, T., Forel, T., Bourgeois, M. T., Jacon, M.: J. Chem. Phys., **72,** 687 (1980).

Me 80b Merbach, A. E., Moore, P., Howarth, O. W., McAteer, C. H.: Inorg. Chim. Acta, **39,** 129 (1980).

Mi 63a Minc, S., Kurowski, S.: Spectrochim. Acta, **19,** 339 (1963).

Mi 63b Minc, S., Kecki, Z., Gulik-Krziwicki, T.: Spectrochim. Acta, **19,** 353 (1963).

Mi 67 Millen, W. A., Watts, D. W.: J. Am. Chem. Soc., **89,** 6858 (1967).

Mi 69 Middelhock, J., Böttcher, C. J. F.: in: Symposium on Molecular Relaxation Processes. The Chemical Society, London, 1969. Special Publication No. 20, p. 69.

Mi 77 Miksche, G., Miksche, H., Murauer, H., Persy, K.: J. Electr. Spectrosc., **10,** 423 (1977).

Mo 63 Moscowitz, A., Wellman, K. M., Djerassi, C.: Proc. Nat. Acad. Sci. U.S.A., **50,** 799 (1963).

Mo 67 Movius, W. G., Matwiyoff, N. A.: Inorg. Chem., **6,** 347 (1967).

Mo 68 Movius, W. G., Matwiyoff, N. A.: J. Phys. Chem., **72,** 3063 (1968).

Mo 71 Moolel, H., Schneider, H.: Z. Phys. Chem. N. F., **74,** 237 (1971).

Mö 58 Mössbauer, R. L.: Z. Phys., **151,** 124 (1958); Naturwiss., **45,** 538 (1958).

Na 70a Narten, A. H.: X-Ray Diffraction Data on Liquid Water in the Temperature Range 4 °C–200 °C. ORBL-4578, 1970.

Na 70b Nagy, G., Dézsi, I.: J. Thermal Anal., **2,** 159 (1970).

Na 70c B. Nagy, S.: Dielektrometria. Műszaki Könyvkiadó, Budapest, 1970.

Na 71 Narten, A. H., Levy, H. A.: J. Chem. Phys., **55,** 2263 (1971).

Na 72 Narten, A. H.: J. Chem. Phys., **56,** 5681 (1972).

Na 77 Naseer Ahmad, Day, M. C.: J. Am. Chem. Soc., **99,** 941 (1977).

No 70 Nozari, M. S., Drago, R. S.: J. Am. Chem. Soc., **92,** 7086 (1970).

Ny 64 Nyberg, S. C., Wood, J. S.: Inorg. Chem., **3,** 468 (1964).

Oe 62 Oehme, F.: Dielektrische Messmethoden. Verlag Chemie, Weinheim, 1962.

Oi 79 Oibrian, D. H., Russel, C. R., Hart, A. J.: J. Am. Chem. Soc., **101,** 633 (1979).

On 79 Ono, K., Konami, H., Murakami, K.: J. Phys. Chem., **20,** 2665 (1979).

Oo 54 Ooshika, Y.: J. Phys. Soc., Japan, **9,** 594 (1954).

Op 33 Oplatka, G.: Physik. Z., **43,** 296 (1933).

Or 71 Ormondroyd, S., Phillpott, E. A., Symons, M. C. R.: Trans. Faraday Soc., **67,** 1253 (1971).

Ös 68 Österberg, R., Sjöberg, B.: Acta Chem. Scand., **22,** 689 (1968); see also Proc. 13th ICCC, Toronto, p. 578.

Pa 68 Partenheimer, W., Epley, T. D., Drago, R. S.: J. Am. Chem. Soc., **90,** 3886 (1968).

Pa 71 Page, D. I., Powles, J. G.: Mol. Phys., **21,** 901 (1971).

Pa 75 Pawelka, Z., Sobczyk, L.: Roczniki Chem., **49,** 1388 (1975).

Pa 76a Pataki, L.: Magy. Kém. Foly., **82,** 271 (1976).

Pa 76b Paul, R., Fuller, G. G.: J. Chem. Phys., **64,** 3809 (1976).

Pa 79a Paul, R.: J. Chem. Phys., **70,** 61 (1979).

Pa 79b Paul, R.: J. Chem. Phys., **70,** 70 (1979).

Pe 62a Perelygin, I. S.: Opt. Spectr. (USSR), **13,** 194 (1962).

Pe 62b Perelygin, I. S.: Opt. Spectr. (USSR), **13,** 198 (1962).

Pe 64 Perelygin, I. S.: Opt. Spectr. (USSR), **16,** 21 (1964).

Pe 66 Pettit, L. D., Bruckenstein, S.: J. Am. Chem. Soc., **88,** 4783 (1966).

Pe 68 Perelygin, I. S., Izosimova, S. V., Kessler, Yu. M.: Zhur. Strukt. Khim., **9,** 390 (1968).

Pe 69 Pelah, I., Ruby, S.: J. Chem. Phys., **51,** 383 (1969).

Pe 73a Petrosyan, V. S., Yashina, N. S., Reutov, O. A., Bryuchova, E. V., Semin, G. K.: J. Organometal. Chem., **52,** 321 (1973).

Pe 73b Petrosyan, V. S., Yashina, N. S., Sacharova, S. G., Reutov, O. A., Rochev, V. Ya., Goldanskii, V. I.: J. Organometal. Chem., **52,** 333 (1973).

Po 54 Pond, T. A.: Phys. Rev., **93,** 478 (1954).

Po 56 Posey, F. A., Taube, H.: J. Am. Chem. Soc., **78,** 15 (1956).

Po 64 Pominov, S. I., Gadziev, A. Z.: Itogovaja Nauchnaja Konferencija Kazanskogo Universiteta za 1963, 31 (1964).

Po 67 Positron Annihilation, Proceedings of the Conference held at Wayne State University, 27–29 July, 1965 (Eds. Stewart, A. T., Roelling, R. O.). Academic Press, New York, London, 1967.

Po 72 Powles, J. G., Dore, J. C., Page, D. I.: Mol. Phys., **24,** 1025 (1972).

Po 79 Popov, A. I.: Pure Appl. Chem., **51,** 101 (1979).

Pr 66 Price, E.: Solvation of Electrolytes and Solution Equilibria. In: The Chemistry of Non-Aqueous Solvents, Vol. I. Academic Press, New York, 1966.

Pr 71 Proc. 2nd Internatl. Conf. Positron Annihilation, Kingstone, Ontario, 1971.

Pr 75 Proc. 3rd Internatl. Conf. Positron Annihilation, Otaniemi, Finland (August 1973) (Eds. Hautojärvi, P., Seeger, A.). Springer Verlag, Berlin, Heidelberg, New York, 1975.

Pr 79 Proc. 5th Internatl. Conf. Positron Annihilation, Lake Yamanata, Japan, 1979.

Pu 58 Pullin, A. D. E., Pollock, J. McC.: Trans. Faraday Soc., **54,** 11 (1958).

Pu 66 Purcell, K. F., Drago, R. S.: J. Am. Chem. Soc., **88,** 919 (1966).

Pu 67 Purcell, K. F., Drago, R. S.: J. Am. Chem. Soc., **89,** 2874 (1967).

Pu 69 Purcell, K. F., Stikeleather, J. A., Brunk, S. D.: J. Am. Chem. Soc., **91,** 4019 (1969).

Pu 65 Pungor, E.: Oscillometry and Conductometry. Publ. Hung. Acad. Sci., Budapest, 1965.

Ra 59 Rampolla, R. W., Miller, R. C., Smith, C. P.: J. Chem. Phys., **30,** 556 (1959).

Ra 73 Ratajczak, H., Orville-Thomas, W. J.: J. Chem. Phys., **58,** 911 (1973).

Re 74 Regis, A., Loupy, J., Corset, J., Josien, M. L.: Abstracts 4th ICNAS (Ed. Gutmann, V.). Vienna, 1974, p. 13.

Re 75 Regis, A., Corset, J.: Chem. Phys. Letters, **32,** 462 (1975).

Ri 77 Richardson, F. S., Riehl, J. P.: Chem. Rev., **77,** 773 (1977).

Ro 67 Roellig, R. O., see, p. 127 in Ref. Po 67.

Ro 68 Rochev, V. Ya.: Candidate's dissertation, Chemical Physical Research Institute of the Academy of Sciences of the Soviet Union, Moscow, 1968.

Ro 70 Rocher, J. P., Van Huog, P.: Chim. Phys., **67,** 211 (1970).

Ro 72 Rockenbauer, A., Budó-Záhonyi, E., Simándi, L. I.: J. Coord. Chem., **2,** 53 (1972).

Ro 76 Rodehüser, L., Schneider, H.: Z. Phys. Chem. N. F., **100,** 119 (1976).

Ru 70 Rumbaut, N., Peeters, A.: Bull. Soc. Chim. Belg., **79,** 45 (1970).

Ru 71 Ruby, S. L., Zabransky, B. J., Stevens, G.: J. Chem. Phys., **54,** 4559 (1971).

Sa 60 Sacconi, L., Cini, R., Ciampolini, M., Maggio, F.: J. Am. Chem. Soc., **82,** 3487 (1960).

Sa 80 Sarvazyan, A. P., Buckin, V. A., Hemmes, P.: J. Phys. Chem., **84,** 692 (1980).

Sc 59 Schleyer, P. E., West, R.: J. Am. Chem. Soc., **81,** 3164 (1959).

Sc 64 Schmulbach, C. D.: J. Inorg. Nucl. Chem., **26,** 745 (1964).

Sc 66 Schneider, H., Strehlow, H.: Z. Phys. Chem. N. F., **49,** 44 (1966).

Sc 68 Schaschel, E., Day, M. C.: J. Am. Chem. Soc., **90,** 503 (1968).

Sc 76 Schneider, H.: Elektrochim. Acta, **21,** 71 (1976).

Se 61 Selbin, J., Bull, W. E., Holmes, Jr., L. H.: J. Inorg. Nucl. Chem., **16,** 219 (1961).

Sh 42 Sheppard, S. E.: Rev. Mod. Phys., **14,** 303 (1942).

Sh 56 Sheka, I. A.: Zh. Obshch. Khim., **26,** 1340 (1956).

Sh 70 Sherry, A. D., Purcell, K. F.: J. Phys. Chem., **74,** 3535 (1970).

Si 67 Siegbahn, K., Nordling, C., Fahlman, A., Nordberg, R., Hamrin, K., Hedman, J., Johansson, G., Bergmark, T., Karlsson, S. E., Lindgren, I., Lindberg, B.: ESCA, Atomic, Electron Spectroscopy. Almquist and Wiksells, Uppsala, 1967.

Si 73 Siegbahn, M., Siegbahn, K.: J. Electron Spectr., **2**, 319 (1973).
Sm 66 Smith, H. L., Daouglas, B. E.: Inorg. Chem., **5**, 784 (1966).
So 64 Soós, J., Szekrényesy, T.: Periodica Polytechnica (Budapest), **8**, 29 (1964).
So 68 Sohár, P., Varsányi, G.: Acta Chim. Acad. Sci. Hung., **55**, 189 (1968).
So 76 Sobczyk, L., Engelhardt, H., Bunzl, K.: in: The Hydrogen Bond (Eds.: Schuster, P., Zundel, G., Sandorffy, C.). North-Holland, Amsterdam, New York, Oxford, 1976. Chapter 20, p. 937.
St 76 Strauss, I. M., Symons, M. C. R.: Chem. Phys. Letters, **39**, 471 (1976).
St 77 Strauss, I. M., Symons, M. C. R.: J. Chem. Soc. Faraday I, **73**, 1796 (1977).
St 78a Strauss, I. M., Symons, M. C. R.: J. Chem. Soc. Faraday I, **74**, 2146 (1978).
St 78b Strauss, I. M., Symons, M. C. R.: J. Chem. Soc. Faraday I, **74**, 2518 (1978).
Sw 62 Swift, T., Connick, J.: J. Chem. Phys., **37**, 307 (1962).
Sy 67 Symons, M. C. R.: J. Phys. Chem., **71**, 172 (1967).
Sy 69 Symons, M. C. R.: Am. Rev. Phys. Chem., **20**, 219 (1969).
Sy 75a Symons, M. C. R., Waddington, D.: J. Chem. Soc. Faraday II, **71**, 22 (1975).
Sy 75b Symons, M. C. R., Waddington, D.: Chem. Phys. Letters, **32**, 133 (1975).
Sy 75c Symons, M. C. R.: Phil. Trans. Roy. Soc., **B272**, 13 (1975).
Sy 80 Symons, M. C. R., Shippey, T., Rastogi, P. P.: J. Chem. Soc. Faraday I, **76**, 2251 (1980).
Ta 58 Tables of Interatomic Distances and Configurations in Molecules and Ions. Chem. Soc. Publ. No. 11, 1958.
Ta 63 Taft, R. W., Price, E., Fox, I. R., Lewic, I. C., Anderson, K. K., Davies, G. T.: J. Am. Chem. Soc., **85**, 709 (1963).
Ta 69 Tao, S. J., Green, H. J.: J. Phys. Chem., **73**, 882 (1969).
Ta 70 Tao, S. J.: J. Chem. Phys., **52**, 752 (1970).
Ta 72 Tao, S. J.: J. Chem. Phys., **56**, 5499 (1972).
Th 64 Thompson, W. K.: J. Chem. Soc. (London), **1964**, 4028.
Th 66a Thwaites, J. D., Sacconi, L.: Inorg. Chem., **5**, 1029 (1966).
Th 66b Thwaites, J. D., Bertini, I., Sacconi, L.: Inorg. Chem., **5**, 1036 (1966).
Th 66c Thomas, S., Reynolds, W. L.: J. Chem. Phys., **44**, 3148 (1966).
Th 70 Thomas, S., Reynolds, W. L.: Inorg. Chem., **9**, 78 (1970).
Ti 63 Timerov, R. K., Lablokov, Yu. V., Ablov, A. V.: Dokl. Akad. Nauk SSSR, **152**, 160 (1963).
To 69 Tong, D. A.: J. Chem. Soc. Dalton, 790 (1969).
Ue 80 Ueji, S.: Tetrahedron Letters, **21**, 475 (1980).
Va 73 Van Gemert, M. J. C., de Loor, G. P., Bordewijk, P., Quickenden, P. A., Suggett, A.: Advances in Molecular Relaxation Processes, **5**, 302 (1973).
Ve 65 Velluz, L., Legrand, M., Grosjean, M.: Optical Circular Dichroism. Academic Press, New York, 1965.
Vé 69 Vértes, A., Burger, K., Suba, M.: Magy. Kém. Foly., **75**, 317 (1969).
Vé 70a Vértes, A.: Acta Chim. Acad. Sci. Hung., **63**, 9 (1970).
Vé 70b Vértes, A., Burger, K., Suba, M.: Acta Chim. Acad. Sci. Hung., **63**, 123 (1970).
Vé 72 Vértes, A., Burger, K.: J. Inorg. Nucl. Chem., **34**, 3665 (1972).
Vé 73a Vértes, A., Burger, K., Molnár, B., Pálfalvi, M.: J. Inorg. Nucl. Chem., **35**, 691 (1973).
Vé 73b Vértes, A., Gaizer, F., Beck, M.: Magy. Kém. Foly., **79**, 310 (1973).
Vé 76 Vértes, A., Nagy-Czakó, I., Burger, K.: J. Phys. Chem., **80**, 1314 (1976).
Vé 78 Vértes, A., Nagy-Czakó, I., Burger, K.: J. Phys. Chem., **82**, 1469 (1978).
Vé 79 Vértes, A., Korecz, L., Burger, K.: Mössbauer Spectroscopy. Elsevier, Amsterdam, 1979.
Vi 71 Vigee, G. S., Ng, P.: J. Inorg. Nucl. Chem., **33**, 2477 (1971).
Vo 70 Vogel, G. C., Drago, R. S.: J. Am. Chem. Soc., **92**, 5347 (1970).
Wa 60 Wallace, P. R.: Positron Annihilation in Solids and Liquids. In: Solid State Physics, Vol. 10 (Eds. Seitz, F., Turnbull, D.). Academic Press, New York, 1960.

Wa 64 Wayland, B. B., Drago, R. S.: J. Am. Chem. Soc., **86**, 5420 (1964).
Wa 70 Walrafen, G. E.: J. Chem. Phys., **52**, 4176 (1970).
Wa 76 Watkins, C. L., Vigee, G. S.: J. Phys. Chem., **80**, 1 (1976).
We 22 Weinland, R., Kiszling, A.: Z. anorg. Chem., **120**, 218 (1922).
We 52 West, P. W., Robichaux, T., Burkhalter, T. S.: Anal. Chem., **23**, 1625 (1951).
We 62 Wells, A. F.: Structural Inorganic Chemistry, 3rd Edition, p. 5. Clarendon Press, Oxford, 1962.
We 65 Wellman, K. M., Briggs, W. S., Djerassi, C.: J. Am. Chem. Soc., **87**, 73 (1965).
We 69 Wertz, D. L., Kruh, R. F.: J. Chem. Phys., **50**, 4013 (1969).
We 71 Weiner, P. H., Malinovskii, E. R.: J. Phys. Chem., **75**, 3971 (1971).
We 72 Westmoreland, T. D., Bhacca, N. S., Wander, J. D., Day, M. C.: J. Organometal. Chem., **38**, 1 (1972).
We 73 Westmoreland, T. D., Bhacca, N. S., Wander, J. D., Day, M. C.: J. Am. Chem. Soc., **95**, 2019 (1973).
Wh 64 White, A. H., Roper, R., Kokot, E., Watermann, H., Martin, R. L.: Austral. J. Chem., **17**, 294 (1964).
Wi 82a Winsor, P., Cole, R. H.: J. Phys. Chem., **86**, 2486 (1982).
Wi 82b Winsor, P., Cole, R. H.: J. Phys. Chem., **86**, 2491 (1982).
Wu 70 Wuepper, J. L., Popov, A. I.: J. Am. Chem. Soc., **92**, 1493 (1970).
Wy 70 Wyss, H. R., Falk, M.: Canad. J. Chem., **48**, 607 (1970).
Yu 78a Yu-Keung Sze, Irish, D. E.: J. Solution Chem., **7**, 395 (1978).
Yu 78b Yu-Keung Sze, Irish, D. E.: J. Solution Chem., **7**, 417 (1978).

6. EFFECT OF THE SOLVENT ON THE STRUCTURES AND STABILITIES OF METAL COMPLEXES IN SOLUTION

From the previous discussion it is clear that in most solutions there is a considerable interaction between the solvent and the solute. Accordingly, the solvent can virtually never be regarded simply as a medium in which various reactions occur. In the course of the reactions of dissolved ions or molecules, the solvent molecules attached to them are almost as important constituents of the reactants as are the ions themselves. It is not surprising that the solvent has an extremely great influence on the chemical processes proceeding in the solution, e.g., on complex formation reactions and on the compositions, structures and stabilities of the complexes in the solution.

The effect of the solvent on a given system may be based on very varied properties. In general, it is not easy to recognize the property of the solvent which has the greatest influence on a given reaction, determining the character of the solvent effect in the system. In different systems, the factor predominantly determining the effect of a particular solvent may be ascribed to different properties. In many cases, the combined effect of a number of properties or processes may appear [Ma 78, 79].

In the following sections, we shall consider the most important processes by which solvents exert their effects on metal complexes in solution.

Effect of the donor strength

Most solvents used in coordination chemistry have donor properties, whereas metal ions are electron pair acceptors; on their dissolution, therefore, the metal ions are solvated by a donor solvent (D). When there are no other ligands in the solution, all of the coordination sites of the metal ion are thus occupied by solvent molecules:

$$M^{n+} + mD \rightleftharpoons MD_m^{n+} \qquad (6.1)$$

Hence, complexing reactions in donor solvents are exchange reactions between the ligands (L) and the solvent molecules coordinated to the acceptor ion:

$$MD_m + nL \rightleftharpoons MD_{m-n}L_n + nD \qquad (6.2)$$

The results of these exchange reactions depend on the number of substituted solvent molecules determined by the relative donor strengths of the solvent molecule and the ligand competing with it, and by their concentrations.

Complexes with an unsaturated coordination sphere are also capable of coordinating solvent molecules:

$$AX_n + mD \rightleftharpoons AX_nD_m \tag{6.3}$$

Solvent coordination (i.e., solvation) in such cases is a typical coordination chemical reaction. Under identical concentration conditions, the decisive factor in the determination of its course is the relative donor strength of the solvent molecule.

As Eq. (6.2) shows, the higher the donor strength of the solvent (i.e., the higher the stability of the solvate complex), the greater is the excess of ligand required for substitution of the solvent molecules situated in the first coordination sphere of the dissolved metal ion. It follows that with an increase in the donor strength of the solvent, the equilibrium stabilities of the complexes in the solution decrease.

The validity of the above correlation has been confirmed by many equilibrium studies, determinations of stability constants, and by data qualitatively reflecting the order of complex stabilities, e.g., polarographic half-wave potentials, etc. [Gu 76, 78].

It can be attributed to the effect of the donor strength of the solvent, for instance, that the equilibrium stabilities of chloro complexes are considerably higher in acetonitrile (which has a lower donicity than water) than in aqueous solution [Ba 63, Ma 65]; that complexes of lower stability are formed in dimethyl sulphoxide than in dimethylformamide, the former solvent having the larger donor strength [Ga 67]; and that nickel bromide is in a fully dissociated state in dimethyl sulphoxide, whereas in dimethylformamide it dissociates only partially [Gu 68a].

For certain compounds, e.g., $Ni(NCS)_2$ and $Ni(CF_3COO)_2$, an almost linear correlation is found between the equilibrium stability constant and the Gutmann donicity of the solvent [Ma 75].

In the following, a number of other examples will be presented that reflect various aspects of the solvent effect, and also point to this regularity.

Effects of steric factors

In addition to the donor strength of a solvent, its space requirement and steric properties, in general, also exert appreciable effects on its coordination, on the stabilities of the solvate complexes and hence on the stabilities and even the compositions of complexes formed with other ligands in solution.

With most donor solvents, the cobalt(II) ion is known to form hexasolvate complexes. On the other hand, it is capable of coordinating only four molecules of hexamethylphosphoramide (HMPA) solvent, with an extremely high donor strength, and the heat of solvation with this solvent was also found to be significantly lower than expected on the basis of the donicity [Gu 69a]. A proton resonance study of the system has shown that owing to the high space requirement of hexamethylphosphoramide, even the four donor atoms in the tetrasolvate cannot approach the central cobalt atom close enough to be strongly bound by it. The central atom has still some mobility within the ligand sphere. This explains why two of the four solvent molecules can be replaced by iodide ion, a ligand of a comparatively low donicity [Gu 69b]. In the resulting mixed complex, the above-mentioned motion within the coordination sphere can no longer be observed.

The absorption spectra and Stokes radii (calculated from the results of conductivity measurements) of solvated complex ions of first transition metals were reported by Abe and Wada [Ab 80] in HMPA. They found that the Stokes radii of these complex ions are sensitive to the configurational change, since the diameter of an HMPA molecule is very large as compared with the crystallographic diameters of central metal ions. Mn(II), Fe(II), Co(II) and Cu(II) ions were found to be tetrahedrally solvated while in the case of Ni(II), equilibria are established among octahedral, square planar and tetrahedral solvated species. Zn(II) forms probably a complex ion solvated octahedrally by HMPA.

Trimethyl phosphate reacts with a number of metal ions to form solvates with appreciably higher stability than expected from its donor strength [Gu 68b, Gu 69c]. The phenomenon has been interpreted by assuming chelate formation:

$$CoI_2 + 2(CH_3O)_3PO \longrightarrow (CH_3O)_2P(O)_2Co(O)_2P(OCH_3)_2 + 2\,CH_3I$$

However, it should be noted that the probability of formation of a four-membered chelate ring is low with oxygen donor atoms.

The donor strength of a solvent containing several different donor atoms also depends on which of the donor atoms is bonded to the metal ion. Dimethyl sulphoxide, for instance, is linked to Pearson's "hard" cations through its oxygen donor atom, and to "soft" ions by its sulphur donor atom. Iodides of the "hard" metals dissociate completely when dissolved in dimethyl sulphoxide, whereas mercury(II) iodide undergoes only slight dissociation in dimethyl sulphoxide solution [Bu 67, Ga 67, Pe 70].

The solvent exchange may also influence the steric properties of some ligands. In the case of macromolecular polypeptides (which are polyfunctional ligands from a coordination chemical point of view), the change in the solvent may cause significant changes in the conformation of the molecule, influencing its coordination chemical behaviour. Corticotropin [Ri 72], for example, shows only slight initial ordering in aqueous solution but assumes the α-helix structure in trifluoroethanol

[Gr 76]. In the solvent mixture water–trifluoroethanol the degree of order varies with the proportions of the solvent components. This change is reflected in the protonation equilibria of the peptide [Bu 79, No 80]. The conformation change also influences its metal coordination.

^{13}C NMR has been used to study the helix-coil transition induced in oligopeptides and polypeptides by trifluoroacetic acid. An attempt was also made by Suzuki et al [Su 80a, b] to separate the influences of conformational change and direct solvent interaction.

Effect of solvation of the anion

The interaction of the solvent molecules with cations is generally greater than with anions, since virtually all of the typical hydrogen-bonding (acceptor) solvents also contain an electron pair donor group, and the strength of the coordinate bonds binding the metal ion to the donor atom is usually larger than that of hydrogen bonds solvating the anion. This is the reason why, in the investigation of the effect of a solvent on complex formation, research workers have dealt primarily with the solvation of the metal ion and the complex formed, and have neglected the effect of the solvation of the anion.

The incorrect nature of this concept has been convincingly demonstrated by the equilibrium studies of Ahrland [Ah 76] in aqueous and in dimethyl sulphoxide solutions. In order to illustrate the effect of the solvation of the anions on the stabilities of their metal complexes, Ahrland compared the stepwise stability constants, determined potentiometrically in water and in dimethyl sulphoxide, of the zinc and cadmium complexes of halide ions, which form hydrogen bonds with strengths decreasing in the sequence $Cl^- > Br^- > I^-$. As can be seen from the data relating to the cadmium complexes in Table 6.1, the absolute values of the equilibrium constants in dimethyl sulphoxide solution were considerably higher than those in water. In addition, the stability sequences for the various halide

Table 6.1. Stepwise stability constants of cadmium(II) halide complexes determined in aqueous and in dimethyl sulphoxide solutions [Ah 76]

Stepwise constants	Ligands					
	in water			in DMSO		
	Cl^-	Br^-	I^-	Cl^-	Br^-	I^-
K_1	38.5	57.0	121.0	16000.0	850.0	150.0
K_2	4.4	3.9	5.0	75.2	71.0	27.0
K_3	1.5	9.5	137.0	430.0	600.0	830.0
K_2	—	2.4	40.0	52.0	44.0	15.0

complexes are reversed. Whereas, in accordance with the rule relating to metals belonging in the Ahrland–Chatt [Ah 58] triangle (group B), the stability sequence of the complexes in aqueous solution is $Cl^- < Br^- < I^-$, in dimethyl sulphoxide it is the opposite.

The phenomenon would be completely incomprehensible if it were to be explained merely on the basis of the interaction between the cation and the solvent. The Gutmann donicity of dimethyl sulphoxide is larger than that of water. Accordingly, lower complex stabilities should be found in dimethyl sulphoxide, since the cation must give solvates of higher stability with a solvent of higher donicity. Hence, in these solvates substitution of the coordinated solvent molecules by some other ligand needs a greater excess of ligand, and this corresponds to smaller equilibrium stability constants.

In reality, however, besides the competition between the solvent and the ligand for the metal ion, the competition between the solvent and the metal ion for the ligand also plays a role in complex formation in solution [Gu 78, Ma 79]. The formation of a coordinate bond between the metal ion and the ligand must involve the desolvation not only of the metal ion, but also of the ligand. Consequently, the stability of the complex formed depends on the stabilities of the solvates of both the metal ion and the ligand.

On the basis of these considerations, the experimental data of Ahrland become perfectly understandable. As a result of its strong hydrogen bond forming property, water is able to solvate the halide ions strongly. The stabilities of these hydrogen-bonded solvates decrease in the sequence of decreasing negativity and increasing radius of the anion, i.e., in the sequence $Cl^- > Br^- > I^-$.

Dimethyl sulphoxide is an aprotic solvent, in which the solvation of the halides is negligibly small. Hence, the sequence of the stabilities of the halide complexes in dimethyl sulphoxide is governed by the magnitude of the affinity between the central ion and the halide ligand. Accordingly, the complex stabilities follow the sequence of the Lewis base strengths of the ligands: $Cl^- > Br^- > I^-$.

In aqueous solution, the solvation of the halide ions diminishes the metal–halide interaction to an increasing extent the more strongly the anion is solvated. This is the reason for both the smaller complex stabilities and the reversal of the sequence of the stability constants, since the stronger the anion is solvated, the greater will be the decrease in the complex stability due to this solvation.

The abnormal ratio of the stepwise constants in dimethyl sulphoxide solution is a result of a stereochemical transformation caused by the complex formation.

Similarly to the aquo complexes, the dimethyl sulphoxide solvates of most metal ions, including zinc and cadmium (studied by Ahrland), have hexacoordinated octahedral structures, and the halide complexes, in both solvents, a tetrahedral structure. However, the octahedral $\leftrightarrow$ tetrahedral transformation associated with the stepwise complex formation with halide ions occurs at different complex compositions in dimethyl sulphoxide and in aqueous solution. Conclusions may be drawn on this transformation from the ratios of the successive stability constants and from the stepwise ΔS and ΔH values [Ah 73, Ah 76].

These data show that in the course of the formation of the zinc halides in dimethyl sulphoxide from the originally octahedral hexasolvate, the tetrahedral complex is formed as a result of the coordination of the second halide. This is the cause of the strikingly low K_1/K_2 value and the strikingly high K_2/K_3 value. In the case of the cadmium complexes, it appears that this transformation occurs during the coordination of the third ligand.

The thermodynamics of the formation of mercury(II) halide and thiocyanate complexes in dimethyl sulphoxide (DMSO) has been also investigated by Ahrland *et al.* [Ah 81]. The order of stability between the halide complexes follows a (b)-sequence, $Cl^- < Br^- < I^-$, though much less marked than in water. The thiocyanate ion behaves in many respects as the iodide ion but also has several traits of its own.

The stereochemical transformation of the mercury halide complexes in the course of the stepwise complex formation could also be followed by means of X-ray examinations [Ga 68]. The results at times cast doubt on the correctness of the conclusions derived from the equilibrium data. In the case of mercury halides, the stability of the third stepwise-formed complex is much higher than that of the fourth, and the value of K_3/K_4 is therefore large. This would indicate that the former complex already has tetrahedral symmetry. According to the X-ray examinations, however, the reason for the anomalous stability sequence is that the second complex is linear, the third is pyramidal and only the fourth has tetrahedral symmetry [Ga 68].

An analogous effect is reflected by the stability constants of other metal-halide complexes determined in dimethyl sulphoxide solution, e.g., indium-, lead- and tin-halide systems [Sa 73].

The anion solvation ability of dimethylformamide (DMF) and propylene carbonate (PC), is similar to that of DMSO, e.g. chloride ion is stronger solvated by water than by DMF. The stability constants of copper(II) chloro complexes were found therefore to be many orders of magnitude higher in N-N-dimethylformamide than in water [El 80].

Spectrophotometric investigation of the UV, visible, and near-infrared regions made possible the determination of the electronic spectra of each individual species and, thus, provided information about the solvation of the copper(II)-chloro complexes in DMF. The matrix rank treatment of more than 1000 spectrophotometric data demonstrates the presence of a minimum of four absorbing species: the free copper ion and three chloro complexes.

The order of the stability of TlX in DMF and PC is found [Sa 81] to be TlCl > TlBr > TlI, while it is TlI > TlBr > TlCl in water. The author concluded that the differences in the order of stability for metal halogeno-complexes in DMF, PC, and water depend mainly on the enthalpy changes for the formation of TlX, which is greatly influenced by the enthalpy changes of the solvation of halide ions.

It is therefore clear that in the study of solvent effects on complex formation it is not possible to neglect the interactions between the ligand and the solvent.

The solvation interaction between the solvent and the dissolved anion can also be characterized quantitatively with the aid of equilibrium constants. For example, Green *et al.* [Gr 69] used NMR and infrared spectrophotometric equilibrium measurements to determine the equilibrium constants of the interactions between 1-phenylethanol and halide ions, and obtained values of 130 for Cl^-, 39 for Br^- and 12 for I^-. It can be seen that even the iodide ion, which displays little tendency to undergo solvation, is solvated by 1-phenylethanol (admittedly to only a slight extent). In the same investigations, it was demonstrated that the bromide ion is more strongly solvated by methanol than by ethanol (equilibrium constants 70 and 40, respectively).

Barcza and Pope [Ba 73] similarly made use of NMR spectroscopic examinations to prove that the perchlorate ion, generally regarded as inert with respect to solvation, does undergo solvation. The equilibrium constants of the complexes formed between the perchlorate ion and some alcohols and phenols in nitrobenzene were determined; these data are given in Table 6.2. With ethylene glycol, which contains two hydroxyl groups, the perchlorate ion can be seen to form a much more stable complex than that with the analogous, monofunctional ethanol. A similar difference is found between the equilibrium constants in the cases of the bifunctional pyrocatechol and the analogous, monofunctional phenol [Ba 74]. This confirms that these bifunctional ligands form hydrogen-bonded chelates with the perchlorate ion.

Table 6.2. Stability constants of complexes of the perchlorate ion with molecules containing hydroxyl groups, in nitrobenzene [Ba 73]

Ligand	K
Pyrocatechol	20.6
Phenol	4.40
Ethylene glycol	3.04
Ethanol	0.36

Solvation due to chelate-type hydrogen-bond formation can also be observed in solutions of organic ligands made with non-aqueous solvents. Propylene glycol, for example, forms chelate-type hydrogen-bonds with deprotonated phenolate oxygen atoms in polypeptides, resulting in a decrease in their protonation constants [Bu 79, No 80] compared with those in water.

It proved readily possible to study the effect of the solvation of the anion on simple dissociation equilibria by means of electromotive force measurements in various non-aqueous solutions of hydrogen chloride in a liquid junction-free cell. In this way, Burger *et al.* [Bu 75], for example, determined the solvation constants of the chloride ion in ethylene glycol and in methanol solutions. They found that the stability of the glycol solvate ($K=0.39$) was about one order of magnitude higher than that of the methanol solvate ($K=0.03$). This can explain why hydrogen chloride is almost completely dissociated in ethylene glycol solution, whereas in methanol the protonation constant of the chloride ion is 5.82. (There is not a great difference between the relative permittivities of the two solvents.) The effect of the solvation of the chloride ion on the dissociation of hydrogen chloride is well shown by the fact that in dimethyl sulphoxide, the relative permittivity of which is considerably larger than those of the former two solvents, the chloride ion is not solvated, and the hydrogen chloride is less dissociated than in ethylene glycol or in methanol. In dimethyl sulphoxide solution, the protonation constant of the chloride ion is 20.1; indeed, even the formation of the ion HCl_2^- can be demonstrated ($K=2.1$). For comparison, it should be noted that in tetrahydrofuran, which similarly does not solvate the chloride ion, but which has a very small relative permittivity, the protonation constant of the chloride ion is 9×10^5, the formation constant of HCl_2^- is 2.6×10^7, and also the dimer H_2Cl_2 is formed ($K = 1.6\times10^8$).

Effect of solvation of the outer-sphere type

Coordination of solvent molecules in the outer sphere may also play a role during the dissolution and reactions of coordinatively saturated complexes. Various spectroscopic and equilibrium measurement methods can be employed to study outer-sphere solvation and some examples will be presented.

Yellin and Marcus [Ma 74, Ye 72, Ye 74] used Raman spectroscopy to examine interactions of this type. They studied the interactions of cadmium(II) and mercury(II) halides and cyanides with water and with various organic solvents, such as alcohols, acetone, dioxane, acetonitrile, formamide, dimethylformamide, N-ethylacetamide, dimethylacetamide and mixtures of these.

It was shown that, in addition to the trihalo complexes that bind the solvent in the inner coordination sphere, tetrahalo complexes with a regular tetrahedral inner coordination sphere are also formed, and that solvent molecules are bound in the outer coordination sphere of the latter complexes. This outer-sphere solvent coordination can also be followed by means of observing its effect on mixed complex formation reactions of the following type [Ye 74]:

$$MX_4^{2-}+Y^- \rightleftharpoons MYX_3^{2-}+X^-$$

where X^- and Y^- are halide ions.

The investigation of this equilibrium isolated from the other equilibria in the system led to the finding that the parent complex and the mixed complex interact with the solvent to different extents.

Raman studies of the solvation of the above systems in binary solvent mixtures provided information on the relative solvating powers of the solvent components in the given system. It was established how the variation of the composition of the solvent mixture affected interactions of the outer-sphere type between the individual solvent components and the dissolved complexes.

In the outer sphere of metal halide complexes, the solvent molecules may be partially hydrogen-bonded to the halide ligand, and partially (as a consequence of their dipole nature) in electrostatic interaction with the complex anion.

Solvent molecules may also be situated in the outer sphere of complex cations. For instance, Fung [Fu 67] used NMR examinations to demonstrate the outer-sphere association between the complex cation tris(ethylenediamine) cobalt(III) and the solvent (deuterated dimethyl sulphoxide–D_2O).

Outer-sphere solvation may exert a considerable effect on the solubility conditions of the system. Thus, even simple solubility examinations may yield information on the process [La 58].

As regards the general knowledge and regularities relating to complex formation of the outer-sphere type which, within certain limits, hold for outer-sphere solvation also, the reader is referred to a review by Beck [Be 68].

Formation of auto-complexes, mixed ligand complexes, DA molecular complexes and polynuclear complexes

Important roles in the determination of the solution structure are played by the various, more complicated equilibria, also, e.g., solvent-induced auto-complex formation and the formation of mixed ligand and polynuclear complexes.

The simplest form of auto-complex formation is illustrated by the equation

$$D + 2\,MX \rightleftharpoons DM^+ + MX_2^-$$

where D is the solvent molecule, M^+ is the metal ion and X^- is the ligand. It must be noted that, depending on the properties of the solvent, auto-complex formation may also occur in addition to solvation, dissociation and complex-formation reactions in the same solution. Different methods of examination are used to distinguish between these processes. Most results in this field have probably been provided by conductometric titration. Valuable information may also be obtained from NMR spectroscopy, potentiometric measurements, polarography and the ultraviolet and visible spectra of the systems.

The role of auto-complex formation can be well seen, for example, in solutions of tin(IV) iodide in non-aqueous solvents [Ma 70]. When donor solvents are added to solutions of tin(IV) iodide in a non-coordinating solvent (e.g., nitrobenzene), the following equilibria may be assumed:

$$SnI_4 + 2\,D \rightleftharpoons SnI_4D_2$$
$$SnI_4D_2 + D \rightleftharpoons SnI_3D_3^+ + I^-$$
$$SnI_4D_2 + 2\,I^- \rightleftharpoons SnI_6^{2-} + 2\,D$$
$$3\,SnI_4D_2 \rightleftharpoons 2\,SnI_3D_3^+ + SnI_6^{2+}$$

The extent to which the individual reactions actually take place depends partly on the properties of the donor solvent and partly on the concentration ratio of the donor solvent and the complex.

In dilute tin(IV) iodide solutions, as a result of the action of high-donicity solvents, it is even necessary to take account of the formation of complex cations, such as $D_4SnI_2^{2+}$, D_5SnI^{3+}, and even D_6Sn^{4+}.

Auto-complex formation, although less marked, can also be detected in iron(III) chloride solutions [Gu 70]; the equilibria are as follows:

$$D + FeCl_3 \rightleftharpoons DFeCl_3$$
$$2\,D + 2\,DFeCl_3 \rightleftharpoons D_4FeCl_2^- + FeCl_4^-$$

It is worthy of special mention that auto-complex formation similarly occurs in the case of metal carbonyls when acted upon by high-donicity molecules [Ch 68, Hi 57a, Hi 57b, Hi 65]. This process is well illustrated by the following equation:

$$4\,Fe(CO)_5 + 6\,py \rightleftharpoons Fe(py)_6^{2+} + Fe_3(CO)_{11}^{2-} + 9\,CO$$

where a polynuclear anion is formed in addition to the solvated cation. The equation also shows that auto-complex formation may be accompanied by a redox reaction. Effects similar to that of pyridine were also exhibited by dimethyl sulphoxide and alcohol solvents in cobalt carbonyl systems [Hi 59, Hi 65].

Very varied mixed ligand and polynuclear complexes may be formed in the course of complexing reactions in non-aqueous solutions. Most non-aqueous solvents have a considerably smaller relative permittivity than water has; these smaller values act against the dissociation processes, and may thus favour the formation of complexes with more complicated compositions.

As demonstrated in the tin(IV) halide and mercury(II) halide systems, mixed complex formation may occur by the interaction of the parent complexes, in accordance with the following general equations:

$$MA_4 + MB_4 \rightleftharpoons 2\,MA_2B_2$$
$$MA_4 + MB_4 \rightleftharpoons MA_3B + MAB_3$$
$$MA_2 + MB_2 \rightleftharpoons 2\,MAB$$

Under anhydrous conditions (to exclude the possibility of hydrolysis of the complexes), or in non-aqueous solutions and solvent mixtures permitting only a minor degree of dissociation of the complexes, this type of mixed ligand formation is favoured [Be 64, Be 74, Ga 71].

Non-aqueous media may similarly favour the formation of heteropolynuclear complexes, such as are formed in dimethylformamide solution in the interaction of mercury(II) iodide and B-metal iodides [Ga 67].

It is known that the solvent may influence the formation equilibria of DA molecular complexes by nonspecific physical effects which depend on solvent polarity and molecular structure as well as by specific interactions. The problem is far from being fully understood. Further development of the theory is hindered by scarcity of data on thermodynamic characteristics of the DA equilibria in different media. In view of the fact that usually small differences in K have to be measured, the best data for comparative studies are those supplied by the same laboratory and derived by the same methods.

Uruska *et al.* [Ur 79a, b, 80] have determined the formation constants of DA molecular complexes (e.g. pyridine and pyridine derivates, with iodine) in a series of organic solvents using spectrometric and calorimetric measurements. The equilibrium constants and other thermodynamic data were found to correlate with solvent properties.

Effect of the structure of the solvent

We have so far discussed the influence of the interactions between the solvent and the solute on the complex equilibria in the solution. However, one cannot neglect the effect of the solute, e.g., complex-forming components, on the interconnection of the solvent molecules. The reason why comparatively few data and references are to be found in the literature with regard to this latter problem is not that it plays a more subordinate role, but rather that its investigation is fairly complicated.

Particularly the molecules of protic solvents are known to form hydrogen-bonded associates with one another, of various sizes and forms [Er 74].

If either electrolytes or non-electrolytes are dissolved in such liquids, the resulting solvation causes some of the hydrogen bonds linking the solvent molecules to break; the associates are progressively degraded, and smaller associates or even monomeric solvent molecules are formed, with properties more or less different from the solvating properties of the original liquid [St 78]. The aspects of this phenomenon in relation to the solvent–non-electrolyte interaction will be discussed in the section dealing with solvent mixtures. For details, the reader is referred to the book by Erdey-Grúz [Er 74].

Because of their much stronger interaction with the solvent molecules, the dissolution of Lewis acids and Lewis bases has a much greater effect on the decomposition of the solvent associates and hence on the changes in the solvent structure than does the dissolution of non-electrolyte molecules. The more concentrated the solution, the more marked is the above action of the dissolved components, which affects the solvating power of the solvent and hence the complex equilibria in the solution.

The changes in the interaction of the solvent molecules that occur as a consequence of complex formation are well reflected by the solvation entropies. From the solvation entropies determined by Parker and Watts [Pa 74] in various solutions, it emerges that in aprotic solvents the solvation entropy generally has a larger negative value, indicative of a higher degree of order, than in protic solvents.

On the above basis, this phenomenon can be explained in that the hydrogen-bonded associates in protic solvents ensure a higher ordering of the solvent. On the dissolution of electrolytes, their solvation causes some of these associates to break down, which results in an increase in the disorder of the system.

In aprotic solvents, on the other hand, which have no similar internal ordering, the coordination of the solvent molecules results in an increase of the order in the system.

Processes analogous to the interaction of the solvent molecules, leading to the formation of hydrogen-bonded associates, are ligand–ligand interactions; these are of importance particularly in systems with low relative permittivity. This is shown well, for example, by the dimerization constants of organic acids that also act as ligands, determined in apolar solvents [Ba 76, Li 69, Li 70]. Although these dimerization equilibria are shifted as a result of complex formation under the action of metal ions, their existence in the solution exerts an influence on the coordination of the ligands.

References

Ab 80 Abe, Y., Wada, G.: Bull. Chem. Soc. Jpn. **53,** 3547 (1980).

Ah 58 Ahrland, S., Chatt, J., Davies, N. R.: Quart. Rev. (London), **12,** 265 (1958).

Ah 73 Ahrland, S.: Struct. Bonding (Berlin), **15,** 167 (1973).

Ah 76 Ahrland, S., Björk, N. O.: Acta Chem. Scand., A **30,** 249, 257, 265, 270 (1976); Coord. Chem. Rev., **16,** 115 (1976).

Ah 81 Ahrland, S., Persson, I., Portanova, R.: Acta Chem. Scand. A **35,** 49 (1981).

Ba 63 Barnes, J. C., Hume, D. N.: Inorg. Chem., **2,** 444 (1963).

Ba 73 Barcza, L., Pope, M. T.: J. Phys. Chem., **77,** 1795 (1973).

Ba 74 Barcza, L., Pope, M. T.: J. Phys. Chem., **78,** 168 (1974).

Ba 76 Barcza, L., Buvári, Á.: Z. Phys. Chem. (Frankfurt), **102,** 25 (1976).

Be 64 Beck, M. T., Gaizer, F.: Acta Chim. Acad. Sci. Hung., **41,** 423 (1964).

Be 68 Beck, M. T.: Coord. Chem. Rev., **3,** 91 (1968).

Be 74 Beck, M. T., Porzsolt, Cs. É., Ling, J.: Reaction Kinetics Catal. Letters, **1,** 125 (1974).

Bu 67 Buckingham, A., Gasser, R. P. H.: J. Chem. Soc. A, **1967,** (1964).

Bu 75 Burger, K., Andrási, E., Barcza, L., Noszál, B.: Unpublished results. See: Andrási, E., Barcza, L., Burger, K.: Proc. Euranal. 2nd Conf., Budapest, 1975, p. 179.

Bu 79 Burger, K., in Metal Ions in Biological Systems (H. Sigel, ed.), M. Deckker, New York, Vol. 9, p. 213, 1979.

Ch 68 Chini, P.: Inorg. Chim. Acta Rev., **2,** 48 (1968).

El 80 Elleb, M., Meullemeestre, J., Schwing-Weill, M. J., Vierling, F.: Inorg. Chem. **19,** 2699 (1980).

Er 74 Erdey-Grúz, T.: Transport phenomena in aqueous solutions. Akadémiai Kiadó, Budapest, 1974.

Fu 67 Fung, B. M.: J. Am. Chem. Soc., **89,** 5788 (1967).

Ga 67a Gaizer, F., Beck, M. T.: Magy. Kém. Foly., **73,** 149 (1967).

Ga 67b Gaizer, F., Beck, M. T.: J. Inorg. Nucl. Chem. **29,** 21 (1967).

Ga 68 Gaizer, F., Johansson, G.: Acta Chim. Scand., **22,** 3013 (1968).

Ga 71 Gaizer, F., Kovács, E., Beck, M. T.: Magy. Kém. Foly., **77,** 343 (1971).

Gr 69 Green, R. D., Martin, J. S., Cassie, W. B. McG., Hyne, J. B.: Canad. J. Chem. **47,** 1639 (1969).

Gr 76 Greff, D., Toma, F., Fermandjian, S., Lőw, M., Kisfaludy L.: Biochim. Biophys. Acta **439,** 219 (1976).

Gu 68a Gutmann, V., Bohunovszky, O.: Monatsh. Chem., **99,** 740 (1968).

Gu 68b Gutmann, V., Fenkart, K.: Monatsh. Chem., **99,** 1452 (1968).

Gu 69a Gutmann, V., Weisz, A., Kerber, W.: Monatsh. Chem., **100,** 2096 (1969).

Gu 69b Gutmann, V., Weisz, A.: Monatsh. Chem., **100,** 2104 (1969).

Gu 69c Gutmann, V., Beer, G.: Inorg. Chim. Acta, **3,** 87 (1969).

Gu 70 Gutmann, V., Wegleitner, K. H.: Monatsh. Chem., **101,** 1532 (1970).

Gu 76 Gutmann, V.: Electrochim. Acta **21,** 661 (1976).

Gu 78 Gutmann, V.: The Donor-Acceptor Approach to Molecular Interactions. Plenum Publ. Corp. New York, 1978.

Hi 57a Hieber, W., Brendel, G.: Z. anorg. allg. Chem., **289,** 338 (1957).

Hi 57b Hieber, W., Werner, R.: Chem. Ber., **90,** 278, 286, 1116 (1957).

Hi 59 Hieber, W., Lipp, A.: Chem. Ber., **92,** 2085 (1959).

Hi 65 Hieber, W., Schubert, E. H.: Z. anorg. allg. Chem., **338,** 37 (1965).

La 58 Larsson, R.: Acta Chim. Scand., **12,** 708 (1958).

Li 69 Liszi, J.: Magy. Kém. Foly., **75,** 158 (1969).

Li 70 Liszi, J.: Magy. Kém. Foly., **76,** 450 (1970).

Ma 65 Manahan, S. E., Iwamoto, R. T.: Inorg. Chem., **4,** 1409 (1965).

Ma 70 Mayer, U., Gutmann, V.: Monatsh. Chem., **101,** 912 (1970).

Ma 74 Marcus, Y., Yellin, N.: Proc. 16th ICCC, Dublin, 1974, p. 233b.

Ma 75 Mayer, U.: Pure Appl. Chem., **41,** 291 (1975).

Ma 78 Mayer, U.: Monatsh. Chem. **109,** 421 (1978).

Ma 79 Mayer, U.: Pure Appl. Chem. **51,** 1697 (1979).

No 80 Noszál, B., Burger, K.: Inorg. Chim. Acta **46,** 229 (1980).

Pa 74 Parker, A. J., Watts, D. W.: Abstracts 4th ICNAS (Ed. Gutmann, V.), Vienna, 1974, p. 14.

Pe 70 Peterson, R. J., Lingane, P. J., Reynolds, W. L.: Inorg. Chem., **9,** 680 (1970).

Ri 72 Riniker, B., Sieber, P., Rittel, W.: Nature New Biol., **235,** 114 (1972).

Sa 73 Samoylenko, V. M., Lyashenko, V. P.: Zhur. Neorg. Khim., **18,** 2402, 2968 (1973).

Sa 81 Sasaki, Y., Takizawa, M., Umemoto, K., Matsuura, N.: Bull. Chem. Soc. Jpn., **54,** 65 (1981).

St 78 Strauss, I. M., Symons, M. C. R.: J. Chem. Soc. Faraday I **74,** 2146, 2518 (1978).

Su 80a Suzuki, Y., Inoue, Y., Chûjô, R.: Macromol. Chem. **181,** 177 (1980).
Su 80b Suzuki, Y., Inoue, Y., Chujo, R.: Macromol. Chem. **181,** 165 (1980).
Ur 79a Uruska, I., Karaczewska, H.: J. Solution Chem. **8,** 108 (1979).
Ur 79b Uruska, I., Inerowicz, H.: Polish Journal of Chemistry **53,** 2579 (1979).
Ur 80 Uruska, I., Inerowicz, H.: J. Solution Chem. **9,** 97 (1980).
Ye 72 Yellin, N., Marcus, Y.: Israel J. Chem., **10,** 919 (1972).
Ye 74 Yellin, N., Marcus, Y.: J. Inorg. Nucl. Chem., **36,** 1325 (1974).

7. EFFECT OF THE SOLVENT ON THE KINETICS AND MECHANISMS OF COORDINATION REACTIONS

So far we have dealt primarily with methods for studying non-aqueous solutions. In discussing the results of these investigations, we presented the effects of the solvents on the compositions and structures of the species formed, and on the constants of the equilibria occurring in solution, such as the stability constants of the complexes.

The solvent effect discussed in connection with measurements of different kinds led to the characterization of the solvents and to a better understanding of their role in different chemical equilibria, in some respect even generally in equilibrium chemistry. It is obvious that this solvent effect is highly significant in most systems. The effect of the solvent is no less marked on the kinetics and mechanism of the various reactions proceeding in the solution.

Almost a century ago, it was established by Menschutkin [Me 90] that reactions cannot be divorced from the medium in which they take place. Indeed, even at that time he attempted to find a correlation between the reaction rate and the structure of the solvent. However, the actual recognition of such a correlation has not yet been achieved: an exact description of general validity characterizing the kinetic effect of the solvents is still unavailable.

The first theory (which still basically holds even today) giving an empirical characterization of the solvent effect was due to Hughes and Ingold [Hu 35]. They described the effect of the solvent on aliphatic substitution and elimination reactions with an electrostatic model of solvation. They demonstrated that the rates of reactions resulting in the appearance or localization of charge increase with an increase in the polarity of the solvent. If the activated complex is more strongly solvated than the reactants, then the larger the difference between the stabilities of these solvates, the lower is the free energy of activation of the reaction. Since the solvation of the ions increases with an increase in the polarity of the solvent, a higher polarity favours the formation of species of more ionic character. Conversely, it follows that the rates of those reactions which are accompanied by the disappearance of charge decrease with an increase in the polarity of the solvent.

The cause of the reaction rate changes induced by the solvent is that solvation will stabilize the activated complex and the reactants to different extents. According to the foregoing, the effect of the solvent on the reaction rate may be characterized quantitatively on the basis of the heats of solvation of the reactants and the activated complex [Ro 59]. This statement is true in any case. However, overestimation of the electrostatic contribution in solvation processes led to the characterization of the stabilities of solvates solely by the relative permittivity of the solvent, and hence the solvent effect was explained exclusively by the dependence of the rate constant on the relative permittivity of the solvent [Am 52, Ki 34, La 56, Sc 32, etc.]. It is natural that such correlations are manifested only in systems where specific interactions are subordinate, and the electrostatic effect is predominant [Da 70]. This condition, however, does not generally hold in complex-formation reactions. For this reason, the specific interactions must also be taken into consideration in a characterization of the solvent dependence of the rates of coordination reactions.

Complex-formation reactions occurring in solution are known to be, in effect, substitution reactions, in which the ligand replaces solvent molecules bound in the solvate sphere of the metal ion. Thus, it is understandable that the rates, kinetics and mechanisms of such reactions are dependent on the solvent. The magnitude of the interaction between the coordinated solvent molecule and the cation (the strength of the coordinate bond) influences the rate of exchange of the coordinated solvent with some other ligand. However, solvents also exert their effects on reactions *via* their different relative permittivities, viscosities, internal cohesions, various chemical properties, acidic or basic natures, hydrogen-bonding abilities, etc. [Am 66].

The solvent may affect the reaction rate without influencing the reaction mechanism or, alternatively, both may be altered at the same time. The reaction mechanism remains unchanged while the reaction rate may change considerably, for example, if the solvent exchange in a diffusion-controlled reaction causes a change only in the viscosity of the solution and hence in the number of effective collisions, but the compositions of the reacting species are not affected. The effect of solvent exchange is similar if there is a change in the relative permittivity and thereby only in the electrostatic forces between the reacting ions, but there is no substantial alteration in the solvation conditions. Such a "simple" solvent effect can be interpreted theoretically, and described quantitatively in mathematical terms. As regards the various theoretical considerations, the reader is referred to the book by Amis [Am 66] and to the literature references given there.

In most cases, however, solvent exchange causes various changes in the systems under examination. This is the reason why the solvent frequently acts jointly on the reaction rate and the reaction mechanism.

As examples, a number of results are presented below that are based on experimental studies of the kinetic effect of the solvent.

The effect of solvation strength (donor or acceptor strength) of solvents

In coordination chemistry, the solvent effect based on the different solvations of the reactants is the most interesting, and accordingly a number of workers have investigated the correlation between the donor strengths or acceptor strengths of solvents and the kinetics of the complex-formation reactions taking place in them.

It has been demonstrated with a T-jump relaxation method, for instance, that in antimony pentachloride solutions, prepared with various donor solvents, there is a linear correlation between the rate of substitution by triphenylchloromethane of the donor solvent molecule bound at the sixth, "free" coordination site of the antimony and its donicity [Gu 71, Gu 72]. Both thermodynamic considerations and experimental data showed an analogous correlation between the rate of formation of $CoCl_4^{2-}$ (by the interaction of $CoCl_3^-$ and Cl^- ions dissolved in various donor solvents) and the Gutmann donicities of the donor solvents [Ma 75].

Dickert *et al.* [Di 71, Di 72, Di 74] employed pressure jump and other relaxation methods to study the kinetics of the formation and dissociation reactions of various nickel complexes. They established that the substitution reactions of some of the nickel(II) complexes (taking place according to the following equation) varied linearly with the donicity of the solvent:

$$NiD_5X^+ + D \rightleftharpoons NiD_6^{2+} + X^-$$

where $X = SCN^-, Cl^-, CF_3COO^-$ and $CH_3HSO_3^-$, and D = the donor solvent [Di 74].

Investigations by Fischer *et al.* [Fi 72], carried out with pressure jump and shock-wave methods, revealed that, in alcoholic solution, the rate-determining step of the formation and dissociation of transition metal halides was the departure of a coordinated solvent molecule from the coordination sphere of the metal.

The acceptor strength of the solvent may be equally important in reactions where the solvation stabilizes an electron pair donor molecule or an anion. Stronger solvation of the reactant ligand or anion inhibits the reaction and the rate will decrease. If the solvated anion is the activated complex or the reaction product, an increase in the stability of the solvate favours the reaction and increases its rate. Examples of each of the two types of effect are presented below.

The ionization reaction of 2-(*p*-methoxyphenyl)-2-methylpropyl *p*-toluenesulphonate is defined so unambiguously by the acceptor strength of the solvent that the logarithm of its rate constant is a parameter that may be used to characterize the acid strength of the acceptor solvent [Sm 61]:

OCH$_3$ | OCH$_3$ | OCH$_3$

$\xrightarrow{k_{ion}}$ [(+) $OTs^{\ominus}$] $\longrightarrow$ + HOTs

CH_3-C-CH_2-OTs | CH_3-C-CH_2 | CH_3 $C{=}CH$

CH_3 | CH_3 | CH_3

The rate of the reaction is governed by the strength of the solvation of the toluenesulphonate anion (OTs^-) in the transition complex. With the increase of the acceptor strength of the solvent, the stability of the solvate increases which, in accordance with the above, leads to an increase in the reaction rate.

Similarly, the solvent dependence of the rates of the following halide or pseudohalide exchange reactions is determined by the acceptor strength of the solvent:

$$Y{:}^- + R_3CX \rightleftarrows [Y{-}CR_3{-}X{:}^-] \rightleftarrows YCR_3 + X{:}^-$$

where Y^- and X^- are the halide or pseudohalide ion.

With an increase in the acceptor strength of the solvent, the stability of the solvate of the halid Y^- increases, which inhibits the formation of the transition product; therefore, the rate of the exchange reaction decreases if the acceptor strength of the solvent increases. This effect is manifested in such a clear-cut manner that the solvent dependence of the rate constant can be described unambiguously by means of the Krygowski–Fawcett [Kr 75] model (cf. Chapter 4).

In kinetically labile simple systems it may be assumed that the regularity reflected by the above examples (i.e., that the rate-determining step is the solvent substitution) is of fairly general validity. The high rates of most of these reactions make their investigation difficult; this is the reason why reliable experimental data have become available for the interpretation of the solvent effect in this field only since the spreading of the fast kinetic methods (stopped flow, T-jump, pressure jump, etc.).

Chattopadhyay *et al.* [Ch 76a, Ch 76b, Ch 77] used a spectrophotometric stopped flow technique to study the kinetics of formation and dissociation of a large number of nickel complexes, with a metal to ligand ratio of 1:1, in various non-aqueous solvents. When their findings were compared with the results obtained earlier in investigations of analogous complexes of other transition metal ions [Ch 73, Ch 74], it was established that in most of the systems the above reaction follows an I_d mechanism. The step determining the reaction rate was found to be the exchange between the solvent bound in the coordination sphere of the metal and the bulk solvent. In reactions with multidentate ligands, the reaction rate is influenced by the solvent-dependent space requirements of these ligands, and in the case of π-acceptor ligands or solvents (acetonitrile) by π-interactions also.

The investigations have shown that the rate of formation or dissociation of the 1:1 complex is correlated linearly with the Gutmann donicities of the solvents only in exceptional cases. A linear correlation appeared more frequently between the activation enthalpy (ΔH) of dissociation of the complex and the Gutmann donicity.

Chattopadhyay *et al.* stated that the first step in the formation of complexes of the nickel(II) ion with monodentate ligands can be described, in accordance with the I_d mechanism, as follows:

$$\mathrm{NiS_6^{2+} + L} \underset{\text{I}}{\overset{K_{12}}{\rightleftharpoons}} \mathrm{S{-}NiS_5^{2+}, L} \underset{k_{32}}{\overset{k_{23}}{\rightleftharpoons}} \mathrm{L{-}NiS_5^{2+}, S} \qquad \text{(II)}$$

where the rate-determining step (II) is preceded by a fast outer-sphere coordination (I), the equilibrium constant of which, K_{12}, can be calculated by the method of Eigen [Ei 54] and Fuoss [Fu 58]. k_{23} and k_{32} are the corresponding rate constants, which could be calculated, with a knowledge of the value of K_{12}, from the experimentally measured overall rate constant.

The kinetics of ligand substitution in bis(N-tert-butylsalicylaldiminato)-copper(II) [Cu(SA=N-t-Bu)$_2$] in various alcoholic media and specially the solvent dependence of this reaction was studied by Elias *et al.* [El 80]. N-ethylsalicylal-dimine (HSA=N-t-Et) served as substituting ligand. Stopped-flow spectro-photometry has been used for these investigations. It was concluded that the mechanism of the solvent effect is made up of the following steps: (i) fast equilibrium Cu(SA=N-t-Bu)$_2$ + ROH $\rightleftharpoons$ Cu(SA=N-t-Bu)$_2$. ROH; (ii) fast proton transfer from bound ROH to the phenolic oxygen of the coordinated tert-butyl ligand; (iii) rearrangement in the coordination sphere with breaking of the Cu-OH (ligand) bond; and (iv) stepwise but fast substitution of the two-tert-butyl ligands by the ethyl ligands. Rearrangement step (iii) is presumably rate determining.

Investigations of the kinetic effect of the solvent have also demonstrated in numerous cases that an important step (and often the most important step) in reactions in solution involving the participation of metal ions is the exchange of the solvent coordinated to the metal with the bulk solvent. Many workers have therefore dealt with the study of the solvent-exchange reactions of metal ions (see, for example, [Ei 65, Gu 74, He 70, Hu 71, Ku 60]). A number of workers have attempted to find a generally valid interpretation of the phenomenon.

Frank and Wen [Fr 57] and later Caldin and Bennetto [Be 71, Ca 73] looked for a correlation between the solvent-exchange kinetic parameters for divalent metal ions and the structural properties of the solvent. Tanaka [Ta 76] demonstrated a correlation between the activation energy of the solvent exchange and the heat of solvation of the metal ion, as well as the heat of vaporization of the solvent.

The latter concept deserves more detailed consideration. Tanaka assumed the following four-step mechanism for the exchange of the solvent molecules coordinated to the metal ion:

(1) The solvent molecule departs from the coordination sphere of the metal ion;
(2) this molecule enters the "bulk" solvent;
(3) another solvent molecule departs from the "bulk" solvent;
(4) the latter solvent molecule enters the coordination sphere of the metal ion.

Steps 1 and 4 depend on the heat of solvation (ΔH_d) of the metal ion, and steps 2 and 3 are proportional to the heat of vaporization (ΔH_v) of the solvent.

In order to calculate the heat of activation $H^{\ddagger}$ of the solvent exchange from these data, Tanaka used the following simple empirical correlation:

$$\Delta H^{\ddagger} = a\Delta H_d + b\Delta H_v$$

where he derived the values of the constants a and b from experimental data measured in various systems. With a knowledge of these, the heat of activation of solvent exchange could also be calculated in other systems.

To illustrate the practical applicability of the model, Table 7.1 shows a comparison of the values of the heat of solvent exchange in various solvents, calculated in the above manner and determined experimentally, for the cobalt(II) ion. The high accuracy of this extremely simplified model in describing the systems is surprising.

Table 7.1. Calculated and measured solvent-exchange heats of activation for the cobalt(II) ion in various solvents [Ta 76]

Solvent	$[\Delta H^{\ddagger}]$, kJ mole^{-1}		
	calculated	measured	difference
Acetonitrile	47.2	47.7	−0.5
Ammonia	49.5	46.8	+2.7
Dimethylformamide	55.9	56.8	−0.9
Dimethyl sulphoxide	53.6	51.0	+2.6
Methanol	57.6	57.7	−0.1
Water	45.0	47.2	−2.2

The examples discussed so far explained the solvent effect in terms of the different equilibrium stabilities or kinetic reactivities of the solvates formed with the different solvents. Solvent exchange is also frequently accompanied by a change in the coordination number of the solvate complex. Solvates which have different coordination numbers, and hence different symmetries, may also have different reactivities. This is illustrated well by the redox kinetic examinations by Gutmann *et al.* [Sc 76] in non-aqueous solutions, carried out with the stopped flow spectrophotometric technique. In connection with the investigation of the kinetics of reduction of iron(III) solvates, Gutmann *et al.* demonstrated that the smaller the coordination number of the solvates formed, the higher is the rate of the reduction. There may even be a difference of several orders of magnitude between the rate constants of the reduction of the hexacoordinated and pentacoordinated iron(III) solvates [Sc 76]. When solvates with the same coordination number are formed with different solvents, the rate of the reduction decreases with the increse in the donicity of the solvent. (The latter trend is a consequence of the higher stability of the solvate of the ion in the higher oxidation state.)

In more complicated systems, a clear interpreration of the solvent effect may also be made difficult by various unexpected side reactions. Solvation, or other complex formation reactions taking place in non-aqueous solution, are often catalyzed by components from which this is scarcely to be expected.

For instance, it has been shown by Beck *et al.* [Rá 76] that the ligand exchange reaction of the kinetically inert hexaquochromium(III) ion in dimethyl sulphoxide solution, leading finally to the formation of the $Cr(DMSO)_6^{3+}$ ion, is catalyzed by the carbon dioxide present in the solution. This effect can readily be seen in Fig. 7.1. Nitrite ion has a similar catalytic effect on the system.

According to Beck *et al.* [Rá 76], this catalytic effect can be explained in that the carbonato or nitrito complex of chromium is formed without splitting of the inert Cr—O bond in the hexaquo complex (in effect, in an oxygen-exchange reaction).

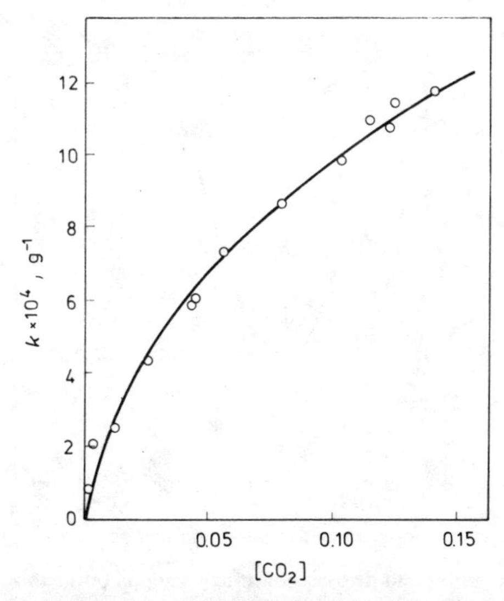

Fig. 7.1. Effect of CO_2 on the rate of the $Cr(OH_2)_6^{3+}$ + DMSO reaction [Rá 76]. [Cr(III)] = 0.025 mole dm^{-3}; [$HClO_4$] = 0.01 mole dm^{-3}; I = 1.0 mole dm^{-3}; T = 298 K

The carbonato or nitrito complex subsequently reacts at a higher rate with dimethyl sulphoxide than does the more inert aquo complex.

Oxyanions do not catalyze the reverse reaction, i.e., the transformation of the dimethyl sulphoxide solvate to the hexaquo solvate, since oxyanions do not participate in an oxygen-exchange reaction with the coordinated dimethyl sulphoxide.

A more intricate solvent effect must be taken into account in complex formation reactions in solvent mixtures. A particularly complicated correlation is to be expected between the reaction rate and the composition of the solvent in systems where both components of the solvent mixture are capable of coordinating to the central ion.

Beck *et al.* [Be 74], for instance, demonstrated that in dioxane–water mixtures the formation and dissociation rate constants of the mixed complex HgBrCN, described by the following equation:

$$HgBr_2 + HgCN_2 \rightleftharpoons 2\,HgBrCN$$

display a non-monotonous change as a function of the composition of the solvent. Figure 7.2 shows the rate constants as a function of the molar fraction of dioxane. The shape of the curve clearly indicates the specific solvation of the two components of the solvent mixture, relating to the corresponding concentrations. The course of the curve cannot be explained by a relative permittivity change due to the change in the solvent composition, since this property exhibits a monotonous decrease with increasing dioxane content of the solvent.

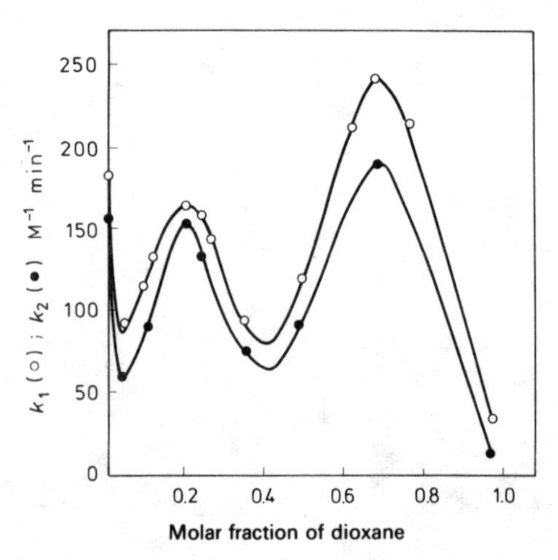

Fig. 7.2. Formation and dissociation rate constants of the mixed complex mercury(II) cyanide bromide as a function of the molar fraction of dioxane in a water–dioxane solvent mixture [Be 74]

A further interesting feature of these investigations is that, despite the great solvent dependence of the rate constants, the equilibrium constant of formation of the mixed complex was almost the same in the different solutions.

Competitive preferential solvation (COPS) theory

For a generally valid description of solvation reactions leading to the formation of a mixture of several solvates, i.e. to the stepwise formation of solvate complexes in solvent mixtures, Nagy *et al.* [Mu 74, Na 76] elaborated their competitive preferential solvation (COPS) theory, which is based on the following conditions:

(1) In the solvent mixture the components j and k compete for the solute i. The species formed as a result of this can be described with the equilibrium constants $K_{i(j)}$.

(2) The composition of the solvate sheath is determined by the concentrations C_j, C_k, ... of the individual solvent components, and by the values of the constants $K_{i(j)}$ relating to them. The higher the value of the product $K_{i(j)}C_j$, the greater is the extent of formation of solvate j (preferential solvation).

(3) The solute is distributed among the various solvates:

$$C_i = \chi_{i(j)} + \chi_{i(k)} + \ldots$$

where

$$\chi_{i(j)} = C_i \frac{K_{i(j)}C_j}{K_{i(j)}C_j + K_{i(k)}C_k + \ldots} C_i P_{i(j)}$$

(4) The effect of solvate formation on the reaction rate is additive:

$$k_{i(j+k)} = P_{i(j)}k_{i(j)} + P_{i(k)}k_{i(k)} + \ldots$$

where $k_{i(j)}$ and $k_{i(k)}$ are the rate constants of solvation with j and k, respectively, and $P_{i(j)}$ and $P_{i(k)}$ are the corresponding distribution factors.

A number of points are neglected in the Nagy scheme: (*a*) Since the stoichiometry of the solvates is not taken into consideration, in the calculation of the solvate concentrations the solvent concentration features to the first power in the $K_{i(j)}C_j$ expression (as if the solvate complexes all had 1 : 1 compositions). This simplification makes the calculation of the distribution factors $P_{i(j)}$ uncertain. (*b*) It does not assume the presence of "free", unsolvated solute. (*c*) "Free" solvent, which is not bound in solvates, features only in the calculation of the solvate concentrations; its solvate-independent effect on the reaction rate is not assumed.

Of the above neglected factors, (*b*) and (*c*) are probably justified, causing scarcely any error, but (*a*) may be a source of serious error. Unfortunately, our limited knowledge relating to the compositions of the solvates prevents the exact formulation of the appropriate equations.

Using several practical examples, Nagy *et al.* [Mu 74, Na 76] have shown that, even in spite of the above simplifications, their concept yields a fairly good description of the solvent dependence of the rates of certain reactions.

The competitive preferential solvation theory [Na 78, Na 79] could also be used for the interpretation of the solvent dependence of the kinetics of organic reactions. It contributed to the understanding of kinetic anomalies, e.g. in the solvent dependence of the n-butylaminolysis of tetrachloro-N-n-butylphthalimide.

Solvent dependence of reactions in organometallic chemistry

Besides those listed above, there is another difficulty in obtaining generalizations, on the basis of literature data, about the effect of the solvents on the kinetics and mechanisms of coordination chemical reactions: most of the simpler reactions have been investigated only in aqueous solutions. Non-aqueous solutions were only used, in general, if the solubility conditions or side reactions due to water (hydrolysis, etc.) hindered work in aqueous solution. Even then, each reaction, or possibly reaction type, was generally studied only in a single solvent.

Data suitable for the interpretation of the solvent effect are found in the field of the chemistry of organometallic complexes. Because of their chemical properties, these compounds are preferably investigated in non-aqueous solutions. Many of their reactions are slow, facilitating kinetic examinations. The difficulty in drawing clear conclusions with regard to the solvent effect here is a consequence of the relatively complicated nature of the systems and their reactions. Some examples of interest in this field are presented below.

Even the very early investigations on the chemistry of organometallic complexes [Zi 29, Zi 30] showed that the course and rate of the reactions of these compounds in solution are influenced considerably by the solvent.

According to Rochow *et al.* [Ro 57, 67], as a result of the interaction between an organometallic compound and a donor solvent (coordination of the donor atom of the solvent to the metal) the polarity of the metal–carbon bond in the organometallic compound decreases, causing a decrease in the reactivity of the compound. Pauson [Pa 68] states that the reactivity of the organometallic compound decreases in proportion to the increase in the donor strength of the solvent.

The opposite view is held by Okhlobystin [Ok 67], according to whom complex formation causes an increase in the polarity of the metal–carbon bond, which must be associated with an increase in the rate of heterolytic substitution and exchange reactions.

In effect, neither point of view can be regarded as holding generally, for the simple reason, among others, that the effect of the solvent on the reactions of organometallic complexes can only exceptionally be interpreted by the solvation of the organometallic molecule alone. The reaction may also be influenced to similar extent by the solvation of the reaction partner, or of the transition complex.

The effect of the former can be seen well, for instance, from a comparison of the reactions between iodine and tetraalkyl lead in benzene and in carbon tetrachloride [Pi 68]. In spite of the facts that both solvents are apolar and have small relative permittivities, and neither of them solvates the tetraalkyl lead molecule, the reaction takes place at a rate 15–20 times higher in benzene than in carbon tetrachloride solution. The explanation of the phenomenon is that with benzene iodine forms a charge-transfer complex of relatively high stability; this causes a greater polarization of the iodine molecule than does the very weak interaction between iodine and carbon tetrachloride.

Many publications have considered the effects of different solvents in the courses of various reactions of organometallic complexes, but there have been few investigations aiming at the interpretation of the solvent effect.

It was shown as early as in Ziegler's classical work [Zi 29, Zi 30] that organolithium compounds were much more reactive in diethyl ether than in solutions prepared with hydrocarbons. This fact has since been confirmed by a number of workers [Ea 63, Sh 66]. It has also emerged that the solvent alters not only the rate but also the mechanism of the reaction. For example, whereas the reaction between n-butyl lithium and benzyl chloride in n-hexane results in the formation of dibenzyl- and n-pentylbenzene, in tetrahydrofuran about 20% *trans*-stilbene is also obtained, in addition to the former two products [Ho 66].

It was pointed out by Schrettas and Eastham [Sc 66] and by Kovrizhnykh and Shatenshtein [Ko 69] that the reactions of organolithium compounds are influenced not only by the nature of the solvent, but also by its amount.

Numerous examinations have shown the effects of solvents in various reactions of the Grignard type [Cu 65, No 63, Za 63, Za 64, Za 65]. Most of these indicate that strongly solvating solvents increase the rates of nucleophilic substitution reactions. However, a number of data point to the reaction rate-decreasing effect of solvation [As 64, Be 63]. For instance, the Grignard reagent reacts with nitriles more slowly in polar solvents than in inert solvents [Be 63].

The apparently contradictory data are indicative of the fact that the effect of solvation is manifested by different mechanisms. According to Wakefield [Wa 66], the following three types of effects must be considered in Grignard reactions:

(1) The steric effect arising as a result of the coordination of the solvent to the transition complex; this influences the further course of the reaction.

(2) The effect of solvation on the strength and reactivity of the carbon–magnesium bond.

(3) The effect of solvation on the reactant.

In Wakefield's view, the solvation of the Grignard reagent makes both the Mg—C and the Mg—X bonds less stable, thereby increasing the reactivity of the carbon atom and favouring the formation of the solvated RMg^+ ion. The latter is more reactive than the undissociated molecule in both electrophilic and nucleophilic reactions. An increase in the solvating ability of the solvent beyond a certain limit, however, leads to the stabilization of the solvate of RMg^+ and thus to its lower reactivity. It can be seen that the above considerations may be employed to explain both a decrease and an increase in reactivity, but they are unsuitable for a prediction of the tendency and nature of the solvent effect.

Solvent effects have also been investigated in numerous reactions of organomercury compounds. For instance, interesting results were obtained in studies dealing with isotope exchange [Be 69, De 61, Ka 64, Re 61, Re 63]; these showed that the energy parameters of organomercury reactions reflect changes in different directions for compounds of different types. Thus, whereas the activation energy and entropy of isotope exchange between phenylmercury acetate and mercury bromide are smaller in aprotic solvents (e.g., pyridine, dimethylformamide, dimethyl sulphoxide) than in protic solvents (aqueous dioxan or ethanol), the energy and entropy values of chemical reactions such as that between benzylmercury chloride and elemental iodine are smaller in protic than in aprotic solvents [Be 67]. In these reactions the solvent may interact not only with the mercury atom of the compounds, but also with the halide or halogen moiety.

In their study of the reactions of organotin compounds, Gielen *et al.* [Gi 62–72] divided the solvent effect into two factors. They showed that whereas in polar solvents (methanol, dimethyl sulphoxide, dimethylformamide, etc.) the reactivity is governed by the steric effects of the substituents on the organotin compound, in apolar solvents (carbon tetrachloride, chlorobenzene, cyclohexane, etc.) inductive effects of these same substituents are manifested. Hence, the reactivity sequence of organotin compounds substituted in various ways is controlled by the solvent. According to this concept, in a polar medium the solvent behaves as a nucleophilic catalyst.

The solvent effect on the reactions of organotin compounds may also be explained by the fact that the nucleophilic solvent molecules solvate the transition complex formed in the course of the reaction [Na 72, Pe 73a, Pe 73b].

Even this brief survey shows that the kinetics and mechanisms of reactions taking place in solution are influenced by the solvents through very varied effects and in many ways. Examples of such effects are those caused by the solvation (with various strengths) of the reactants, of the transition or activated complexes and of the reaction products (in which process the solvent may be an electron pair donor or acceptor, a former of hydrogen bounds, or simply a polar molecule in a dipole–dipole or possibly only in electrostatic interaction), by the change of the dielectric properties of the system, or possibly of its viscosity or other physical properties, etc. The differentiation of these effects and the establishment of the predominant effect may be expected from a varied experimental investigation of the individual systems. In analogous systems, on the other hand, the solvent dependence may be described by empirical or semi-empirical correlations constructed on the basis of models that are of necessity simplified (see also Chapter 4).

References

Am 52 Amis, E. S.: J. Chem. Ed., **29,** 337 (1952).

Am 66 Amis, E. S.: Solvent Effects on Reaction Rates and Mechanisms. Academic Press, New York, London, 1966.

As 64 Ashby, E. C., Smith, M. B.: J. Am. Chem. Soc., **86,** 4363 (1964).

Be 63 Becker, E. I.: Trans. N. Y. Acad. Sci., **2,** 25, 513 (1963).

Be 67 Beletskaya, I. P., Fetisova, T. P., Reutov, O. A.: Izv. Akad. Nauk SSSR, Ser. Khim., **1967,** 990.

Be 69 Beletskaya, I. P., Zakharycheva, I. I., Reutov, O. A.: Zh. Org. Khim., **5,** 2087 (1969).

Be 71 Bennetto, H. P., Caldin, E. F.: J. Chem. Soc. *A*, **1971,** 2198.

Be 74 Beck, M. T., Porzsolt, E. Cs., Ling, J.: Reacton Kinetics Catal. Letters, **1,** 125 (1974).

Ca 73 Caldin, E. F., Bennetto, H. P.: J. Solution Chem., **2,** 217 (1973).

Ch 73 Chattopadhyay, P. K., Coetzee, J. F.: Inorg. Chem., **12,** 113 (1973).

Ch 74 Chattopadhyay, P. K., Coetzee, J. F.: Anal. Chem., **46,** 2014 (1974).

Ch 76a Chattopadhyay, P. K., Kratochvil, B.: Inorg. Chem., **15,** 3104 (1976).

Ch 76b Chattopadhyay, P. K., Kratochvil, B.: Canad. J. Chem., **54,** 2540 (1976).

Ch 77 Chattopadhyay, P. K., Kratochvil, B.: Canad. J. Chem., **55,** 1609, 3449 (1977).

Co 74 Coetzee, J. F., Chattopadhyay, P. K.: Abstracts IV. ICNAS, Vienna, 1974, p. 23.

Cu 65 Cuvigny, T., Normant, H.: Bull. Soc. Chim. France, **1964,** 2000; **1965** 1872, 1881.

Da 70 Dack, M. R. J.: Chem. Brit., **6,** 347 (1970).

De 61 Dessy, R. E., Lee, J. K., Kim, J.: J. Am. Chem. Soc., **83,** 1163 (1961).

Di 71 Dickert, F., Hoffmann, H.: Ber. Bunsenges. phys. Chem., **75,** 1320 (1971).

Di 72 Dickert, F., Wank, R.: Ber. Bunsenges. phys. Chem., **76,** 1028 (1972).

Di 74 Dickert, F., Hoffmann, H., Janjic, T.: Ber. Bunsenges. phys. Chem., **78,** 712 (1974).

Ea 63 Eastham, J. K., Gibson, G. W.: J. Am. Chem. Soc., **85,** 2171 (1963).

Ei 65 Eigen, M., Wilkins, R. G.: Adv. Chem. Ser. **49,** 55 (1965).

El 80 Elias, H., Fröhn, U., Irmer, A., Wannowius, K. J.: Inorg. Chem. **19,** 869 (1980).

Fi 72 Fischer, P., Hoffmann, H., Platz, G.: Ber. Bunsenges. phys. Chem., **76,** 1060 (1972).

Fr 57 Frank, H. S., Wen, W. Y.: Discuss. Faraday Soc., **24,** 133 (1957).

Fu 58 Fuoss, R. M.: J. Am. Chem. Soc., **80,** 5059 (1958).

Gi 62–64 Gielen, M., Nasielski, J. *et al.*: Bull. Soc. Chim. Belg., **71,** 32 (1962); **73,** 214, 293 (1964)

Gu 71 Gutmann, V., Schmid, R.: Mh. Chem., **102,** 1217 (1971).

Gu 72 Gutmann, V.: Topics in Current Chemistry **27,** 98 (1972).

Gu 74 Gutmann, V., Schmid, R.: Coord. Chem. Rev., **12,** 263 (1974).

He 70 Hewkin, D. J., Prince, R. H.: Cord. Chem. Rev., **5,** 45 (1970).

Ho 66 Hoeg, D. F., Lusk, D. L.: Organometal. Chem., **5,** 1 (1966).

Hu 35 Hughes, E. D., Ingold, C. K.: J. Chem. Soc., **1935,** 244

Ka 64 Kalyavin, V. A., Smolina, T. A., Reutov, O. A.: Dokl. Akad. Nauk SSSR, **156,** 95 (1964).

Ki 34 Kirkwood, J. G.: J. Chem. Phys., **2,** 351 (1934).

Ko 69 Kovrizhnykh, E. A., Shatenshtein, A. I.: Usp. Khim., **38,** 1836 (1969).

Kr 75 Krygowski, T. M., Fawcett, W. R.: J. Am. Chem. Soc., **97,** 2143 (1975).

Ku 70 Kustin, K., Swinehart, J.: Inorg. Reaction Mechanisms. Vol. 13 (Ed. Edwards, J. O.), Interscience, New York, 1970, p. 107.

La 56 Laidler, K. J., Landskroener, P. A.: Trans. Faraday Soc., **52,** 200 (1956).

Ma 75 Mayer, U.: Pure Appl. Chem., **41,** 291 (1975).

Me 90 Menschutkin, N.: Z. Phys. Chem., **5,** 589 (1890); **6,** 41 (1890).

Mu 74 Mukanawa Muanda, B. Nagy, J., B. Nagy, O.: Tetrahedron Letters, **38,** 3421 (1974).

Na 72 Nasielski, J.: Pure Appl. Chem., **30,** 3 (1972).

Na 76 B. Nagy, O., B. Nagy, J.: Environmental Effects on Molecular Structure and Properties. D. Reidel Publ. Co., Dordrecht, 1976, pp. 179–203.

Na 78 B. Nagy, O., Wa Muanda, M., B. Nagy, J.: J. Chem. Soc. Faraday I, **74,** 2210 (1978).

Na 79 B. Nagy, O., Wa Muanda, M., B. Nagy, J.: J. Phys. Chem., **83,** 1961 (1979).

No 63 Normant, J.: Bull. Soc. Chim. France, **1963,** 1868, 1876, 1888.

Ok 67 Okhlobystin, O. Yu: Usp. Khim., **36,** 34 (1967).

Pa 69 Parker, A. J.: Chem. Rev., **69,** 1 (1969).

Pa 68 Pauson, P. L.: Organometallic Chemistry. Edward Arnold, London, 1968.

Pe 73a Petrosyan, V. S., Reutov, O. A.: J. Organometal. Chem., **52,** 307 (1973).

Pe 73b Petrosyan, V. S., Yashina, N. S., Reutov, O. A.: J. Organometal. Chem., **52,** 315 (1973).

Pi 68 Pilloni, G., Tagliavini, G.: J. Organometal. Chem., **11,** 557 (1968).

Rá 76 Rábai, Gy., Bazsa, Gy., Beck, M.: Magy. Kém. Foly., **82,** 60 (1976).

Re 61 Reutov, O. A., Sokolov, V. I., Beletskaya, I. P.: Izv. Nauk SSSR, Ser. Khim., **1961,** 1213, 1217, 1247, 1561.

Re 63 Reutov, O. A., Sokolov, V. I., Beletskaya, I. P.: Izv. Akad. Nauk. SSSR, **1963,** Ser. Khim. 965, 967.

Ro 59 Robinson, R. E., Heppolette, R. L., Scott, J. W.: Canad. J. Chem., **37,** 803 (1959).

Ro 57 Rochow, E. G., Hurd, D. T., Lewis, R. N.: The Chemistry of Organometallic Compounds. Wiley, New York, 1957.

Ro 67 Rochow, E. G.: Organometallic Chemistry, Reinhold, New York, 1967.

Sc 32 Scatchard, G.: Chem. Rev., **10,** 229 (1932).

Sc 66 Schrettas, G. G., Eatsham, J. F.: J. Am. Chem. Soc., **88,** 5668 (1966).

Sc 76 Schmid, R., Sapunov, V. V., Gutmann, V.: Ber. Bunsenges. phys. Chem., **80,** 456, 1302, 1307 (1976).

Sh 66 Shatenshtein, A. I., Kovrizhnykh, E. A., Basmanova, V. M.: Kinet. Katal., **7,** 957 (1966).

Sm 61 Smith, S. G., Fainberg, A. H., Winstein, S.: J. Am. Chem. Soc., **83,** 618 (1961).

Ta 76 Tanaka, M.: Inorg. Chem., **15,** 2325 (1976).

Wa 66 Wakefield, B. J.: Organometal. Chem. Rev., **1,** 131 (1966).

Za 63 Zakharkin, L. I., Bilevitch, K. A., Okhlobystin, O. Yu.: Dokl. Akad. Nauk SSSR, **152,** 338 (1963).

Za 64 Zakharkin, L. I., Okhlobystin, O. Yu., Bilevitch, K. A., J. Organometal. Chem., **2,** 309 (1964).

Za 65 Zakharkin, L. I., Okhlobystin, O. Yu., Bilevitch, K. A.: Tetrahedron **21,** 881 (1965).

Zi 29 Ziegler, K., Grossmann, F., Schöfer, O.: Justus Liebigs Ann. Chem., **473,** 1 (1929).

Zi 30 Ziegler, K., Colonius, H.: Justus Liebigs Ann. Chem., **479,** 135 (1930).

8. INTERACTIONS IN SOLVENT MIXTURES AND THEIR INVESTIGATION

In previous chapters, mainly solvent effects occurring in pure non-aqueous solvents have been discussed. This survey has shown that, even in the simplest systems, where a single solute is present in a given homogeneous solvent, it is necessary to take account of very complicated interactions, which are frequently difficult to follow. The higher the number of components in a system, the more complicated will be these interactions and the more difficult will be their interpretation.

During coordination chemical equilibrium and kinetic studies, particularly with complexes involving organic ligands, the solubility conditions of the complex or the ligand may require that the examinations be carried out not in a single pure solvent, but in a mixture of (generally two) solvents. One of the two solvents is usually water. The interpretation of studies in such solvent mixtures is made especially difficult by the fact that the system is considerably influenced not only by the interaction between the dissolved complex and the solvent, but also by the interactions between the individual components of the solvent mixture itself. Accordingly, in such cases the study of the solvent effect demands a knowledge of the interactions between the solvent components, too.

Apart from the interactions between the components of the solvent mixture, the picture of the system is made more complicated by the complexity of the solvation of the dissolved ions. Even in the case of a solvent mixture consisting of solvents with similar properties, one cannot be sure that the dissolved ions will be solvated to the same extent by the components of the solvent (in accordance with their concentration proportions). Special, involved and laborious examinations are required to establish the extent of solvation of an ion or a complex by the individual components of the solvent mixture. For this reason, the results achieved in this field are rather limited and in many cases also contradictory.

In addition to the specific interactions between the two components of the solvent mixture, and between the solvent and the solute, great effects are exerted on the equilibria in the solution by the dielectric and other properties, determined by the composition of the solvent mixture (viscosity, density, etc.).

Interaction between the components of the solvent mixture

With the aid of various methods of investigation, over a long period Erdey-Grúz *et al.* [Er 58–73] dealt with the study of interactions between the individual components of solvent mixtures. They found that, even in mixtures of chemically related substances, it was necessary to take into account interactions of various types, occurring to at least a slight extent.

The study of the *viscosities* of solvent mixtures, and of the composition- and concentration-dependent *viscosity changes*, led to fundamental results. The viscosities of methanol–water [Er 58], ethanol–water and propanol–water [Er 59a, Er 68a] solvent mixtures display maxima that depend on the alcohol concentration. The longer the carbon chain in the alcohol, the higher is the value of the maximum in the viscosity curve. In polyalcohol–water systems, on the other hand, viscosity increases monotonously with increase in the alcohol concentration [Er 59b, Er 68b], and no viscosity maximum appears.

This difference in the effects of the two types of alcohol on the viscosity permits conclusions concerning the structures of the two types of solvent mixture to be drawn [Ag 67]. Water molecules in pure water are known to be linked by hydrogen bonds to form various associates. Each water molecule is capable of forming two independent hydrogen bonds. Accordingly, if it is assumed that the associates are open chains, or contain open chains, these terminate in free hydroxyl groups. *Via* the latter, further water molecules may be linked to the chain-terminal water molecules; alternatively, these chain-terminal water molecules may readily undergo cleavage and/or exchange with other water molecules. This breaking-down and building-up process presumably has a viscosity-decreasing effect.

Monohydric alcohols can form only one hydrogen bond, since they contain a single hydroxyl group; hence their attachment to the end of an open-chain water complex means the termination of that chain. This hinders the above-mentioned breaking-down and rebuilding processes. The larger associates formed in this way result in an increase in the viscosity of the solution. In a pure alcohol, on the other hand, the associates present display an appreciably lower degree of polymerization (generally being dimeric), and have lower viscosities than the above water–alcohol formations. Assuming that the latter have definite compositions, Erdey-Grúz *et al.* estimated the compositions of these formations on the basis of the compositions of the solvent mixtures corresponding to the viscosity maxima. For example, they showed that in solvent mixtures of water with methanol, ethanol or propanol, species containing eight water molecules, terminated by alcohol molecules at their two ends, predominate at room temperature at the positions of the viscosity maxima. With elevation of the temperature, these progressively break down, and smaller complexes are formed.

Since they are capable of forming more than one independent hydrogen bond, polyalcohols naturally do not terminate the water chains, and therefore a maximum does not appear in the viscosity curve of such a system as a function of the alcohol concentration.

However attractive the above explanation may be, it is only a hypothesis and not a proved fact; particular attention was drawn to this by Erdey-Grúz [Er 74]. According to another concept for instance that of Mitchell and Wynne-Jones [Mi 53] and Mihajlov [Mi 61], these phenomena can also be interpreted by means of the filling of the structural cavities in water.

When the alcohol is present in low concentration in the solvent mixture, the alcohol molecules are situated in the structural cavities of water, deforming but not breaking down the liquid structure of the water. If the alcohol concentration is increased to such an extent that there is no longer room for all of the alcohol molecules in the structural cavities, a new type of solution structure develops, which influences the properties of the solvent in a different way from the previous effect that caused the deformation of the structural cavities. In effect, properties varying in accordance with the different extremes can be interpreted with this theory.

It should be noted that the two theories do not exclude each other. It is conceivable that the two effects (linkage of the alcohol molecules to the water associates by means of hydrogen bonds, and the filling of the structural cavities) are exerted simultaneously, and jointly influence the properties of the solvent.

The strong interaction between alcohols and water is also well reflected by *conductivity measurements* on electrolyte solutions, e.g., hydrochloric acid in water–methanol solvent mixtures. It was demonstrated by Erdey-Grúz and Majthényi [Er 58] that in this solvent mixture (at a methanol content of ca. 90 mole-%) the equivalent conductivity of the hydrogen ion is surprisingly low, hardly exceeding that of the chloride ion. This shows that the predominant part of the conductivity of the hydrogen ion in this solvent mixture originates from the hydrodynamic migration of the hydrogen ion; this may be explained by the methanol presumably cleaving the hydrogen bonds holding together the water associates, thereby greatly reducing the possibility of proton transfer from one water molecule to another and thus prototropic conductivity. With a further increase in the methanol concentration (above 90 mole-%), there is an increase in the number of hydrogen ions transferred, which indicates that in methanol with a low water content the proton transfer occurs with the mediation of the methanol molecules. In water with a low methanol concentration, on the other hand, the reason why the number of hydrogen ions transferred increases in the presence of the alcohol is that this transfer is promoted as the structural cavities in the water are filled up by alcohol molecules, which may result in strengthening of the liquid structure [Er 59c, Er 67, Er 71a, We 61]. The above correlations are supported by studies on the *auto-diffusion* of methanol-containing solvent mixtures [Er 71b].

Measurement of the auto-diffusion coefficient, and the activation energies of auto-diffusion, have also yielded valuable information with regard to the interaction between the two components in water–dioxane solvent mixtures [Er 71c]. It was shown experimentally that with a water content in the range 0–60 mole-% the solution structure is hardly changed by the water; it is only in mixtures with higher water contents that the liquid structure based on hydrogen bonding between the water molecules can develop. Analogous conclusions could be drawn from conductivity measurements made in the same solvent mixtures [Er 71a].

The significant nature of the interaction between dioxane and water molecules is also shown by the fact that the *dipole moment* of water molecules dissolved in dioxane is larger than that of free water molecules. This indicates that the dielectric polarization of the water molecule is greater in dioxane solution than in the gaseous state, owing to hydrogen bonding between the water and the dioxane molecules [To 61, Wi 30]. The linkage between water and dioxane is also proved by viscosity measurements [Sk 68], on the basis of which it was concluded that, in a solution with a dioxane of molar fraction 0.2, four molecules of water are coupled to one molecule of dioxane. The formation of these complexes results in a strengthening of the liquid structure and hence an increase in viscosity. At higher dioxane concentrations, the water content is no longer sufficient for the development of these associates, and a decrease in viscosity is observed.

These considerations together reflect the occurrence of the water–dioxane interaction; indeed, they have proved suitable for the interpretation of the changes in certain physical properties of the solvent mixture as a function of the composition. It can further be seen that the combined evaluation of the results of various methods of investigation may be expected to yield a closer understanding of the structures of the solvents.

An analogous conclusion was drawn by Ratkovics *et al.* from a series of studies on the properties of alcohol–amine mixtures. In systematically selected binary model systems, composed of a primary, secondary or tertiary amine and a member of the homologous series of alcohols, these authors examined the compositions and concentrations of the associates formed in solution. They found that numerous physico-chemical characteristics of the mixtures and their components were in such a close correlation with the association phenomena that an understanding of the complicated equilibria might be hoped for on the basis of their joint investigation.

Hence, they carried out measurements on the vapour–liquid equilibria [Ra 73a] and the heats of mixing [Ra 73b, Ra 74a, Ra 76d], and studied the dielectric properties of the systems [Li 74, Ra 76a], their viscosities [Ra 74b, Ra 74c, Ra 76b, Ra 77c] and the electric conductances of the mixtures [Ra 74d, Ra 76c, Ra 77a, Ra 77b].

As a result of these many-sided examinations, they confirmed the formation of various alcohol–amine mixed associates and auto-associates of the components in the systems. They were able to demonstrate the effects of the compositions and structures of the components and of their concentration on the sizes and structures of the associates and on the association processes. In certain cases it was even possible to draw conclusions on the form of the associates (chain or ring).

This work has the great value that it demonstrates from many aspects the complicated association equilibria that must be considered in comparatively simple binary systems, which are mixtures of two dipole molecules (electrolytically neutral, containing at most one donor atom each, and a few acidic hydrogen atoms). On the basis of these investigations, it may be conceived how much more complicated the system becomes if the coordination reactions of some metal ion also play a role in such or similar solvent mixtures.

Infrared spectra of acetone in the C═O stretching region have been studied as a function of the composition of mixed protic-aprotic solvents [Sy 80]. For methanolic systems, equilibria between hydrogen-bonded and non hydrogen-bonded acetone dominate the spectral changes, but for aqueous systems at least four distinct components need to be considered.

The reaction conditions ensured by one of the components may be such (e.g., low relative permittivity), that, besides the species formed as a result of association between the different components of the solvent mixture, it is necessary also to take into consideration the auto-association reactions of the other component: its dimerization, and possibly its polymerization. With *infrared spectrophotometric* measurements, for instance, Náray and Liszi [Na 73] showed that cyclohexanol undergoes dimerization even at low concentrations in apolar solvents. A series of examinations in a wider concentration range, based on the measurement of dielectric properties, also pointed to the formation of cyclic tetramers and chain associate [Li 73].

In the case of highly polar molecules there is not even any need for coordinative interactions (e.g., hydrogen bonding) for the formation of the associates. In polar solvents, therefore, the electrostatic interactions between the molecules may also lead to dipole associates. In the associates the monomer molecules may be situated in either a parallel or an antiparallel arrangement. In the case of an antiparallel arrangement, the dipole moments of the monomer molecules are in opposite directions, whereas in a parallel arrangement they are in the same direction. If the arrangement is fully antiparallel, the directional polarization becomes zero; the associated dipoles mutually counteract the effect of one another, and the polarization will be equal to the sum of the electronic and atomic polarizations.

Liszi [Li 75] studied the phenomenon of dipole association in nitrobenzene–carbon tetrachloride, nitrobenzene–n-heptane and nitrobenzene–benzene mixtures, i.e., mixtures in which the second component is an apolar solvent. With a knowledge of the densities, he used the dielectrometrically (at a constant frequency of 3 MHz) determined relative permittivities of the solutions to calculate the molar polarizations of the mixtures, and the dependence of these on the composition. In this way, conclusions were drawn on the formation of the dipole auto-associates. These data were used, with the aid of the Sugden equation [Su 34], to calculate the permanent dipole moment of nitrobenzene. It was concluded from the examinations that the nitrobenzene molecules form dipole associations with an antiparallel arrangement. All of these findings lose their validity, however, in mixtures in which the component accompanying the nitrobenzene is also a polar solvent.

In addition to the methods of investigation mentioned so far, which may be regarded as simple "classical" methods, even more extensive use is being made of the various relaxation procedures to study the interactions (primarily hydrogen bonding) between the components of solvent mixtures. Among others, such investigations in systems of water–alcohol [Go 69, Oa 73], water–acetone [Re 70] and methanol–apolar solvent [In 75] are worthy of mention.

Attempts are also made to calculate the different parts of the thermodynamic properties of mixing. Thus, Liszi [Li 76a] dealt with the calculation of that part of the internal energy change, the excess free energy change and the excess entropy change of mixing which arise from the dipolar interactions. With a modified procedure based on the Onsager theory [On 36] relating to polar liquids, he derived correlations for binary mixtures, but also presented the means of generalization for multi-component systems. Similar interest is attached to the considerations of Liszi [Li 76b, Li 76c] on the calculation of the molar polarizations and the partial molar polarizations of binary liquid mixtures.

The interactions between the components of a solvent mixture can also be followed by means of quantum chemical calculations. In this way, Hinton [Hi 74] found strong evidence suggesting the existence of various hydrogen-bonded associates of the two components in water–formamide mixtures. At present this method can be employed to only a limited extent. The results of the calculations may be accepted as definitely correct only after experimental confirmation.

Solvation processes in solvent mixtures

Solvation processes occurring in a solvent mixture, i.e., the interactions between the solvent and the solute, are even more involved than the interactions between the solvent components themselves. Padova [Pa 68] stated that it is the partial molar free enthalpy of solvation of the components that determines which of the components of the solvent mixture will solvate the dissolved ion. Naturally, this is true in general, but it hardly gives a factual basis for the interpretation of the special interactions.

To a first approximation it is also true that in solvent mixtures containing water, the formation of the aquo complex is favoured over solvation with the other component. Many exceptions are found, however. For example, by means of NMR examinations Covington *et al.* [Co 74a] demonstrated that, whereas the sodium ion is indeed solvated more strongly by the water in a water–acetonitrile mixture, in an ethylenediamine–water mixture solvation by ethylenediamine is preferred. When caesium nitrate is dissolved in a dimethyl sulphoxide–water solvent mixture, both the cation and the anion interact more strongly with the dimethyl sulphoxide molecule than with the water molecule; this is particularly surprising in the case of the anion, since dimethyl sulphoxide is a very weak acceptor molecule, forming no hydrogen bond. In an acetonitrile–water solvent mixture, the nitrate ion undoubtedly interacts more strongly with the water molecules. These findings were deduced by Covington *et al.* [Co 74a] from the dependence of the NMR shift of the dissolved ion on the composition of the solvent mixture; they are also supported by electron-excitation spectrophotometric measurements.

^{23}Na-chemical shifts for the $NaClO_4$ solute was shown to depend markedly upon the composition of binary solvent mixtures of THF with amines (pyridine, peperidine, pyrrolidine, aniline, propylamine, and isopropylamine, respectively) [De 81] indicating the preferential solvation of the sodium ion by the amines.

Costa *et al.* [Co 74b] employed viscosity measurements to study the effects due to the different solvating powers of the components of solvent mixtures. In lithium chloride solutions prepared with water–alcohol mixtures of various compositions, they demonstrated the dependence of the viscosity of the solution on the composition of the solvent mixture. From the nature of the viscosity curve and from the position of the extreme in the curve, conclusions were drawn on the different solvating abilities of the components of the solvent. Information appearing to be especially promising is that which may be obtained from the dependence of constant B of the Jones–Dole equation on the composition of the solvent mixture.

Many studies of complex equilibria are carried out in water–dioxane solvent mixtures, since these may be examined with the customary pH-metric methods, even in the case of comparatively high dioxane contents. Early considerations set out from the assumption that in these solvent mixtures the dissolved ions are solvated virtually exclusively by the water. It was shown by Grunwald [Gr 62], however, that the formation of dioxane complexes of the cations must also be taken into consideration. Dioxane may coordinate to the cations to a significant extent through its oxygen donor atom. The energy of the cation–solvent dipole interaction in the case of water did not prove more negative than for dioxane. In the solvation of the anions involving hydrogen bonding, the role of the dioxane is naturally a subordinate one. It should be noted that as the dioxane content of the solvent mixture is raised, the water associates progressively break down [Er 73], resulting in an increase in the concentration of monomeric water molecules, which have a stronger tendency to coordination.

In water–organic solvent mixtures the decrease in the water concentration is generally accompanied by degradation of the water associates and hence by the formation of smaller associates, which finally leads to the liberation of monomeric water molecules. It must be noted, however, that this process does not result in enhancement of the donor properties of the water in every system. For example, Moreau and Douhéret [Mo 74] reported that, in a water–acetonitrile solvent mixture, the breakdown of the water structure as a consequence of the addition of acetonitrile is not associated with an increase in the solvation of the proton.

A complete understanding of the solvation conditions in solvent mixtures is made extremely difficult by the fact that, in addition to the parent solvates formed with the two solvent components, mixed solvates simultaneously containing both solvents are also formed: in fact, in systems with low relative permittivities the central metal ion may be capable of binding not only the two kinds of solvent molecule, but also the anion, in its first coordination sphere.

NMR spectroscopic examinations of cobalt(II) perchlorate and cobalt(II) nitrate carried out in a water–acetone solvent mixture by Ablov *et al.* [Ab 72, Gu 73] revealed that it is necessary to take account of the above, extremely complex interactions even in systems that seem to be comparatively simple.

^{1}H- and ^{13}C-NMR studies by Dickert [Di 77] in a dimethylformamide–methanol solvent mixture showed that the nickel(II) ion coordinates first of all (most strongly) two dimethylformamide molecules. Occupation of the other coordination sites takes place in a proportion corresponding to the composition of the solvent mixture. A point of special interest is the finding that the coordination of the dimethylformamide increases the rate of exchange of the methanol molecules bound to the nickel.

The solvation number of the cobalt(II) ion in aqueous solutions [Ma 68a] and in solutions prepared with polar solvents in general [Mu 64a, Lu 64b, Ma 67, Th 67] is known to be six. Few data are available with respect to how the coordination numbers of ions change if the pure solvent is diluted with some other solvent miscible with it.

By means of the resonance absorptions of the ^{13}C atom of acetone and the ^{35}Cl of the perchlorate anion, Gulja *et al.* [Gu 73] found that, depending on the acetone content, mono- and diacetone solvates are formed in solutions of cobalt(II) perchlorate in a water–acetone solvent mixture. In the diacetone solvate one of the coordination sites of the cobalt(II) ion is occupied by the perchlorate ion, and the other three coordination sites by water molecules.

When these results were compared with the data from an analogous investigation on cobalt(II) nitrate solutions, where the strength of the interaction between the cobalt and the nitrate was followed via the ^{14}N resonance absorption, it could be demonstrated that the increase in the acetone content of the mixture favoured the coordination of the nitrate ion [Ab 72]. On the basis of these studies, therefore, it is clear that, as a consequence of the elevation of the acetone content of the solvent mixture, an enhancement is observed not only in the coordination of the acetone to the cobalt, but also in the coordination of the anion.

The dilution with acetone of aqueous solutions containing aluminium ion showed that the acetone is not coordinated to the aluminium in solutions with low acetone contents [Fr 69, Ma 68b]. Ethylene glycol and propylene glycol have been shown to have a solvating power in different solvent mixtures [Bo 79, Ka 80, No 80].

In simpler cases, solubility measurements in solvent mixtures may also provide information on the solvating effects of the solvent components. On the basis of such solubility measurements in solutions of HgI_2 in a dimethylformamide–dimethyl sulphoxide solvent mixture, for instance, Gaizer and Beck [Ga 67] have shown that HgI_2 forms a mixed solvate with the two solvents. In an analogous study, they drew attention to the phenomenon that in a dimethylformamide–water system mixed solvates HgI_2–DMF–H_2O are formed.

Complex formation in solvent mixtures

In most complex-equilibrium investigations in solvent mixtures, the choice of the solvent was determined by the solubilities of the components of the system. The aim of the investigations, therefore, was not the detection of the solvent effect, but the determination of the compositions and stabilities of the complexes formed in solution, and possibly some of the factors governing these. Relatively little attention has been devoted to the study of the solvent effect in solvent mixtures.

Selection of the composition of the solvent mixture in accordance with the above meant that, in most examinations, one of the components of the mixture was water, and the other was an organic solvent serving to increase the solubility of the complex. The preparation of such solvent mixtures can be regarded as the dilution of the aqueous solution with some organic solvent miscible with water. This process results in shifts in the association–dissociation equilibria existing in solution. As mentioned in the preceding section, there may be various reasons for this: the water structure breaks down, new solvation equilibria arise, the decrease in the water activity causes decreases in the concentration and stability of the aquo complex, the change in the dielectric properties of the medium results in variations in the activity coefficients of the ions participating in the equilibria, etc.

Solvent dependence of complex stability

Published results of equilibrium studies in solvent mixtures show that in water–organic solvent mixtures an increase in the concentration of the organic component is generally accompanied by an enhancement of the complex stability. A number of different explanations may be given for this.

In systems where only water molecules are coordinated to the metal ion (but not the organic component), the cause of the higher complex stability is the lower stability of the aquo complex, which is due to the reduced water activity resulting from dilution by the organic solvent [Be 70]. A similar effect may be achieved if the water activity of an aqueous solution is diminished by the dissolution of an inert salt, as shown by the complex stability measurements of Burger *et al.* [Bu 68] in concentrated alkali metal perchlorate solutions.

In a system containing a ligand strongly solvated by water, a decrease in the water activity is accompanied by a decrease in the solvation of the ligand, which again results in a higher complex stability.

In solvent mixtures in which the organic component also is coordinated, the higher complex stability in the solvent mixture may be due to the lower stability of the resulting complex in comparison with the aquo complex.

Naturally, in systems where the organic solvent forms a more stabile solvate complex with the metal ion than water does, an increase in the concentration of the organic component must cause a reduction in the complex stability. Few such data are known in the literature. Similarly, in systems containing an organic solvent that solvates the ligand more strongly than does water, an increase in the concentration of the organic solvent favours the advance of the dissociation equilibrium, and hence a decrease in the stability of the complex.

With the exception of formamide, the relative permittivities of organic solvents miscible with water are smaller than that of water. An increase in the concentration of the organic component of a solvent mixture therefore causes a decrease in the relative permittivity of the system; in the case of the association of charged species, this favours their electrostatic interaction and hence an increase of the stability of the complex.

It may be seen that, on the basis of the above qualitative considerations, the dependence of the stability of the complex on the composition of the solvent mixture can often be predicted, or at least explained. However, the situation is, in general, too complicated to permit differentiation of the above factors from one another, and thus separate evaluation of their effects.

With proton-acceptor ligands, in addition to the factors mentioned earlier, a significant effect is also exerted on the complex stability by the solvent dependence of the protonation of the ligand. It has been demonstrated that, whereas with ligands containing a carboxylate donor group the protonation constant of the ligand usually increases with increase in the concentration of the organic component of the solvent mixture (similarly to the complex stability constants), this regularity is not manifested with ligands containing an amino group [Bu 79, Fa 70], and the opposite trend could also be observed [Bu 81a], namely a decrease in the protonation constants of the carboxylate groups and an increase in those of the amino groups due to the increased concentration of the organic component in water–organic solvent mixtures. Since the solvent dependence of the protonation constants plays a role in the solvent dependence of the complex stability constants, a knowledge of this is indispensable for an interpretation of the solvent effect.

To a first approximation, the solvent dependence of the protonation constants is determined by two factors: the basicity of the solvent molecules, and the relative permittivity, which governs the interactions of charged species.

In addition to these factors, various specific interactions between the solvent and the ligand, and between the solvent and the stepwise-formed complexes, may also influence the solvent dependence of the stability constants and structure of the complex in the various systems. In most cases, therefore, it is very difficult to find a simple correlation between the complex stabilities and some physical or chemical property of the solvent mixture or its components. The complexity of this question is well illustrated by the equilibrium studies by Gaizer *et al.* [Ga 74a], in which the stability of the pyridine-2-carboxylic acid complex of cobalt(II) was determined in 50% and 75% dioxane, 75% acetone and 75% acetonitrile containing water–organic solvent mixtures. The logarithms of the protonation constants of the ligand in the various solvent mixtures and the relative permittivities of the solutions are listed in Table 8.1; the logarithms of the complex stability constants are given in Table 8.2.

Table 8.1. Logarithms of the protonation constants (K) of pyridine-2-carboxylate in various solvents, and the relative permittivities (ε) of the solutions [Ga 74a]

Solvent	log K	ε
Water	5.12	81.0
50% v/v dioxane	5.44	37.0
75% v/v dioxane	5.91	15.9
75% v/v acetone	5.68	38.2
75% v/v acetonitrile	5.95	48.5

(The latter ones have been determined by spectrophotometric measurements at four different wave numbers. The data in the table show the accuracy of measurement and evaluation.)

The tabulated data show that the protonation constants measured in the solvent mixtures (the reciprocals of the acid dissociation constants) are larger than those measured in aqueous solutions. However, their sequence in increasing order does not follow the decreasing sequence of the relative permittivities of the solutions, as would be expected on the basis of electrostatic considerations. For example, the value of the protonation constant measured in 75% v/v acetonitrile (which has a high relative permittivity of $\varepsilon = 48.5$) agrees, within the limits of experimental error, with that measured in 75% v/v dioxane (which has a low relative permittivity of $\varepsilon = 15.9$).

Table 8.2. Logarithms of the stability constants (β) of the pyridine-2-carboxylate complex of cobalt(II) in solvent mixtures [Ga 74a]

Solvent	ν nm	$\log \beta_1$	$\log \beta_2$	$\log \beta_3$
Water	335	6.69	10.36	14.51
	340	6.48	10.70	14.59
50% v/v dioxane	335	5.93	10.98	15.58
	340	5.89	11.02	15.56
75% v/v dioxane	335	5.74	10.95	15.60
	340	5.69	10.95	15.57
75% v/v acetone	335	5.55	10.92	15.68
	340	5.81	10.85	15.57
	350	5.70	11.12	15.73
	360	5.68	11.16	15.69
75% v/v acetonitrile	335	5.54	10.42	15.22
	340	5.55	10.49	15.18

The solvent dependence of the stepwise complex stability constants exhibits some differences (see Table 8.2). The value of $\log \beta_1$ is larger in aqueous solution than in the solvent mixtures, whereas the values of $\log \beta_2$ and $\log \beta_3$ are larger in the solvent mixtures (as expected on the basis of the protonation constants). The values of the stability constants measured in the solvent mixtures display scarcely any dependence on the composition of the solvent, and do not follow either the sequence of the relative permittivities of the solutions or the sequence of the protonation constants of the ligand.

The low solvent dependence of the stability constants, and their independence from the protonation constants, indicate that the stability of the complex in this system is controlled primarily by the strength of the bonding between the metal and the nitrogen of the pyridine, the effect of the coordination of the carboxylate oxygen being a secondary one.

With the aid of the equilibrium constants determined in the different solutions, Gaizer *et al.* calculated how the concentrations of the complexes in the solutions vary with the change in the concentration of the free ligand. Figure 8.1 was constructed from the results. It can be seen that, whereas the complexes of composition MA and MA_3 predominate over a wide free ligand concentration range in aqueous solution, the species MA_2 playing a subordinate role, in all of the solvent mixtures the concentrations of MA_2 and MA_3 increase significantly at the expense of MA.

These investigations reveal that, even in this comparatively simple system, it is not possible to give a clear interpretation of the effects of solvents on the stabilities of complexes.

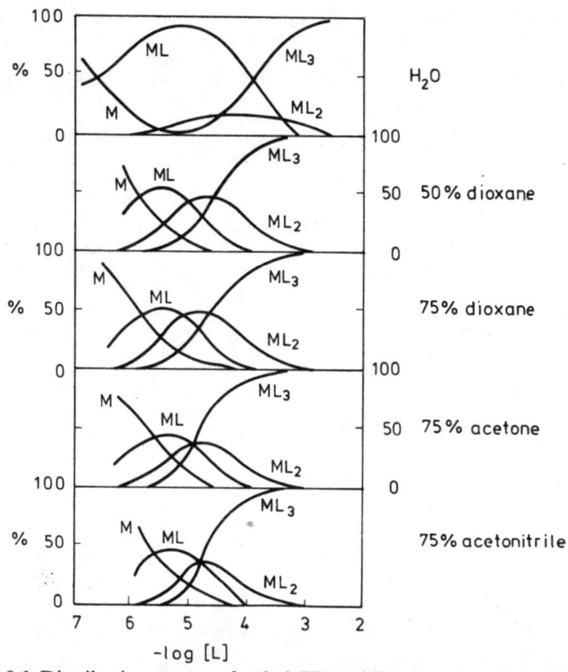

Fig. 8.1. Distribution curves of cobalt(II)-pyridine-2-carboxylic acid complexes in water and in various solvent mixtures [Ga 74a]

The solvent dependence of coordination equilibria is even more complicated in systems in which the solvent change also influences the conformation of the organic ligand. The protonation constants of two corticotropin fragments (of the N-terminal four amino acid-containing $ACTH_{1-4}$ and of the N-terminal 32 amino acid-containing $ACTH_{1-32}$ polypeptides) measured in water and in trifluoro-ethanol–water and propylene glycol–water solvent mixtures (of 1:1 v/v composition) are shown in Table 8.3 [Bu 81b, No 80].

By comparing the data, conclusions can be drawn on the effect of solvents on the protonation processes. Since $ACTH_{1-4}$ does not undergo structural changes due to the solvent effect its protonation constants reflect the variation in the basicity of its functional groups (carboxylate, amino and phenolic hydroxyls) on changing the solvent.

Terminal carboxyl groups show higher protonation constants in solvent mixtures (log $K = 3.88$ and 3.89, whereas log $K = 3.20$ was obtained in water), which can be attributed to the lower relative permittivity of solvent mixtures favouring the formation of neutral particles ($RCOO^- + H^+ \rightleftharpoons RCOOH$).

Table 8.3. Protonation constants of ACTH fragments in aqueous solutions and solvent mixtures (logarithmic values) [Bu 81b, No 80]

Functional group	Aqueous solution		Trifluoroethanol–water		Propylene glycol–water	
	$ACTH_{1-4}$	$ACTH_{1-32}$	$ACTH_{1-4}$	$ACTH_{1-32}$	$ACTH_{1-4}$	$ACTH_{1-32}$
Terminal COO^-	3.20	3.54	3.88	4.08	3.84	4.00
Glutamic acid COO^-		3.59		4.17		4.21
		4.10		4.73		4.46
Aspartic acid COO^-		4.25		4.80		5.02
		5.03		5.47		5.29
Histidine N		6.43		6.13		6.06
Terminal NH_2	7.27	7.47	6.77	6.77	6.87	6.82
Lysine NH_2		9.70		8.41		8.56
		10.13		9.49		9.73
		10.45		9.75		9.84
		10.82		9.82		9.84
Tyrosine OH	10.70	10.94	9.95	9.84	10.04	9.84
		10.98		9.87		9.84

The constants log $K = 6.77$ and 6.87 belonging to the terminal amino group are lower than those obtained in water (log $K = 7.27$), indicating changes in solvation conditions of unprotonated and protonated amino groups favouring the unprotonated state. In $ACTH_{1-32}$, similar statements can be made also with respect to the imidazole group. In fact, the solvent mixture favours the electrically neutral state of these groups also.

The lower constants for phenolic hydroxyls (compared with aqueous media) are probably due to different effects in the two solvent mixtures. In solutions containing propylene glycol, the interaction of phenolate oxygen with propylene glycol, being suitable for forming a hydrogen bond of a chelate nature with the deprotonated phenolate, is responsible for this phenomenon, whereas in solutions containing trifluoroethanol the specific solvation effect of the latter on the proton (the autoprotolysis constant of this mixture is $pK_w = 11.04$) causes the lower protonation constants.

In $ACTH_{1-32}$, the difference in the solvent dependence of the protonation constants from those mentioned above can be explained by modifications in the secondary structure of the polypeptides. It is known that a decrease in the water activity of corticotropin solutions results in the transformation of the unordered globular structure of the polypeptide in aqueous solutions into the ordered α-helix structure [Gr 76].

In both solvent mixtures, the lowest value of the constants assigned to the ε-amino groups of $ACTH_{1-32}$ shows a decrease compared to the corresponding value obtained in aqueous medium, while the constants belonging to the five most basic groups show a much smaller deviation from each other and the mean value than in water. According to this, the helical arrangement that occurs with a decrease in the concentration of water favours the development of one hydrogen bond, producing an extremely low stability constant; however, it hinders the formation of further hydrogen bonds with the participation of ε-amino or phenolic OH groups yielding the globular structure characteristic of the aqueous solutions.

Equilibrium studies in water–dioxane solvent mixtures

Since the solubilities in water of organic ligands and their complexes are generally low, but they dissolve well in water–dioxane solvent mixtures and such solutions are suitable for potentiometric equilibrium measurements, equilibrium studies on numerous complex systems have been carried out in such mixtures. A number of researchers have also examined the effects of the composition of the solvent mixture on the stabilities of the complexes.

In most of the systems, both the protonation constants and the complex stability constants increase with an increase in the dioxane content of the solvent, i.e., with a decrease in the relative permittivity. The literature contains reports [Ir 56, Va 53] that there is a linear correlation between the molar fraction of the dioxane and the logarithms of the complex stability constants. As early as 1920, Born [Bo 20] published an equation describing the correlation between the relative permittivity of the solvent and the dissociation constant of a weak electrolyte dissolved in it.

Nevertheless, if a comparison is made of the literature data on equilibrium measurements in dioxane–water solvent mixtures, more data are to be found which contradict the above conclusions than those which support them. In order to investigate this question, Gaizer *et al.* [Ga 74b] determined the stability constants of the iodide mixed complexes obtained from three cobalt(III) dioxime parent complexes, in solvent mixtures with a dioxane content of 10–75%. The three dioximes were dimethylglyoxime, cyclohexanedione dioxime (Nioxime) and furil dioxime. The results are given in Table 8.4, together with the stoichiometric constants determined in solutions of ionic strengths of 0.1 M, and also the thermodynamic constants obtained with the activity coefficients calculated for the various solvent mixtures (with a knowledge of their relative permittivities). A comparison of the data obtained in the different solvent mixtures is possible only in the case of the

Table 8.4. Constants of equilibria between the iodide ion and dioxime parent complexes, measured in dioxane–water solvent mixtures [Ga 74b]

Parent complex*	Dioxane content, %	Stoichiometric constants		Thermodynamic constants		x	a
		$\log \beta_1$	$\log \beta_2$	$\log \beta_1$	$\log \beta_2$		
$Co(dmg)_2^+$	10	1.60	—	1.88	—		
	25	1.73	—	2.09	—	4.20	−64.3
	50	1.97	—	2.68	—	4.56	−63.4
	75	2.14	—	4.43	—	6.53	−51.5
$Co(nx)_2^+$	10	1.66	—	1.94	—		
	25	1.61	—	1.97	—	0.60	−16.8
	50	1.71	—	2.42	—	2.74	−38.8
	75	2.14	—	4.43	—	6.38	−50.3
$Co(fdo)_2^+$	10	1.30	2.01	1.58	2.29		
	25	1.51	1.66	1.87	2.02	3.80	−88.6
	50	1.52	1.25	2.23	1.96	3.71	−51.5

* dmg = dimethylglyoximate ion
nx = nioximate ion
fdo = furil dioximate ion

latter constants (calculated with the activity coefficients). The reason is that the activity coefficients of the iodide ion and the parent complex cation depend to a great extent on the dioxane content of the solution.

To study the correlation between the composition of the solvent mixture and the equilibrium constants, the value of x was calculated in the following relation between the molar fraction of the dioxane (M) and the logarithms of the thermodynamic equilibrium constants (A):

$$A = xM + B$$

where B is the value measured in 10% dioxane [Tr 72].

With the same aim, the value of a was calculated for the Born equation [Bo 20]:

$$\log K = \log K_0 - a\left(\frac{1}{D} - \frac{1}{D_0}\right)$$

where D and D_0 are the relative permittivities of the solution and of 10% dioxane (serving as the reference), respectively. In addition to the stability constants, the values of x and a are included in Table 8.4.

The data indicate that in the concentration range investigated, the dimethylglyoxime and Nioxime parent complexes coordinate a single iodide ion each, while the furil dioxime parent complex is able to coordinate two iodide ions. The $\log \beta_1$ values increase with an increase in the dioxane content of the solvent, i.e., with a decrease in the relative permittivity of the solution. It is striking that the log β_2 values decrease with increase in the dioxane content.

The scatters in the values of both x and a are fairly high, showing that in the systems in question the correlation between the solvent and the equilibrium constants is more complicated than would be expected on the basis of the literature data referred to [Ir 56, Va 53].

The fact that these systems behaved in contrast to expectations could be explained on the basis of a spectrophotometric study of the parent complexes.

The shapes of the absorption spectra of the parent complexes in the various solvent mixtures are analogous in character. However, the magnitudes of the molar absorptivities (extinction coefficients) definitely depend on the solvent. Table 8.5 gives the molar absorptivities of the parent complexes, measured at the positions of the absorption maximum of the iodide mixed complexes (the coefficients were refined by computer during the evaluation of the equilibrium data).

Table 8.5. Solvent dependence of the molar absorptivities (ε) of dioxime parent complexes [Ga 74b]

Parent complex	Solvent	ε	Wawelength, nm
$Co(dmg)_2^+$	10% dioxane	360	443
	25% dioxane	370	443
	50% dioxane	230	443
	75% dioxane	170	443
$Co(fdo)_2^+$	10% dioxane	1278	464
	25% dioxane	1346	464
	50% dioxane	1448	464
$Co(nx)_2^+$	10% dioxane	420	445
	25% dioxane	440	445
	50% dioxane	690	445
	75% dioxane	620	445

The dependence of the molar absorptivity on the composition of the dioxane–water solvent mixture indicates that in these mixtures both solvents are capable of coordinating to the cobalt(III) parent complex; thus, the system may contain different solvates (and possibly mixed solvates), the absorbances of which are not identical. The concentrations of these solvates depend on the composition of the solvent mixture. In this way the composition of the solvent mixture influences the value of the absorbance. On the other hand, the stability of the iodo complex examined is governed by the stabilities of those solvates from which the mixed solvate is formed in the given system by means of substitution of the coordinated solvent molecules by iodide. Hence, it is understandable that the correlation between the composition or relative permittivity of the solvent mixture and the equilibrium constants measured in it is too complicated to be described with the above simple relations.

An analogous conclusion was reached by Gergely *et al.* [Ge 74]. They investigated the protonation equilibria of certain amino acids and the formation of their copper(II) complexes in water–dioxane solvent mixtures of various concentrations, and established that neither the protonation constants nor the parent complex formation constants varied linearly as functions of the reciprocal of the relative permittivity, thus indicating the effect of specific solvation processes in the system.

Equilibrium studies in solvent mixtures with varying compositions

McBryde *et al.* attempted to interpret the solvent effect appearing in solvent mixtures on the basis of a large amount of experimental data [Kw 74, Mc 74]. They determined the stability constants of the ethylenediamine and glycine complexes of nickel(II), zinc(II) and manganese(II) ions, and the acid dissociation constants of the ligands in water–methanol, water–dioxane, water–acetonitrile and water–dimethylformamide solvent mixtures, in which the concentration of the organic solvent was varied between 0 and 80%. The aim of the investigations was to establish how the complex stabilities are influenced by different parameters (the basicity of the ligand, the relative permittivity of the system, the solvating abilities of the components of the solvent mixture, etc.).

The examinations leading to the acid dissociation constants showed that, in accordance with the literature data [Ir 56, Ro 71], in all four solvent mixtures the carboxylate group and the amino group behaved fundamentally differently from one another. Whereas the acid dissociation constant of the carboxylate group of glycine increases more or less linearly with increasing concentration of the organic solvent, the dissociation constant of the hydrogen of the protonated amine varies according to a curve with a maximum. Thus, the dissociation of the carboxylate group is described to a reasonably good approximation by the Born equation, whereas that of the protonated amino group is not; this was attributed to McBryde *et al.* to a solvation interaction between the proton and the organic component of the solvent mixture in the latter system.

The stabilities of the complexes of the different metals display changes of a similar nature in the individual solvent mixtures. The values of the stability constants generally decrease in the order

$$\text{dioxane} > \text{acetonitrile} > \text{methanol} > \text{water}\,.$$

Dimethylformamide cannot be fitted into this sequence: its effect on the complex stability proved to be different for the different ligands and in solvent mixtures with different compositions. It is conceivable that the abnormal behaviour of dimethylformamide was caused by products of basic nature produced during the decomposition of the solvent, which promoted the hydrolysis of the metals.

It can be seen from the above that the complex stabilities do not follow the sequence of the relative permittivities unambiguously. Dioxane, with the lowest permittivity, shows the most marked complex stability-increasing effect, and water, with the highest permittivity, shows the weakest such effect; however, acetonitrile precedes methanol in the series denoting the magnitude of the solvent effect, although the relative permittivity of the former solvent is higher. On the other hand, the sequence of the Reichardt solvation parameters derived for pure non-aqueous solvents [Re 65] corresponds to the stability sequence observed.

McBryde *et al.* interpreted the solvent effect not by the dielectric properties of the system, but primarily by the solvation of the metal ion by the components of the solvent mixture. In accordance with this, they defined the equilibrium constant of complex formation as the equilibrium constant of the substitution reaction of the coordinated solvent molecules by the ligands. This equation system also contains the water activity and the organic solvent activity or concentration. With the aid of data relating to solvent mixtures of various compositions, they attempted to establish the cases in which the organic component played a role in the solvent effect by coordinating itself to the metal ion, and those in which it affected the concentration of the aquo complex only by altering the water activity of the system.

Rossotti *et al.* [Fa 70] found that the ratio of stability constants obtained by equilibrium measurements in solvent mixtures similar to the above was independent of the composition of the solvent. McBryde *et al.* showed that in most of the systems they examined, the ratio of the first two stepwise complex stability constants did not depend on the solvent. In the case of the stepwise acid dissociation constants this correlation was approximately true only for ethylenediamine; for glycine the ratio of the protonation constants increase with increase in the concentration of the organic component. However, if the formation of the glycine complexes is defined as a reaction between the protonated ligand and the metal ion, accompanied by the liberation of a hydrogen ion, the ratio of the equilibrium constants is approximately independent of the composition of the solvent mixture.

Gaizer and Gilbert [Ga 80] found that the stability constants of zinc chloride complexes in DMSO–water solvent mixtures of varying composition are higher than expected from the values measured in pure DMSO and pure water.

The experimental data discussed above do not confirm in every case the assumptions of McBryde *et al.*, which are otherwise based to some extent on the earlier concept of Bjerrum and Jørgensen [Bj 53]; nevertheless, it appears beyond doubt that the interpretation of the solvent effect in solvent mixtures should be soluble with the aid of equilibrium constants which also include the solvent activities [Ma 70]. However, the determination of such constants is a very difficult task.

Equilibrium studies of complex formation in solutions of solvent mixtures with varying compositions may lead to seemingly anomalous formation functions too.

If the molar fractions of mononuclear complexes formed in stepwise processes are plotted as a function of the free ligand concentration, in general a single extreme is observed. Exceptions from this fairly general regularity occur in systems where the ratio of the components of the solvent mixture varies during the equilibrium measurement. For example, Vértes *et al.* [Vé 73] employed Mössbauer examinations to follow the interaction of tin tetraiodide and dimethylformamide in rapidly frozen solutions in carbon tetrachloride as inert solvent, and showed that a plot of the sum of the molar fractions of the species SnI_6^{2-} and $Sn(DMF)_6^{4+}$ as a function of the dimethylformamide concentration exhibited two extremes. The phenomenon was explained by Nagypál and Beck [Na 75] in that a double effect occurs in the solution with an increase in the dimethylformamide concentration:

(1) The relative permittivity of the medium increases considerably, which causes decreases in the stabilities of the complexes formed in the solution.

(2) The concentration of the dimethylformamide ligand increases, which increases the stability of the dimethylformamide solvate complex.

The former effect may cancel out or even exceed the latter, i.e., the stability decrease caused by the increase in the relative permittivity may overcompensate the stability increase caused by the higher concentration of the donor solvent.

Using the simplified model system described below, Nagypál and Beck carried out calculations for the quantitative description of this phenomenon. They set out from the following conditions:

(1) In an inert solvent, the metal ion M^+ reacts with the ligand L^- to form the parent complexes ML and ML_2^-.

(2) On the action of a donor solvent (S), the associations MS^+, MS_2^+ and MLS are also formed in the system.

(3) The inert and the donor solvent mix ideally, and the logarithms of the complex stability constants vary linearly as a function of the reciprocal of the relative permittivity of the solvent mixture.

With the aid of assumed equilibrium constants, Nagypál and Beck calculated the concentration distributions of the various complexes formed in this assumed system, as functions of the donor solvent concentration and the relative permittivity determined by this. Figure 8.2 presents the distribution curves constructed from these data. It can be seen that the distributions of several ionic species display anomalous behaviour. This is particularly well reflected by the two maxima and one minimum appearing in the distribution curve of the complex MS. In the case of each complex the anomalous behaviour occurs within the same donor solvent concentration range, indicating that in this range the increase of the solvate concentration accompanying the increase in the concentration of the donor solvent is equaled, or even exceeded by the effect of the lower stability constant resulting from the increase in the relative permittivity of the system.

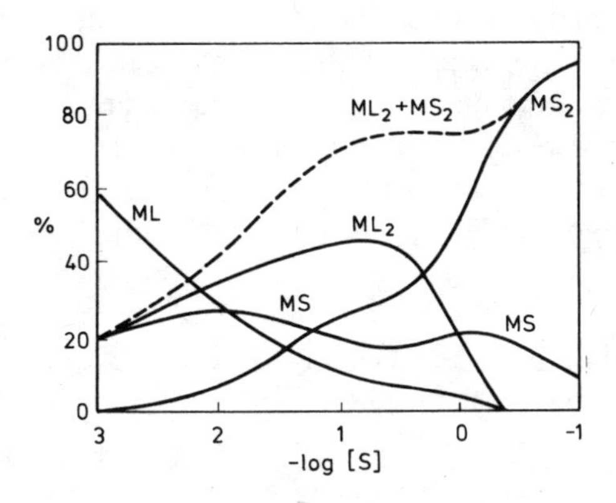

Fig. 8.2. Distribution curves of the complexes ML, ML_2, MS and MS_2, calculated by Nagypál and Beck [Na 73]

Figure 8.2 also shows that the sum of the concentrations of the complexes ML_2^- and MS_2^+ exhibits two extremes, as was observed in the tin(IV) iodide–dimethylformamide system studied by Vértes *et al.* [Vé 73].

The model calculations by Nagypál and Beck [Na 75] convincingly illustrate the effects which may be exerted on the complex equilibria in solution by a change in the ratio of the components of the solvent mixture. Unfortunately, the exact quantitative description even of such a comparatively simple system encounters almost insuperable difficulties. In the above case, e.g., the concentration distributions of the complexes formed in the solution have been determined by sixteen parameters, and to establish the values of these completely, or even merely to take them into consideration, is an extremely difficult task.

Equilibrium measurements in non-aqueous solvent mixtures

Relatively little attention has been paid to the study of complex formation in solvent mixtures that do not contain water. Since little is known about the chemical properties of such solvent mixtures, the interactions between their components, etc., the interpretation of the equilibrium conditions existing in these systems is even more uncertain than for the systems discussed so far.

Archipova *et al.* [Ar 72] carried out an equilibrium study of copper–picoline complexes in alcohol–benzene, alcohol–carbon tetrachloride and alcohol–dimethylformamide mixtures, with a view to finding a correlation between the composition of the solvent mixture and the complex stability constants. It was found that the stability constants were the highest in 100% alcohol. The values of the stability constants slightly decrease with increase in the carbon tetrachloride content of the solvent mixture, and to a greater extent with increase in the proportion of benzene. In solutions containing dimethylformamide, the stability of the picoline complex decreases up to a dimethylformamide content of 25% and then increases with further increase in the dimethylformamide content. Relying on literature data, Archipova *et al.* explained this apparently anomalous behaviour of dimethylformamide-containing solvent mixtures partly by the formation of mixed solvates [Ga 68], and partly by the interaction between the alcohol and dimethylformamide [Br 70]. The effect of the solvation of the ligand has also been taken into consideration in each system.

Calorimetric equilibrium studies

The application of calorimetric equilibrium measurements appears to be a highly promising possibility for investigating the solvent effect in solvent mixtures. Some investigations of this kind are considered below.

In a calorimetric study of the protonation of 8-hydroxyquinoline-5-sulphonate, Gutnikov and Freiser [Gu 68] established that the heat of the protonation of the ligand-oxygen is smaller in a 50% dioxane–water mixture than in water, whereas that of the quinoline-nitrogen is scarcely affected. A study of the nickel(II), copper(II), and zinc(II) complexes of this ligand showed that the stability increase in the water–dioxane mixture manifests itself in a change in both the enthalpy and the entropy.

Hill and Irving [Hi 69] used calorimetric measurements and Rao and Mathur [Ra 71] studied the temperature dependence of the equilibrium constants; both groups found that the heat of the protonation of acetylacetone was higher in water–dioxane mixtures than in water. The heats of formation of the copper and nickel complexes, determined by the latter workers, reflect a similar tendency.

In an equilibrium study of the nickel(II)-glycine complex in water and in a 50% dioxane–water mixture, Li *et al.* [Li 56] found that the stability increase resulting from the increase in the dioxane concentration was due only to a change in the entropy.

Gergely and Kiss [Ge 77] performed a calorimetric investigation of the thermodynamic data of the formation of copper(II) complexes of alanine, phenylalanine and tyrosine in water–dioxane solvent mixtures. In connection with the protonation constants of the ligands, they showed that a change in the solvent primarily influences the carboxylate-proton and phenolate-proton interactions, which are predominantly of electrostatic character. The increase in the protonation constants with a decrease in the relative permittivity is expressed mainly in the entropy factor, although an increase can also be observed in the $-\Delta H$ value.

A similar change was observed by Grenthe and Williams [Gr 67] in a study of the protonation of the acetate ion in water and in a 95% methanol–water mixture. The heat of protonation of the acetate ion changed from $\Delta H = -0.22$ (in water) to $\Delta H = -2.12$ in the methanol–water mixture, and the entropy change increased from $\Delta S = 6.0$ to $\Delta S = 7.4$. The change in the medium had barely any effect on the protonation of the amino group.

By means of calorimetric measurements reflecting the dependence of the stability constants of metal complexes on the composition of the solvent mixture, Gergely and Kiss have shown that the stability increase resulting from a decrease in the relative permittivity shows up primarily in an increase in the $-\Delta H$ values, whereas the ΔS values decrease or remain unchanged compared with those measured in aqueous medium. A comparison of the ΔH values for aliphatic and aromatic amino acid complexes of copper(II) has revealed that the change in the medium acts to a lesser extent on the ΔH values of the aromatic amino acid complexes. This is probably due to the presence of the hydrophobic phenyl group in the aromatic ligands, as a consequence of which the extent of hydration even in aqueous medium is lower than in the case of alanine.

Together with the results of other complex equilibrium studies in analogous solvent mixtures, and primarily the work of McBryde *et al.* [Mc 74], the investigations by Gergely and Kiss [Ge 77] support the assumption that the changes in the thermodynamic quantities of coordination reactions in the various organic solvent–water mixtures can be attributed primarily to changes in the solvent properties in the immediate environment of the solute species; the effects of the macroscopic properties of the solvent are probably subordinate.

References

Ab 72 Ablov, A. V., Gulja, A. P.: Zh. Neorg. Khim., **17**, 12 (1972).

Ag 67 Ageno, M., Frontall, C.: Proc. Nat. Acad. Sci. USA, **57**, 856 (1967).

Ar 72 Archipova, N. V., Muftahov, A. G., Nachimov, C. R.: Zh. Neorg. Khim., **17**, 2952 (1972).

Be 70 Beck, M. T.: Chemistry of Complex Equilibria, Van Nostrand, London, 1970, pp. 165–170.

Bj 53 Bjerrum, J., Jørgensen, C. K.: Acta Chem. Scand. **7**, 951 (1953).

Bo 20 Born, M.: Z. Physik, **1**, 45, 221 (1920).

Bo 79 Bose, K., Kundu, K. K.: Indian J. Chem., **17A**, 122 (1979).

Br 70 Brodskii, A. I.: Uspekhi Khimii, **39**, 753 (1970).

Bu 68 Burger, K., Pintér, B., Papp-Molnár, E., Nemes-Kósa, S.: Acta Chim. Acad. Sci. Hung., **57**, 363 (1968).

Bu 79 Burger, K.: in: Metal Ions in Biological Systems (Ed.: H. Sigel), Vol. 9, p. 227 (1979).

Bu 81a Burger, K., Zay, I., Gaizer, F.: Inorg. Chim. Acta, **55**, L 23 (1981).

Bu 81b Burger, K., Noszál, B.: in: Advances in Solution Chemistry. Plenum Press, New York, London, 1981, p. 139.

Co 74a Covington, A. K., Thain, J. M.: Abstracts IV ICNAS (Ed.: Gutmann, V.), Vienna, 1974, p. 11.

Co 74b Costa, J. M., Miro, J.: Abstracts IV ICNAS (Ed.: Gutmann, V.), Vienna, 1974, p. 15.

De 81 Delville, A., Detellier, Ch., Gerstmans, A., László, P.: Helvetica Chimica Acta, **64**, 547 (1981).

Di 77 Dickert, F. L.: Z. Phys. Chem. Neue Folge, **106**, 155 (1977).

Er 58a Erdey-Grúz, T., Kugler, E., Reich, A.: Acta Chim. Acad. Sci. Hung., **13**, 429 (1958).

Er 58b Erdey-Grúz, T., Majthényi, L.: Acta Chim. Acad. Sci. Hung., **16**, 417 (1958).

Er 59a Erdey-Grúz, T., Kugler, E., Hidvégi, J.: Acta Chim. Acad. Sci. Hung., **19**, 89 (1959).

Er 59b Erdey-Grúz, T., Kugler, E., Hidvégi, J.: Acta Chim. Acad. Sci. Hung., **19**, 363 (1959).

Er 59c Erdey-Grúz, T., Majthényi, L.: Acta Chim. Acad. Sci. Hung., **20**, 73, 175 (1959).

Er 67 Erdey-Grúz, T., Majthényi, L., Nagy-Czakó, I.: J. Inorg. Nucl. Chem., **53**, 28 (1967).

Er 68a Erdey-Grúz, T., Kugler, E., Majthényi, L.: Electrochim. Acta, **13**, 947 (1968).

Er 68b Erdey-Grúz, T., Kugler, E.: Acta Chim. Acad. Sci. Hung., **57**, 301 (1968).

Er 71a Erdey-Grúz, T., Nagy-Czakó, I.: Acta Chim. Acad. Sci. Hung., **67**, 283 (1971).

Er 71b Erdey-Grúz, T., Fodor-Csányi, P., Lévay, B., Szilágyi-Győri, E.: Acta Chim. Acad. Sci. Hung., **69**, 423 (1971).

Er 71c Erdey-Grúz, T., Lévay, B.: Acta Chim. Acad. Sci. Hung., **69**, 215 (1971).

Er 73 Erdey-Grúz, T., Lévay, B.: Acta Chim. Acad. Sci. Hung., **79**, 401 (1973).

Er 74 Erdey-Grúz, T.: Transport phenomena in aqueous solutions. Hilger, London, 1974.

Fa 70 Faraglia, G., Rossotti, F. J. C., Rossotti, H. S.: Inorg. Chim. Acta, **4**, 488 (1970).

Fo 71 Fowler, F. W., Katritzky, A. R., Rutherford, R. J. D.: J. Chem. Soc. **B, 1971**, 460.

Fr 69 Fratiello, A., Lee, R. E.: Inorg. Chem., **8**, 69 (1969).

Ga 67 Gaizer, F., Beck, M. T.: J. Inorg. Nucl. Chem., **29**, 21 (1967).

Ga 74a Gaizer, F., Buxbaum, P., Papp-Molnár, E., Burger, K.: J. Inorg. Nucl. Chem., **36**, 859 (1974); Magy. Kém. Foly., **78**, 420 (1973).

Ga 74b Gaizer, F., Tran, T. B., Burger, K.: J. Inorg. Nucl. Chem., **36**, 1601 (1974).

Ga 80 Gaizer, F., Silber, H. B.: J. Inorg. Nucl. Chem., **42**, 1317 (1980).

Ge 74 Gergely, A., Nagypál, I., Kiss, T., Király, R.: Magy. Kém. Foly., **80**, 181 (1974).

Ge 77 Gergely, A., Kiss, T.: J. Inorg. Nucl. Chem., **39**, 109 (1977); See also Gergely, A., Kiss, T.: Magy. Kém. Foly., **81**, 15 (1975) and **82**, 89 (1976).

Go 68 Golub, A. M., Golovuriskin, V. I.: Zh. Fiz. Khim., **42**, 1902 (1968).

Go 69 von Goldammer, E., Zeidler, M. D.: Ber. Bunsenges. phys. Chem., **73**, 4 (1969).

Gr 62 Grunwald, E.: in: Electrolytes (Ed.: Pesce, B.). Pergamon Press, Oxford, 1962, p. 62.

Gr 67 Grenthe, I., Williams, D. R.: Acta Chem. Scand., **21**, 341 (1967).

Gr 76 Greff, D., Toma, F., Fermandjian, S., Löw, M., Kisfaludy, L.: Biochim. Biophys. Acta, **439**, 219 (1976).

Gu 68 Gutnikov, G., Freiser, H.: Anal. Chem., **40**, 39 (1968).

Gu 73 Gulja, A. P., Cherbakov, V. A., Ablov, A. V.: Dokl. Akad. Nauk SSSR, **269**, 851 (1973).

Hi 69 Hill, J. O., Irving, R. J.: J. Chem. Soc. **A, 1969**, 2759.

Hi 74 Hilton, J. F.: Abstracts IV ICNAS (Ed.: Gutmann, V.), Vienna, 1974, p. 42.

In 75 Inlow, R. O., Joesten, M. D., Van Wazer, J. R.: J. Phys. Chem., **79**, 2307 (1975).

Ir 56 Irving, H., Rossotti, H. S.: Acta Chem. Scand., **10**, 72 (2956).

Ka 80 Kalidas, C., Sivaprasad, P.: Indian J. Chem., **18**, 532 (1980).

Kw 74 Kwan-Kit Mui, McBryde, W. A. E.: Canad. J. Chem., **52**, 1821 (1974).

Li 56 Li, N. C., White, J. M., Yoest, R. L.: J. Am. Chem. Soc., **78**, 5218 (1956).

Li 73 Liszi, J., Náray, M.: Magy. Kém. Foly., **79**, 298 (1973).
Li 74 Liszi, J., Salamon, T., Ratkovics, F.: Acta Chim. Acad. Sci. Hung., **81**, 467 (1974).
Li 75 Liszi, J.: Acta Chim. Acad. Sci. Hung., **84**, 125 (1975).
Li 76a Liszi, J.: Magy. Kém. Foly., **82**, 338, 427 (1976).
Li 76b Liszi, J.: Magy. Kém. Foly., **82**, 375 (1976).
Li 76c Liszi, J.: Magy. Kém. Foly., **82**, 396 (1976).
Lu 64a Luz, Z., Meiboom, S.: J. Chem. Phys., **40**, 1058 (1964).
Lu 64b Luz, Z.: J. Chem. Phys., **41**, 1748 (1964).
Ma 67 Matwiyoff, N. A., Hooker, S. V.: Inorg. Chem., **6**, 1127 (1967).
Ma 68a Matwiyoff, N. A., Darley, P. E.: J. Phys. Chem., **72**, 2659 (1968).
Ma 68b Matwiyoff, N. A., Darley, P. E., Movius, W. G.: Inorg. Chem., **7**, 2173 (1968).
Ma 70 Marshall, W. L.: J. Phys. Chem., **74**, 346 (1970).
Mc 74 McBryde, W. A., Mui, K. K., Nieboer, E.: Abstracts IV ICNAS (Ed.: Gutmann, V.), Vienna, 1974, p. 10.
Mi 53 Mitchell, A. G., Wynne-Jones, W. F. K.: Disc. Faraday Soc., **15**, 161 (1953).
Mi 61 Mihajlov, B. A.: Zh. Strukt. Khim., **2**, 677 (1961).
Mo 74 Moreau, C., Douhéret, G.: Abstracts IV ICNAS (Ed.: Gutmann, V.), Vienna, 1974, p. 11.
Ná 73 Náray, M., Liszi, J.: Magy. Kém. Foly., **79**, 279 (1973).
Na 75 Nagypál, I., Beck, M. T.: Magy. Kém. Foly., **81**, 117 (1975).
No 80 Noszál, B., Burger, K.: Inorg. Chim. Acta, **46**, 229 (1980).
Oa 73 Oakes, J.: J. Chem. Soc. Faraday Trans., **69**, 1311 (1973).
On 36 Onsager, L.: J. Am. Chem. Soc., **58**, 1486 (1936).
Pa 68 Padova, J.: J. Phys. Chem., **72**, 796 (1968).
Ra 71 Rao, B., Mathur, H. B.: J. Inorg. Nucl. Chem., **33**, 2919 (1971).
Ra 73a Ratkovics, F., Liszi, J., László, M.: Acta Chim. Acad. Sci. Hung., **79**, 387 (1973).
Ra 73b Ratkovics, F., László, M.: Acta Chim. Acad. Sci. Hung., **79**, 395 (1973).
Ra 74a Ratkovics, F., Guti, Zs.: Acta Chim. Acad. Sci. Hung., **83**, 63 (1974).
Ra 74b Ratkovics, F., Salamon, T., Domonkos, L.: Acta Chim. Acad. Sci. Hung., **83**, 53 (1974).
Ra 74c Ratkovics, F., Salamon, T., Domonkos, L.: Acta Chim. Acad. Sci. Hung., **83**, 71 (1974).
Ra 75 Ratkovics, F., László-Parragi, M.: Acta Chim. Acad. Sci. Hung., **84**, 45 (1975).
Ra 76a Ratkovics, F., Domonkos, L.: Acta Chim. Acad. Sci. Hung., **89**, 325 (1976).
Ra 76b Ratkovics, F., Salamon, T.: Acta Chim. Acad. Sci. Hung., **89**, 331 (1976).
Ra 76c Ratkovics, F., László, M., Salamon, T.: Acta Chim. Acad. Sci. Hung., **89**, 245 (1976).
Ra 76d Ratkovics, F., Salamon, T.: Acta Chim. Acad. Sci. Hung., **91**, 165 (1976).
Ra 77a Ratkovics, F., László, A.: Magy. Kém. Foly., **83**, 25 (1977).
Ra 77b Ratkovics, F., László, A.: Magy. Kém. Foly., **83**, 43 (1977).
Ra 77c Ratkovics, F., Salamon, T.: Magy. Kém. Foly., **83**, 112 (1977).
Re 65 Reichardt, C.: Angew. Chem. Int. Edn., **4**, 29 (1965).
Re 70 Reeves, L. W., Yue, C. P.: Canad. J. Chem., **48**, 3307 (1970).
Ro 71 Rorabacher, D. B., MacKellar, W. J., Shu, F. R., Sister-Bonavita, M.: Anal. Chem., **43**, 561 (1971).
Sh 56 Shedlovsky, T., Kay, R. L.: J. Phys. Chem., **60**, 151 (1956).
Sk 68 Skodin, A. M., Levitskaya, H. F., Lozhnikov, V. A.: Zh. Strukt. Khim., **38**, 1006 (1968).
Su 34 Sugden, S.: Nature, **133**, 415 (1934).
Sy 80 Symons, M. C. R., Eaton, G., Shippey, A., Harvey, M.: Chem. Phys. Letters, **69**, 344 (1980).
Th 67 Thomas, J., Reynolds, W. L.: J. Chem. Phys., **46**, 4164 (1967).
To 58 Tourky, A. R., Mikhail, S. Z.: Egypt. J. Chem., **1**, 1, 13 (1958).
To 61 Tourky, A. R., Rizk, H. A., Girgis, Y. M.: J. Phys. Chem., **65**, 40 (1961).

Tr 72 Tran, T. B.: Equilibrium measurements in non-aqueous solutions. Thesis (in Hungarian) for the degree of Candidate of Sciences, Budapest, 1972.

Va 53 Van Uitert, L. G., Hass, C. G., Fernelius, W. C., Douglas, B. E.: J. Am. Chem. Soc., **75**, 455 (1953).

Vé 73 Vértes, A., Gaizer, F., Beck, M.: Magy. Kém. Foly., **79**, 310 (1973).

We 61 Wear, J. O., McNully, C. V., Amis, E. S.: J. Inorg. Nucl. Chem., **19**, 278 (1961).

Wi 30 Williams, J. W.: J. Am. Chem. Soc., **52**, 1838 (1930).

9. GENERAL SUGGESTIONS FOR THE CHARACTERIZATION OF NON-AQUEOUS SOLUTION SYSTEMS

In earlier chapters it has been shown that an extremely large number of experimental methods have been employed to study the solvent effect, and that any given method may be used in many different ways. A survey of the results of these investigations also reflects that the processes taking place in solution are influenced by the solvents used in very many, and often very different, ways.

Depending on the type of solvent and the nature of the interaction to be studied, it may be necessary to utilize different experimental methods for the investigation of the solvent effect. In fact, the method of application of the same technique may vary considerably according to the properties of the system in question. This is the explanation of the extremely diverse nature of the investigations in this field, and why it is so difficult (and perhaps impossible) to develop a general method or common procedure for the characterization of solution systems.

With a knowledge of these limitations, the purpose of this chapter is merely to review general aspects that serve as a basis for further specific examinations of the individual systems.

Analytical examination of the solution

A primary requirement in all solvation studies is the analytical examination of the solvent and the solute. Since water, owing to its amphoteric nature, behaves as a base in acidic solvents and as an acid in basic solvents, strongly solvating both Lewis acids and Lewis bases as a consequence of its ability to donate an electron pair and to form hydrogen bonds, the first and perhaps the most important task of the analyst is to determine the moisture contents of the solvent, of the components to be dissolved, and finally of the solution. This is followed by the detection, and if necessary the determination, of any possible decomposition products of the solvent.

The impurities to be analyzed differ, depending on the problem under investigation. In the course of the study of coordination chemical systems, the results are distorted considerably by any electron pair donor or acceptor molecule, even at low concentrations, whereas the presence of impurities that are "inert" as regards complex formation may be permissible in much larger amounts. In the study of processes accompanied by the formation of hydrogen bonds, appreciable disturbances result from the presence of molecules that either provide the pillar atom or serve as the hydrogen donor for hydrogen bridges; at the same time, small amounts of apolar impurities cause scarcely any interference. In optical studies, impurities absorbing in the spectral range of interest are additional sources of disturbance.

It is a general rule that all examinations should be carried out in systems of the maximum possible purity, containing only the components required for the process to be investigated. The weaker the interactions that must be studied, the purer the solvent required.

In most cases checking off the purity of a solvent does not necessitate tests for the various impurities; instead, various solvent characteristics, physical constants sensitive to the presence of impurities are determined (conductivity, refractive index, boiling point, dielectric constant, infrared spectrum, ^{1}H NMR spectrum, etc.). In the case of non-aqueous solutions or solvents, the most important, and often the only specific analytical task, is the determination of the moisture content. In addition, the quality control of the solvent is based on the determination of various physical constants and the recording of characteristic spectra.

Determination of moisture content

Several physical and chemical procedures have been elaborated for the determination of small amounts of water in solvents [Ha 72, Mi 61], but in most cases the Karl Fischer titrimetric method [Fi 35] has proved the most suitable. The essence of the method is that, in methanolic solution containing pyridine as proton binder, the reaction between sulphur dioxide and iodine in the presence of water takes place quantitatively in accordance with the equation

$$SO_2 + I_2 + H_2O \rightleftharpoons SO_3 + 2\,HI$$

or, if the formation of pyridine associates and the solvation effect of methanol are also taken into consideration

$$C_5H_5N \cdot SO_2 + C_5H_5N \cdot I_2 + H_2O + C_5H_5N + CH_3OH \rightleftharpoons$$
$$\rightleftharpoons 2\,C_5H_5N \cdot HI + C_5H_5N \cdot HOSO_2OCH_3$$

Accordingly, one molecule of water ensures the oxidation of one molecule of sulphur dioxide by one molecule of iodine. As the reaction is quantitative, it can be utilized for the determination of the water content of the system. If a solution containing moisture is titrated with a methanolic Karl Fischer titrant solution containing sulphur dioxide, iodine and pyridine, the iodine will be reduced in accordance with the above reaction as long as water is available in the system. The appearance of iodine, which is generally observed instrumentally ("dead-stop" method), indicates the end-point of the titration.

The method has been successfully employed for the determination of the moisture contents of very varied organic compounds, e.g., saturated and unsaturated hydrocarbons, alcohols, organic acids, acid anhydrides, esters, ethers, amines and amides [Mi 61]. It can be seen that the possible water content in the majority of non-aqueous solvents can be controlled by this means.

The measurement is subject to interferences from compounds that react with one or the other component of the Fischer reagent. Examples are molecules containing active carbonyl groups, which form acetals and ketals with methanol; reducing agents, such as ascorbic acid and organic mercaptans, which reduce iodine; or oxidizing agents, such as quinones and peroxides, which oxidize iodide [Mi 48]. The interfering effect of inorganic compounds is of less interest when the solvents are examined, but very important in the determination of the moisture contents of solutions [Br 41]. In addition to the oxidizing or reducing inorganic salts, interference is also caused by metal oxides and hydroxides, and by the salts of alkali and alkaline earth metals formed with weak acids, since these react with the Karl Fischer reagent to yield water, and this naturally leads to false analytical results.

The investigation of systems containing interfering components has in many cases led to modifications suitable for the elimination of the effect. For example, if the active carbonyl group is converted with cyanide to the cyanohydrin, it no longer affects the water determination [Br 40]. The interfering effect of reducing mercaptans can be eliminated by allowing the mercaptan to react with an active olefin, e.g., octene, in the presence of boron trifluoride catalyst. Replacement of methanol by another solvent also frequently leads to the elimination of disturbing side reactions. For instance, anhydrous acetic acid may serve as a solvent in place of methanol in the presence of basic amine compounds [Mi 48]. In cases when the disturbing side reaction is stoichiometric, e.g., some reactant produces or consumes water in an equivalent amount, or similarly oxidizes or reduces iodine, a knowledge of the concentration of this component permits the correction of the experimental result. For example, one sulphide ion consumes one molecule of iodine. With a knowledge of the sulphide concentration, therefore, the volume of the Karl Fischer titrant solution consumed in the oxidation of sulphide can be taken into account in the measurement of the moisture content of a sulphide-containing solution [Mi 49].

At present, the biamperometric (dead-stop) method is used virtually exclusively for indicating the end-point in Karl Fischer titrations. In this way the water content may be measured with appropriate accuracy down to a concentration limit of 0.001%. Such a sensitivity is unquestionably necessary, as a water content of 0.01%, for instance, corresponds to a water concentration of 5.56×10^{-3} M. In a spectroscopic examination of the solvent effect, the concentration of the solvate may also be of this order only. To avoid appreciable interference from water present as an impurity, its concentration must be at least one order of magnitude lower than those of the reactants.

Other methods have been also worked out for the determination of moisture in different samples.

A new spectrophotometric procedure was elaborated by Khalaf and Rimpler [Kh 80] for the determination of the water content of basic solvents (e.g., pyridine, DMSO, DMFA) based on the reaction of 5-isothiocyanato-1,3-dioxo-2-*p*-tolyl-2,3-dihydro-1H-benz(de)isochinoline with water. The reaction results in the formation of the 5-amino product of this reagent, causing the appearance of a new absorption band in the vicinity of 430 nm wavelength. The intensity of this new band was shown to be in linear correlation with the water content of the sample between 0.3–0.02 vol%. The reproducibility was found to be $\pm 2.2\%$.

A new simple procedure for the determination of water (even in trace amounts) in organic solvents was elaborated by Langhals [La 81]. The solvatochromism of the pyridiniumphenol betaine, determined by a simple UV-absorption measurement, together with a two-parameter equation, permits an exact determination. The procedure is rapid and is, therefore, an alternative to the Karl Fischer titration.

Determination of other impurities

Besides the discussed above, gas chromatographic analysis has also been successfully employed for the determination of moisture contents [Se 71]. This method is most preferred for the selective detection and determination of other impurities. Paper, thin-layer and, most recently, high-performance liquid chromatography are also used to check the purities of solvents. However, in general, they are employed only if some special impurity is to be detected or determined. Otherwise, it is sufficient to determine some characteristic, impurity-sensitive, physical property of the carefully purified system.

Purification of solvents

The initial step in investigations of the solvent effect is always the careful purification of the solvent. Since a considerable proportion of non-aqueous solvents are hygroscopic, a very important step in the purification is the removal of the moisture content.

Some solvents are prone to decomposition, and in view of this, they usually contain some stabilizer. (As an example, commercial chloroform is stabilized with ca. 1% of ethanol to inhibit phosgene formation.) However, the stabilizers are not inert, and may alter the various chemical interactions occurring in the system. Ethanol, for instance, not only displays a strong ability to form hydrogen bonds, but also has good donor properties by virtue of the free electron pair on its oxygen atom. Hence, it may enter into strong interactions with reactants that are either Lewis acids or Lewis bases. A solvent used for investigations of solvent effects, therefore, must be purified not only from impurities present as a consequence of the manufacturing method, or impurities produced as a result of its decomposition, but also from stabilizers. Errors caused by any decomposition of the solvent can be avoided if freshly purified solvent is used for the examinations.

The most general means of purifying solvents is distillation at a controlled temperature and pressure. Fortunately, in many cases the boiling point of the solvent to be purified differs substantially from those of the accompanying materials and impurities. Hence, appropriate fractional distillation may be effective for purification. Its effectiveness may be increased even further if, prior to distillation, the solvent is treated with a reagent binding the accompanying or contaminating substances. Certain solvents may undergo decomposition at the boiling temperature, and these can be purified by vacuum distillation.

Removal of the moisture content

In a fortunate case, distillation aimed at purification may also result in the drying of the solvent. Distillation may be particularly effective in the removal of moisture if azeotropic mixtures with low boiling points are formed. For instance, the first step in the dehydration of ethanol is distillation after the addition of benzene, when water is removed in the volatile ternary azeotrope. Other solvents may also be dehydrated by distillation, e.g., benzene, chloroform, carbon tetrachloride, ethylene dichloride, heptane, hexane, toluene and xylene. In distillations with the aim of dehydration, the apparatus must be fitted with moisture traps (containing calcium chloride, silica gel or some other drying agent). It must be borne in mind that many anhydrous organic solvents are hygroscopic.

Dehydration by distillation is generally only the first step in the removal of the moisture content. As a rule, the water concentration attained in this way is still too high to permit the study of the solvent effect. The distillation is therefore followed by binding of the residual water with various drying agents. These agents can be classified into three groups:

(a) Compounds binding water reversibly (anhydrous salts which bind the water in the form of water of crystallization, e.g., sodium sulphate, magnesium sulphate, calcium chloride; oxides and hydroxides forming hydrates with water, e.g., calcium oxide, phosphorus pentoxide; concentrated acids, e.g., sulphuric acid).

(b) Compounds reacting irreversibly with water (metal amalgams, e.g., magnesium amalgam; metal hydrides, e.g., lithium aluminium hydride; alkali metals, e.g., sodium or potassium).

(c) Molecular sieves, which are the most modern drying agents offering great advantages [Je 77].

Table 9.1. Solvent drying agents

Type of solvent	Drying agent*
Organic acids	$CaSO_4$, $MgSO_4$, Na_2SO_4
Alcohols	CaO, $CaSO_4$, $MgSO_4$, K_2CO_3 and subsequently metallic Mg and I_2, m.s.
Alkyl halides	$CaCl_2$, $CaSO_4$, $MgSO_4$, P_2O_5, Na_2SO_4, m.s.
Amines	BaO, CaO, KOH, Na_2CO_3, $NaOH$, m.s.
Aryl halides	$CaCl_2$, $CaSO_4$, $MgSO_4$, P_2O_5, Na_2SO_4
Esters	$MgSO_4$, K_2CO_3, Na_2SO_4, m.s.
Heterocyclic bases	$MgSO_4$, K_2CO_3, $NaOH$, m.s.
Hydrocarbons	$CaCl_2$, $CaSO_4$, $MgSO_4$, P_2O_5, m.s., Na (not for olefins)
Ketones	$CaSO_4$, $MgSO_4$, K_2CO_3, Na_2SO_4, m.s.
Nitro compounds and nitriles	$CaCl_2$, $MgSO_4$, Na_2SO_4, m.s.

* m.s. = molecular sieve with a pore size of 0.4 nm

Agents used for the drying of some important solvent types are listed in Table 9.1. However, despite the existence of an arsenal of desiccants, the presence of water in some systems continues to be a problem. Recommended agents for the removal of water and polar impurities from other solvents may not be suitable, e.g. for amines. The radiotracer method for water assay developed by Burfield *et al.* [Bu 77] can be applied to obtain quantitative data on the drying of some representative solvents. In the case of amine type solvents, e.g. only molecular sieves and CaH_2 were shown to have efficient siccative effect [Bu 81].

In the selection of the drying agent, care must always be taken that it does not attack the solvent itself. For example, in dehydrations with lithium aluminium hydride, it is not only water which reacts with the drying agent; the hydride also reduces aldehydes, ketones and esters to alcohols, and nitriles, amides and aldimides to amines. It is naturally not suitable, therefore, for the dehydration of solvents having such chemical composition.

Molecular sieves have the great advantage that the occurrence of such side reactions need not be considered. Owing to their excellent utilizability, they will be discussed below in some detail.

Molecular sieves and their application

Molecular sieves are crystalline zeolites (sodium and calcium aluminosilicates) which lose their water of hydration on heating, producing cavities of molecular magnitude in their crystal lattices. These uniform-sized cavities are able to admit and bind small molecules, in particular water, but at the same time larger molecules are not bound. This reversed "sieve" effect (separation of the smaller molecules from the larger ones) leads to the binding of the water content of a solvent. Although molecular sieves are sensitive to strong acid, they can be employed in the comparatively wide pH range of 5–11.

The pore size of a molecular sieve can be modified by exchange of the cations incorporated into the aluminosilicate skeleton. In general, molecular sieves of small pore size (ca. 4 Å; 0.4 nm) are the most suitable for dehydration (e.g., the Linde 0.4 nm product of Union Carbide, or products equivalent to this). In addition to water molecules, this molecular sieve is suitable for the binding of carbon dioxide, hydrogen sulphide, sulphur dioxide, ammonia, methanol and other molecules of similar size. For chemical reasons, however, its affinity for water is so much higher than that for the other molecules of similar size that it may even be used for the dehydration of methanol. As an example of the effectiveness of this sieve, it can reduce the water content of ethanol from 0.5% to 10 ppm.

One reason why dehydration with molecular sieves is popular is that an exhausted molecular sieve can be regenerated simply by heating between 150 and 300°C. It is preferable to carry out this process in a current of dry nitrogen, and then to leave the dried sample to cool in a desiccator.

The literature data and our own experience indicate that drying with a molecular sieve can be employed with good results for the following solvents: acetone, acetonitrile, benzene, butane, butyl acetate, carbon tetrachloride, dichloroethane, cyclohexane, diethyl ether, dimethylformamide, dimethyl sulphoxide, ethanol, heptane, isopropanol, di-isopropyl ether, pyridine, toluene and xylene.

Portions of not more than 1 litre of the solvent to be dehydrated are shaken for 24 h with about 30–50 g of the molecular sieve; after sedimentation of the molecular sieve, the liquid is decanted into a distillation flask and distilled with the exclusion of atmospheric moisture. It is advisable to store hygroscopic liquids over a molecular sieve, and to distil them only immediately before use.

Purification procedures for some solvents

In the purification of solvents it is, of course, necessary to remove not only the moisture content, but also any other impurities, such as decomposition products, originating from, and perhaps not removed during, the production of the solvent.

Certain purification procedures may themselves bring about partial decomposition of the solvent, thus giving rise to new contamination.

The present author's researches, dealing with equilibrium measurements in non-aqueous solutions and with various (primarily Mössbauer) spectroscopic studies, required in every case the use of meticulously dehydrated and purified solvents. An account is given below of the methods that proved satisfactory for the purification of some of the more important solvents, with reference to any disturbing factors. For the individual solvents, the boiling point (b.p.), melting point (m.p.) and refractive index (n_D^t) at temperature t are given, where appropriate.

Acetic acid (glacial) (b.p.: 118 °C; m.p.: 16.6 °C; n_D^{25}: 1.36995)

Apart from water, the most frequent impurity is acetaldehyde. The water content is eliminated by reaction with acetic anhydride. Acetaldehyde can be oxidized to acetic acid with chromium(VI) oxide or potassium permanganate. The simplest purification procedure consists of refluxing glacial acetic acid containing acetic anhydride for 1 h with chromium(VI) oxide, followed by fractional distillation.

Acetone (b.p.: 56.2 °C; n_D^{25}: 1.35609)

Because of its hygroscopic nature, acetone is virtually always contaminated with water. Several of the customary drying agents cannot be used for its dehydration; with silica gel or aluminium hydroxide, and even (to a lesser extent) on treatment with phosphorus pentoxide or sodium amalgam, water is formed in the system as a result of aldol condensation. Hence the water content of improperly "dried" acetone may be higher than that of the starting material. Acetone can be dried well with a molecular sieve having a pore size of 0.4 nm. According to experience in this laboratory, the moisture content of acetone distilled off this molecular sieve may be as low as 0.001–0.003%.

Acetonitrile (b.p.: 81.6 °C; n_D^{25}: 1.34163)

Acetonitrile is hygroscopic and also prone to hydrolytic decomposition. Hence, in addition to water, it virtually always contains acetamide, ammonia and possibly ammonium acetate as impurities. It cannot be dried with potassium hydroxide, since this would catalyze the hydrolysis. A molecular sieve with a pore size of 0.4 nm has proved effective for the drying of acetonitrile. Following treatment with this, the solvent is subjected to fractional distillation. Application of the molecular sieve makes unnecessary the previously employed treatment with calcium hydride and phosphorus pentoxide.

Benzene (b.p.: 80.1 °C; n_D^{25}: 1.49790)

Frequent impurities in benzene are thiophene, cyclohexane and sometimes toluene. Removal of the thiophene content is of particular importance and can be effected by shaking with concentrated sulphuric acid. Benzene is then freed from sulphuric acid by washing with dilute sodium hydroxide solution, and the residual alkali is removed by washing with water. The solvent is dried by treatment on a molecular sieve, and the purification process is completed by fractional distillation.

Chloroform (b.p.: 61.2 °C; n_D^{15}: 1.44858)

The commercial product is usually stabilized with 1% of ethanol, which can be removed by washing with distilled water. Most of the water dissolved in the chloroform can then be bound with anhydrous potassium carbonate or calcium chloride. In order to remove the residual traces of water, the solvent is refluxed with phosphorus pentoxide and then distilled.

Pure chloroform undergoes partial decomposition yielding phosgene, particularly on exposure to light. Accordingly, all work involving the use of chloroform should be carried out with the freshly purified solvent. Even during short storage periods purified chloroform must be protected from light. Metallic sodium should never be used for the dehydration of chloroform, as it causes considerable decomposition.

1,2-*Dichloroethane* (b.p.: 83.4 °C; n_D^{15}: 1.44759)

The commercial product usually contains about 1% of ethanol as a stabilizer. For the removal of this, the solvent is extracted with concentrated sulphuric acid, then washed with 5% sodium carbonate solution and finally with water. Since dichloroethane forms an azeotropic mixture containing 8.9% of water (b.p. at normale pressure 77 °C compared with 83.4 °C for the pure solvent), the bulk of the water content may be removed by simple distillation. Dichloroethane is then refluxed with phosphorus pentoxide to bind the residual moisture content, and finally fractionally distilled.

Diethyl ether (b.p.: 34.6 °C; n_D^{20}: 1.35272)

The most dangerous impurity is diethyl peroxide, which can be removed by refluxing for 1–2 h with tin(II) chloride. The water content may then be bound with calcium chloride. The purification is completed by distillation from phosphorus pentoxide.

N,N-Dimethylformamide (b.p.: 76 °C at 52 mbar, 153.0 °C at 1.01 bar; n_D^{25}: 1.4269)

This solvent is employed widely in analytical and coordination chemistry. At its boiling point it undergoes partial decomposition to yield dimethylamine and carbon monoxide. The decomposition is catalyzed by various substances, particularly those with acidic or basic properties; this must be taken into consideration in the selection of the material used for drying. Under no circumstances may this solvent be refluxed with, for example, potassium hydroxide, sodium hydroxide or calcium hydride. Dimethylformamide can be dehydrated most advantageously with a molecular sieve of pore size 0.4 nm; however, calcium sulphate, magnesium sulphate or silica gel may also be employed. After dehydration, the solvent may be purified by vacuum distillation.

Dimethyl sulphoxide (b.p.: 75.6 °C at 16 mbar; 190 °C at 1.01 bar; m.p.: 18.0–18.5 °C)

This solvent may be purified in an analogous manner to dimethylformamide. If a considerable amount of water is present, the dimethyl sulphoxide may be frozen out of the mixture prior to the purification procedure.

1,4-*Dioxane* (b.p.: 101.3 °C; n_D^{25}: 1.42025)

In addition to water, this solvent is almost invariably contaminated with peroxide and acetaldehyde. For removal of the peroxide impurity, dioxan is treated with tin(II) chloride and then distilled. Subsequently, the above process is repeated with anhydrous potassium hydroxide for removal of the acetaldehyde. Finally, a molecular sieve with a pore size of 0.4 nm may be used to bind traces of water. This procedure is followed again by distillation. In this way the purification of dioxane may be carried out in three steps, with triple distillation. For simplification, refluxing with lithium aluminium hydride may be attempted, followed by distillation which frees the dioxane from all three impurities in one step.

Ethanol (b.p.: 78.3 °C; n_D^{25}: 1.35941)

Many procedures have been developed for the preparation of anhydrous ethanol. In our own experience, this solvent may be purified most simply by drying with molecular sieve, followed by distillation. However, drying with metallic calcium or calcium hydride has also proved effective.

Isopropanol (b.p.: 82.5 °C; $n_D^{25.8}$: 1.3739)

Of the possible impurities, the presence of peroxide is particularly disturbing. This contaminant can be reduced by refluxing with tin(II) chloride. For the removal of ammonia or basic impurities, the isopropanol is distilled from sulphanilic acid. The distillate may be freed from water most simply with a molecular sieve. Isopropanol forms an azeotropic mixture with water (alcohol content 91%, boiling point 80.3 °C); hence, for samples with higher water contents, the bulk of the water may be removed by azeotropic distillation.

Methanol (b.p.: 64.5 °C; n_D^{25}: 1.32663)

In our experience, in addition to moisture, commercial methanol may also contain of organic impurities such as acetone, formaldehyde, ethanol, methyl formate and even acetaldehyde. The water content can be reduced to less than 0.01% by fractional distillation. Further drying may be effected with calcium hydride or calcium sulphate, but most efficiently with metallic sodium. Experiments in this laboratory showed that treatment with sodium, twice repeated, decreased the water content to 5×10^{-5}%. If a larger amount of acetone is present as an impurity, it can be precipitated first in the form of iodoform, by treatment with iodine and sodium hydroxide. Methanol is then distilled off from the precipitate, and the dehydration process is performed afterwards.

Nitrobenzene (b.p.: 84–86.5 °C at 9–11 mbar, 210.8 °C at 1.01 bar; n_D^{20}: 1.55257)

The main impurities are nitrotoluene, dinitrophenol, dinitrobenzene and aniline, which can be removed by steam distillation from a mixture with dilute sulphuric acid. The water content of the nitrobenzene may then be bound by calcium chloride treatment. Finally, the purification is completed by vacuum distillation from phosphorus pentoxide. The purified sample should be stored in a dark bottle.

Pyridine (b.p.: 115.6 °C; n_D^{20}: 1.51021)

Various purification procedures have been proposed, and in our experience pyridine with a purity sufficient for most solvent studies may be obtained by dehydration of analytical-reagent grade material (Merck, p.a.) with a molecular sieve, followed by fractional distillation.

Tetrahydrofuran (b.p.: 65.4 °C; n_D^{25}: 1.4040)

Apart from water, the most frequent impurity in this solvent is the peroxide, but it may also contain other organic impurities. Refluxing with lithium aluminium hydride and subsequent distillation may be employed for simultaneous dehydration and purification in one step.

The above procedures are suitable for the purification of relatively large amounts of solvent (in the litre range). Preparative gas chromatography has proved the most suitable method for the preparation of small portions of solvents (50–500 mg) of special purity. However, the resulting extremely pure solvents are required only for the solution of special problems on the micro scale.

As regards the purification of solvents not listed above, the reader is referred to the book by Perrin *et al.* [Pe 66].

Characterization of solvents by their physical parameters

The physical characteristics of solvents are influenced to various extents by the presence of impurities, and hence are more or less suitable for checking the purity of a solvent, although the impurity remains unidentified. Here the physical parameters suitable for the detection of impurities are listed in decreasing order of their sensitivity; the various types of impurities that can thus be detected are also indicated.

(1) Conductivity is extremely sensitive to the presence of ions. Hence, conductivity measurements are particularly suitable for the indication of all decomposition processes (hydrolysis, oxidation) accompanied by the liberation of hydrogen ions. If the high mobility of the hydrogen ion is also taken into account, the special sensitivity of this method in these processes is readily understood. In addition, it may be utilized to follow the production of other ionic formations, e.g., to indicate the hydrolytic decomposition of halogenated hydrocarbons.

(2) The relative permittivity and the dielectric properties in general are sensitive to impurities more polar than the solvent examined. By this means, therefore, the presence of water, aliphatic or aromatic amines, heterocyclic molecules, alcohols, etc., can usually be readily detected in apolar solvent.

(3) The refractive index is suitable for indicating the presence of components having different optical densities. Although the refractive index is changed to only a slight extent by such components, this property can be measured with such an accuracy (particularly by interferometry) that it may even be employed to indicate the presence of an impurity at a concentration of less than 1%.

(4) The boiling point is fairly insensitive to the presence of a small amount of impurity, especially if the boiling point of the contaminant is not very different from that of the solvent in question. Naturally, a larger amount of impurity will result in elevation of the boiling point. However, since the boiling ranges of solvents are mostly 2–3 °C wide, observation of this effect is somewhat uncertain. More information is provided by the differences between the boiling points of the various fractions during fractional distillation.

(5) Density is another physical constant which is not very sensitive to impurities. Although the densities of liquids can be determined with very high accuracy by pycnometry, the measurement of density is not suitable for the detection of impurities occurring in micro amounts.

(6) Viscosity is likewise a fairly insensitive parameter.

Special mention must be made of the spectroscopic methods appropriate for the characterization of solvents and solutions. All organic molecules, including those of a solvent, exhibit characteristic vibration spectra; protic solvents have, in addition, characteristic proton resonance spectra. The presence of foreign molecules in the system under examination is indicated less sensitively by the infrared spectrum, but more sensitively by the NMR spectrum, if these impurities contain infrared-active or NMR-active groups that are not present in the solvent molecule.

In aliphatic or aromatic hydrocarbons, alcohols may be detected sensitively on the basis of the strongly infrared-active OH bands, as may aldehydes or ketones on the basis of the similarly high-intensity CO bands; the presence of organic amines is revealed by the NH bands, and that of nitriles by the CN bands. These same bands may be used, although with lower sensitivity, for the determination of aldehyde impurities in alcohols, or alcohol impurities in nitriles. The appearance of new bands in the infrared spectrum does not always lead to identification of the impurity component, but it definitely draws attention to the contaminated nature of the solvent.

The proton resonance spectrum allows differentiation between protons which are in different chemical environments, such as in different bonds. By appropriate modification of the sensitivity of the instrument, it is possible to detect impurities present in very small amounts beside the main component, if they contain functional groups with protons whose bands appear in the spectrum far enough from the bands of the protons in the main component. For instance, OH, NH and SH protons may readily be seen in the presence of CH protons; aromatic protons may be detected in the presence of the protons of aliphatic compounds, and *vice versa.*

New possibilities for such purity examinations have opened up with the development of ^{13}C-resonance spectroscopy. By means of this method it is now possible to distinguish between carbon atoms in different chemical environments. In accordance with the above considerations, this may be utilized for qualitative investigations. It should be noted, however, that the sensitivity of the carbon-resonance method is lower than that of proton-resonance spectroscopy. Owing to the low abundance of the ^{13}C isotope, the measurements must be carried out in more concentrated solutions. At the same time, the resolution of carbon-resonance spectroscopy is higher than that of proton-resonance spectroscopy, hence the former can be used for the detection of impurities (when present in sufficiently high concentrations for detection) which are chemically little different from the solvent. Whereas the contamination of benzene by toluene, for example, can scarcely be demonstrated by the proton-resonance method, under optimal conditions this problem can be solved with carbon-resonance spectroscopy.

Electron excitation spectroscopy is less generally utilizable for similar examinations. On the other hand, in a system in which the light absorption of the impurity appears in a range where the main component does not absorb, this method has a higher sensitivity than the previously mentioned methods. For instance, aromatic impurities in aliphatic solvents can be detected most simply by UV spectroscopy.

Characterization of the solvating powers of solvents

Before investigations of the solvent effect, it is advisable to characterize the properly purified solvent also chemically, completing its characterization by the physical parameters. Determination of the solvating power appears to be the most suitable for this purpose.

As clearly shown by the examples in Chapters 2 and 3, the solvating power of a solvent is the resultant of a combination of several specific and non-specific interactions. It is difficult to differentiate these from one another. This is the reason why so many different types of empirical solvent scales have been proposed for the characterization of the solvating power.

In studies of the solvent effect, great importance is attached to the specific donor and acceptor interactions. Hence, for the general characterization of solvents, the solvent scales that seem to be the most suitable are those which distinguish between donor and acceptor solvation.

Selection from among the various donor and acceptor strength scales available for the characterization of a solvent depends on the aim of the intended investigation. In our experience, the Gutmann donicity scale [Gu 66] and the Gutmann–Mayer acceptor strength scale [Ma 75] can be used most generally. Accordingly, we propose the determination of these for the characterization of the solvating powers of solvents. (These parameters are already known for the solvents most widely used in practice.)

The Gutmann donicity may be determined by calorimetric measurement of the heat of reaction. A solution of the reference acceptor, antimony pentachloride, in dichloroethane is mixed in a suitable calorimeter with a dichloroethane solution of the solvent under investigation. Under adiabatic conditions, the change in temperature of the reaction mixture is proportional to the heat of reaction. (Naturally, the value of the heat of dilution must also be taken into consideration.) For more detail, the reader is referred to the book by Gutmann [Gu 68].

The Gutmann–Mayer acceptor number is determined with the aid of phosphorus-resonance spectroscopy. Triethylphosphine oxide is used as reference donor. If the solvent dependences of the ^{31}P-NMR chemical shifts measured in the various solvents are referred to the chemical shift of the complex formed between antimony pentachloride and the reference donor (the latter value is taken as 100), the acceptor number values are obtained. A description of the methodology can be found in the original publication by Mayer [Ma 75].

It cannot be ignored that most solvents have both donor and acceptor properties. This also holds for those solvents which are regarded as typical donor or typical acceptor solvents; this merely means that one of the properties is much more subordinate [Gu 78]. It is particularly true, however, for amphiprotic solvents such as water and alcohols, the oxygen of which acts as a donor, while the hydroxyl proton (by forming hydrogen bonds) behaves as an acceptor.

It follows that the placing of a solvent on a single given solvent scale is not equivalent to its chemical characterization. In order to be able to draw conclusions on the chemical behaviour of a solvent, and primarily its solvating power, both its donor strength and acceptor strength must be characterized. In each case, therefore, it is advantageous to give both the Gutmann donicity and the Mayer acceptor number.

Our above proposal does not imply that the other solvent scales described in Chapter 4 have no justification. The solvent scale which will best characterize the solvent effect in the system in question is decided by the nature of the task to be studied. The available experimental means may also have a considerable influence on the choice of the empirical solvent scale to be employed in a given examination series. It is of no use for the Gutmann–Mayer acceptor number to appear advantageous if a NMR instrument suitable for measurement of the phosphorus-resonance spectrum is not available. For the experimental determination of the Reichardt and Kosower solvent parameters based on the shifts in the charge-transfer bands (see Chapter 4), the only instrument required is a spectrophotometer that measures in the UV and visible spectral ranges, this being available in most laboratories. At the same time, these two empirical scales are also suitable for the characterization of the acceptor strength. For a complete characterization of the solvent effect (solvating power) in composite systems the two and multi parameter equations discussed at the end of Chapter 4 (p. 76–90) have to be used.

Characterization of solvent mixtures

If examinations are to be carried out in a solvent mixture, the latter is always prepared by mixing in an exact ratio the carefully purified and controlled components. Hence, the characterization of such mixtures might in effect be traced back to the characterization of the components. However, in the mixture these components are involved in interactions of various strengths with one another. Futher, the interactions existing between the molecules in the pure solvents are changed by the act of mixing. Accordingly, the chemical and physical properties of a solvent mixture may differ appreciably from those of the components.

As a direct result of the above interactions, in many cases the mixing of two solvents is accompanied by a volume change. This volume change is the larger the stronger are the interactions. Because of the volume change, even the composition of the solvent mixture cannot be given simply on the basis of the volumes and densities of the components mixed; the density of the final mixture must always be determined, and the actual composition can be calculated by taking this into consideration.

Next, for characterization of the solvent mixture, use may be made of the same physical parameters and spectra with which the components were characterized. In every case, however, the data obtained for the mixture must be compared with the corresponding data for the pure components. These examinations usually yield information on the interactions between identical and different molecules in the mixture. Thus, changes in the infrared and proton-resonance spectra as a consequence of the mixing of the components give a guide to the formation or breakdown of hydrogen-bonded associates, etc. Numerous examples of the methods of evaluating the data are presented in the appropriate parts of Chapter 5 and in Chapter 8.

The chemical characterization of a solvent mixture is much more difficult because it is not the mixture, but its components, that take part in the solvation reactions. However, the solvating powers of the components are affected by their interactions with one another. Depending on the chemical properties of the system, the solvating powers of the individual components may differ widely. Hence, the effects of the components in forming a solvate sheath will not be proportional to their ratio in the mixture. An understanding of the system is made even more complicated by the fact that, besides solvates that contain only one or the other solvent, the solute can also form mixed solvates in the solvate spheres of which both solvents are present together, possibly in various ratios. This topic is treated in greater detail in Chapter 8.

It follows that the empirical solvent parameters which are based on measurement of specific interactions of given molecules cannot be used for the characterization of a solvent mixture. If, for example, the Gutmann donicity of a solvent mixture is measured calorimetrically, the resulting ΔH value, under optimal conditions, is equal to the sum of the heats of reaction of the interactions (taken in appropriate proportions) between the components and antimony pentachloride, but a contribution may possibly be made by a term originating from the interaction of the components of the mixed solvent. This equally holds for most of the solvent parameters discussed in Chapter 4.

The exact characterization of the solvating power of a solvent mixture with the aid of the empirical solvent parameters can be achieved only in a mixture where one of the components is inert towards the system examined. Despite these considerations, a number of workers have employed very varied solvent parameters to characterize solvent mixtures [La 82] also in cases that do not satisfy this condition; the chemical content of such data is uncertain.

In the present author's view, greater assistance is provided towards an understanding of the processes that occur in solvent mixtures by the solvent parameters relating to the pure components, the chemical background of which is well known, than by the parameters determined for the mixture, as the latter reflect overall processes the components of which are inseparable from one another, and hence can scarcely interpreted, even qualitatively.

References

Br 40 Bryant, W. M. D., Mitchell, J. Jr., Smith, D. M.: J. Am. Chem. Soc., **62,** 3504 (1940).

Br 41 Bryant, W. M. D., Mitchell, J. Jr., Ashby, E. C.: J. Am. Chem. Soc., **63,** 2924, 2927 (1941).

Bu 77 Burfield, D. R., Lee, K. H., Smithers, R. H.: J. Org. Chem., **42,** 3060 (1977).

Bu 81 Burfield, D. R., Smithers, R. H., Sui Chai Tan: J. Org. Chem. **46,** 629 (1981).

Fi 35 Fischer, K.: Angew. Chem., **48,** 394 (1935).

Gu 66 Gutmann, V., Wychera, E.: Inorg. Nucl. Chem. Letters, **2,** 257 (1966).

Gu 68 Gutmann, V.: Coordination Chemistry in Non-Aqueous Solutions. Springer, Vienna, New York, 1968.

Gu 78 Gutmann, V.: The Donor-Acceptor Approach to Molecular Interactions. Plenum Press, London, 1978.

Ha 72 Harris, C.: Talanta, **19,** 1523 (1972).

Je 77 Jezorek, T. R., Manly, C. F.: Anal. Chem., **49,** 1874, (1977).

Kh 80 Khalaf, H., Rimpler, M.: Fresenius Z. Anal. Chem., **302,** 204 (1980).

La 81 Langhals, H.: Fresenius Z. Anal. Chem., **305,** 26 (1981).

La 82 Langhals, H.: Nouv. J. Chim. **6,** 285 (1982); See also Langhals, H.: Z. phys. Chem. N.F. **127,** 45 (1981).

Ma 75 Mayer, U.: Pure Appl. Chem., **41,** 291 (1975).

Mi 48 Mitchell, J. Jr., Smith, D. M.: Aquametry, Application of the Karl Fischer Reagent to Quantitative Analyses involving Water. Interscience, New York, London, 1948.

Mi 49 Milberger, E. C., Uhrig, K., Becker, H. C., Levin, H.: Anal. Chem., **21,** 1192 (1949).

Mi 61 Mitchell, J. Jr.: in: Treatise on Analytical Chemistry (Eds.: Kolthoff, I. M., and Elving, P. J.), Part II, Vol. I. Chap. 4. Interscience, New York, 1961.

Pe 66 Perrin, D. D., Armarego, W. L. F., Perrin, D. R.: Purification of Laboratory Chemicals. Pergamon Press, Oxford, 1966.

Se 71 Sellers, P.: Acta Chem. Scand., **25,** 2295 (1971).

10. RESEARCH TRENDS IN THE FUTURE

In an examination of the literature dealing with the study of non-aqueous solutions, it is striking that there is virtually no experimental procedure for the investigation of matter that has not been employed to study solvation or other processes involving solvent effects. Scarcely does a new method appear in the arsenal of the analyst than it is applied to this field of solution chemistry. This tendency is well shown by the example of typical methods, developed for the investigation of solid substances, such as Mössbauer spectroscopy and ESCA. The application of these to the study of solvation processes was made possible by the elaboration of the technique of quenching solutions by rapid freezing.

In spite of the many-sided use of the various methods of examination, involving small or large instruments, it cannot be said that the understanding of the solvent effect in the various systems is simple. In non-aqueous solutions the conditions are perhaps even more complicated than in aqueous solutions. In most non-aqueous solvents the interaction between the solvent and the solute (the solvation) is not substantially less than in water. At the same time, the interactions of the components of the solutes with one another, the association reactions favoured by the lower relative permittivity (ion pair and complex formation, oligomerization and polymerization) are much more pronounced in non-aqueous solutions than in water. Non-aqueous solvents favour in particular the formation of the species with more complicated compositions (e.g., mixed ligand, polynuclear and outer-sphere type complexes). This also shows up in the more involved mechanisms of the reactions occurring in such systems. Nevertheless, as may be seen from the subject matter surveyed in this volume, more authors have dealt with the study of the simple formations, primarily mononuclear parent complexes and simpler mixed complexes, and with the reactions in which they participate, than with the study of the more complicated systems, which occur more frequently in reality and which are of greater practical importance.

This tendency can be explained by the fact that the more complicated the system to be examined, the more sensitive and refined are the experimental methods and the higher the performance of the evaluation procedures that are required. The introduction and rapid spread of electronic computers led to the development of methods suitable for the simultaneous evaluation of series of large numbers of experimental data of various types, necessary for the study of the most complex systems in both coordination chemistry and analytical chemistry. Naturally, the applicability of these methods does not depend on the nature of the solvent, but the latter can be an important factor in deciding the efficiency of the method applied, and particularly the accuracy and reproducibility of the measurements.

The different procedures based on the measurement of electromotive force that may be applied in many varied ways in the equilibrium chemistry of aqueous solutions have a much more restricted use in non-aqueous solutions. It is difficult to construct measuring cells that have small and readily reproducible diffusion potentials or that are without a liquid junction. The glass electrode, perhaps the most extensively used type in aqueous solutions, does not function at all in some non-aqueous solutions, and with very low accuracy in others. Evaluation of the data obtained with its use is hampered by the limits of the acidity scales employed in such systems. The most promising type of ionselective membrane electrodes, the liquid ionic exchange membrane electrode, virtually cannot be employed in non-aqueous solutions.

Similarly, the other electroanalytical methods, primarily the various types of voltammetry and conductometry, are also of limited applicability in systems with low relative permittivities.

In most non-aqueous solvents, the solubility conditions of the inorganic components restrict the equilibrium investigations to a fairly narrow concentration interval, causing uncertainties in the study of the more complicated systems in particular.

The clear evaluation of the experimental results is also hindered by the difficulties encountered in the perfect purification of non-aqueous solvents. Several of them are hygroscopic, and even an extremely low water content may cause fundamental changes in the chemical properties of numerous solvents. As an electron-pair donor, the water molecule may behave as a ligand, and as a consequence of the ability of its hydrogen atoms to form hydrogen bonds, it may also act as an acceptor. This may lead to the occurrence of unexpected side reactions. In acidic solvents water behaves as a base, and in basic solvents as an acid, thereby disturbing the courses of the reactions to be investigated. The removal of trace amounts of water and the performance of work under anhydrous conditions is a difficult task.

From what has been said so far, it emerges that the most important task in future studies of non-aqueous solutions is the investigation of the more complicated systems, e.g., mixed ligand and polynuclear complexes and outer-sphere type associates, and reactions involving their participation. This may lead to the elaboration of methods suitable for the characterization and control of solution systems important in industrial practice, and to a more complete understanding of the regularities that hold for non-aqueous solutions.

Conditions for the success of such research work are the selection or development of the appropriate experimental methods, and the attainment of the maximum accuracy possible in the measurements. Many results of research being carried out in this field at present are already indicative of this trend.

Experiments are under way for the production of ion-selective membrane electrodes that can also be used in non-aqueous solutions, and studies are being made of the possibilities of application of liquid junction-free potentiometry in non-aqueous solutions. High-performance computer evaluation procedures permit the employment of spectrophotometric equilibrium measurements in the study of complex systems. Spectrophotometric measurements are not prone to greater errors in non-aqueous solutions than in water.

Calorimetric investigations have already contributed appreciable results towards the understanding of non-aqueous solutions. This is the area in particular in which many valuable new results may be expected in the future. In addition to the determination of equilibrium constants, equilibrium studies combined with calorimetric measurements also contribute to the recognition of the driving forces of chemical processes, *via* the measurement of thermodynamic data (enthalpy and entropy).

The more complicated the system under examination, the more comprehensive its study must be if a full understanding is to be achieved. This is particularly so for cases such as the investigation of non-aqueous solutions, in which it has been seen that the accuracy of measurement is automatically lower than in aqueous solutions.

In accordance with the above, it is even more desirable in non-aqueous solutions than in aqueous systems that the equilibrium studies should be combined with the results obtainable by structural examination methods. Determination of the electronic structures and the structures and symmetries of the species in solution by means of some independent method is of substantial help in the evaluation of the data from equilibrium measurements. It contributes to the correct choice of the mathematical model necessary for the calculations to be performed on the direct experimental data, and in addition it assists in recognizing the factors which govern the processes occurring in solution.

Many valuable new results may be expected in this respect from the wider ranging application of electron paramagnetic resonance spectroscopy, NMR spectroscopy and solution X-ray examinations, and from the introduction of neutron diffraction and other diffraction methods in the chemistry of non-aqueous solutions. New possibilities in this field, which have by no means been fully exploited, are Mössbauer and ESCA investigations of rapidly frozen solutions.

Capillary Mössbauer Spectroscopy (CMS), the study of liquid samples trapped in the pores of "thirsty" glass, seems to open a new field in the study of solvation phenomena.

Other new experimental methods, such as those provided recently by positronium chemistry, may contribute to the fuller understanding of non-aqueous complex systems in just the same way as certain classical procedures which have undergone a revival in recent years, e.g., autodiffusion measurement.

With a knowledge of the factors determining the solvent effect, one may expect many new results from the investigation of the kinetics and mechanisms of reactions taking place in non-aqueous solutions. The use of different procedures suitable for the investigation of fast reactions (stopped flow and relaxation methods) may assist towards the solution of problems previously considered unapproachable. The most recent relaxation procedures, such as the microwave temperature jump and laser temperature jump methods, are also applicable to the examination of non-aqueous solutions.

The understanding of reactions that occur in non-aqueous solutions is important not only because of their theoretical interest, the deeper knowledge of the theory of solutions and the investigation of systems that do not exist or cannot be examined in aqueous solutions, but also owing to the practical significance of non-aqueous solutions in the industrial practice. The last reason is the primary explanation of the great interest with which research workers throughout the world are devoting themselves to the examination of non-aqueous systems.

SUBJECT INDEX

Numbers in italics refer to Tables,
bold-face numbers to solvent purification procedures